Methods in Molecular Biology

Series Editor
John M. Walker
School of Life and Medical Sciences
University of Hertfordshire
Hatfield, Hertfordshire, UK

For further volumes:
http://www.springer.com/series/7651

For over 35 years, biological scientists have come to rely on the research protocols and methodologies in the critically acclaimed *Methods in Molecular Biology* series. The series was the first to introduce the step-by-step protocols approach that has become the standard in all biomedical protocol publishing. Each protocol is provided in readily-reproducible step-by-step fashion, opening with an introductory overview, a list of the materials and reagents needed to complete the experiment, and followed by a detailed procedure that is supported with a helpful notes section offering tips and tricks of the trade as well as troubleshooting advice. These hallmark features were introduced by series editor Dr. John Walker and constitute the key ingredient in each and every volume of the *Methods in Molecular Biology* series. Tested and trusted, comprehensive and reliable, all protocols from the series are indexed in PubMed.

Expressed Protein Ligation

Methods and Protocols

Edited by

Miquel Vila-Perelló

ProteoDesign, Barcelona, Barcelona, Spain

Humana Press

Editor
Miquel Vila-Perelló
ProteoDesign
Barcelona, Barcelona, Spain

ISSN 1064-3745 ISSN 1940-6029 (electronic)
Methods in Molecular Biology
ISBN 978-1-0716-0433-5 ISBN 978-1-0716-0434-2 (eBook)
https://doi.org/10.1007/978-1-0716-0434-2

This Humana imprint is published by the registered company Springer Science+Business Media, LLC, part of Springer Nature.
The registered company address is: 1 New York Plaza, New York, NY 10004, U.S.A.

Preface

Expressed protein ligation (EPL) is a powerful method to generate site-specifically modified proteins [1, 2] and has become a key technology in the toolbox of researchers to study protein structure and function. Methodologically, EPL requires the combination of molecular biology techniques for recombinant protein expression, with chemical strategies of peptide synthesis to generate precisely modified proteins. This combination of skills and techniques is often hard to come by. Hence, one of the key motivations in preparing this volume was to provide readers all the required technical detail, so that researchers without strong background in recombinant protein production, and/or peptide synthesis, would be able to successfully implement these protocols in their own research.

EPL has been demonstrated to be an extremely versatile technology to generate precisely modified proteins, without any limitation on protein size, or identity of the modifications to be introduced. This versatility is exemplified by the great wealth of modified proteins, obtained by EPL, for a variety of purposes: from post-translationally modified proteins to study the intricacies of their biological function and bioconjugates with therapeutic potential [3, 4]. As a result of all this work, EPL is currently a technology with great potential in key areas of biology and biomedicine, and this book aims to make it readily available to a broad audience of interdisciplinary researchers.

EPL was developed in the late 1990s, by the labs of Phil Cole and Tom Muir [1], based on the chemoselective reaction between a peptide thioester and an N-terminal Cys containing peptide (termed native chemical ligation (NCL)) [5]). The NCL reaction using synthetic peptides provides access to fully synthetic polypeptides and proteins with absolute control on their chemical structure. However, fully synthetic methods have limitations regarding the size of the proteins that can be accessed. The application of the NCL principle to larger proteins required the ability to obtain the necessary starting materials, particularly protein thioesters, via recombinant methods. The discovery of inteins in 1990 and all the ensuing mechanistic work on protein splicing [6] allowed researchers to identify protein thioesters as intermediates on that process. Engineering of inteins to promote accumulation of these protein thioesters was harnessed in the first report on EPL to expand the NCL reaction beyond fully synthetic proteins to semi-synthetic ones, thus opening the door to the precise modification of large proteins without the size limitations imposed by fully synthetic methods.

This book provides an overview of EPL, including general protocols, methods for the discovery and preparation of the key required reagents, EPL variations and combinations with other technologies, as well as specific protocols for the preparation of particularly challenging modified proteins.

The first part of the book contains reviews by Philip A. Cole and Zhipeng Wang (Chapter 1), and Marc Ribó et al. (Chapter 2) on the development of EPL, and protein splicing and inteins, respectively. These introductory chapters are followed by chapters by Neel Shah and Adam Stevens (Chapter 3) and Kenneth Mills et al. (Chapter 4) on methods for the discovery and characterization of inteins, as key tools for semi-synthetic protein ligation strategies. JP Pellois and Joshua Diaz (Chapter 5) describe a series of general EPL protocols that include the preparation of all required reagents, as well as the performance of ligation reactions under native and denaturing conditions. Yiming Li et al. (Chapter 6) and

Juan B. Blanco-Canosa and Judith Palá-Pujadas (Chapter 7) describe protocols for the preparation of synthetic peptide thioesters using the hydrazide and the acyl urea methods, respectively. These two strategies have become two of the most versatile and widely used methods to prepare peptide thioesters, which are key for the generation of semi-synthetic proteins bearing modifications at the N-terminus. Next, the book contains a series of chapters on variations of EPL and combination with related technologies. Galia Debelouchina et al. (Chapter 8) describe methods to generate complex modified proteins using engineered fused split-inteins combined with amber suppression; Silvia Frutos (Chapter 9) covers methods of using split inteins as masked thioesters for the modification of antibodies and generation of site-specific antibody-drug conjugates; Yael David et al. (Chapter 10) present methods combining peptide ligation and protein trans-splicing to label proteins inside living cells, and Eric Shusta and Benjamin Umlauf (Chapter 11) provide protocols for the site-specific modification of proteins produced in yeast using EPL. In the final part of the book, methods for the semi-synthesis of different types of modified proteins are provided. Louise Clemmensen et al. (Chapter 12) describe EPL-based methods for the semi-synthesis of phosphorylated proteins; Beat Fierz and Nora Guidotti (Chapter 13) describe a semi-synthetic methodology for the preparation of asymmetrically modified nucleosomes; Champak Chatterjee and Caroline Weller (Chapter 14) provide protocols for their recently developed strategy to obtain semi-synthetic ubiquitylated peptides using a ligation auxiliary. Matthew Pratt and Ana Galesic (Chapter 15) describe a method for the preparation of the glycosylated α-synuclein protein to investigate its role in Parkinson's disease. Julio Camarero et al. (Chapter 16) provide protocols for the synthesis of cyclic proteins using EPL. In the last chapter of the book, Christian Becker et al. (Chapter 17) describe semi-synthetic methods, which combine desulfurization with selective Cys side chain protection to obtain precisely modified proteins.

Overall, the book provides a complete overview of the principles of EPL, its potential for biological and biochemical research as well as a great wealth of protocols and methods to successfully implement it on any project that could benefit from, or would require, access to site-specifically modified proteins.

Finally, I would like to wholeheartedly thank all the authors for their great contributions and their commitment to the project from the very beginning, without their enthusiastic participation this book would have not been possible. I would also like to thank Dr. John M. Walker for his guidance and support throughout the process.

Barcelona, Spain *Miquel Vila-Perelló*

References

1. Muir TW, Sondhi D, Cole PA (1998) Expressed protein ligation: a general method for protein engineering. Proc Natl Acad Sci USA 95:6705–6710
2. Muir TW (2003) Semisynthesis of proteins by expressed protein ligation. Annu Rev Biochem 72:249–289
3. Flavell R, Muir T (2008) Expressed Protein Ligation (EPL) in the Study of Signal Transduction, Ion Conduction, And Chromatin Biology. Acc Chem Res. 42:107-116
4. Gentilucci L, Tosi P, Bauer A, Marco RD (2016) Modern tools for the chemical ligation and synthesis of modified peptides and proteins. Future Med Chem 8:2287–2304
5. Dawson PE, Muir TW, Clark-Lewis I, Kent SB (1994) Synthesis of proteins by native chemical ligation. Science 266:776–779
6. Paulus H (2000) Protein splicing and related forms of protein autoprocessing. Annu Rev Biochem 69:447–496

Contents

Contributors

CHRISTIAN F. W. BECKER • *Faculty of Chemistry, Institute of Biological Chemistry, University of Vienna, Vienna, Austria*

RAYMOND F. BERKELEY • *Department of Chemistry and Biochemistry, University of California, San Diego, La Jolla, CA, USA*

JUAN B. BLANCO-CANOSA • *Department of Biological Chemistry, Institute for Advanced Chemistry of Catalonia—CSIC (IQAC-CSIC), Barcelona, Spain*

JULIO A. CAMARERO • *Department of Pharmacology and Pharmaceutical Sciences, School of Pharmacy, University of Southern California, Los Angeles, CA, USA; Norris Comprehensive Cancer Center, Keck School of Medicine, University of Southern California, Los Angeles, CA, USA*

MARIA JOSE CAMPBELL • *Department of Pharmacology and Pharmaceutical Sciences, School of Pharmacy, University of Southern California, Los Angeles, CA, USA*

JESSICA CASTRO • *Laboratori d'Enginyeria de Proteïnes, Departament de Biologia, Facultat de Ciències, Universitat de Girona, Girona, Spain*

CHAMPAK CHATTERJEE • *Department of Chemistry, University of Washington, Seattle, WA, USA*

LOUISE S. CLEMMENSEN • *Department of Drug Design and Pharmacology, Center for Biopharmaceuticals, University of Copenhagen, Copenhagen, Denmark*

PHILIP A. COLE • *Division of Genetics, Department of Medicine, Brigham and Women's Hospital, Harvard Medical School, Boston, MA, USA; Department of Biological Chemistry and Molecular Pharmacology, Harvard Medical School, Boston, MA, USA*

YAEL DAVID • *Tri-Institutional PhD Program in Chemical Biology, New York, NY, USA; Chemical Biology Program, Memorial Sloan Kettering Cancer Center, New York, NY, USA; Department of Pharmacology, Weill Cornell Medicine, New York, NY, USA; Department of Physiology, Biophysics and Systems Biology, Weill Cornell Medicine, New York, NY, USA*

GALIA T. DEBELOUCHINA • *Department of Chemistry and Biochemistry, University of California, San Diego, La Jolla, CA, USA*

JOSHUA DIAZ • *Department of Biochemistry and Biophysics, Texas A&M University, College Station, TX, USA*

ZHENMING DU • *Department of Chemistry, Georgia State University, Atlanta, GA, USA*

BEAT FIERZ • *École Polytechnique Fédérale de Lausanne (EPFL), Lausanne, Switzerland*

SILVIA FRUTOS • *ProteoDesign, Barcelona, Spain*

ANA GALESIC • *Department of Chemistry, University of Southern California, Los Angeles, CA, USA*

VERÓNICA GORDO • *Laboratori d'Enginyeria de Proteïnes, Departament de Biologia, Facultat de Ciències, Universitat de Girona, Girona, Spain*

NORA GUIDOTTI • *École Polytechnique Fédérale de Lausanne (EPFL), Lausanne, Switzerland*

YI-MING LI • *School of Food and Biological Engineering, Hefei University of Technology, Hefei, China*

BYUNG JOON LIM • *Department of Chemistry and Biochemistry, University of California, San Diego, La Jolla, CA, USA*

SANA MA • *Department of Drug Design and Pharmacology, Center for Biopharmaceuticals, University of Copenhagen, Copenhagen, Denmark*
MARIA MATVEENKO • *Faculty of Chemistry, Institute of Biological Chemistry, University of Vienna, Vienna, Austria*
KENNETH V. MILLS • *Department of Chemistry, College of the Holy Cross, Worcester, MA, USA*
SOMNATH MUKHERJEE • *Faculty of Chemistry, Institute of Biological Chemistry, University of Vienna, Vienna, Austria*
JUDITH PALÀ-PUJADAS • *Institute for Research in Biomedicine (IRB Barcelona), The Barcelona Institute of Science and Technology (BIST), Barcelona, Spain*
JEAN-PHILIPPE PELLOIS • *Department of Biochemistry and Biophysics, Texas A&M University, College Station, TX, USA; Department of Chemistry, Texas A&M University, College Station, TX, USA*
MATTHEW R. PRATT • *Department of Chemistry, University of Southern California, Los Angeles, CA, USA; Department of Biological Sciences, University of Southern California, Los Angeles, CA, USA*
NICHOLAS A. PRESCOTT • *Tri-Institutional PhD Program in Chemical Biology, New York, NY, USA; Chemical Biology Program, Memorial Sloan Kettering Cancer Center, New York, NY, USA*
MARC RIBÓ • *Laboratori d'Enginyeria de Proteïnes, Departament de Biologia, Facultat de Ciències, Universitat de Girona, Girona, Spain*
ALEJANDRO ROMERO-CASAÑAS • *Laboratori d'Enginyeria de Proteïnes, Departament de Biologia, Facultat de Ciències, Universitat de Girona, Girona, Spain*
NEEL H. SHAH • *Department of Chemistry, Columbia University, New York, NY, USA*
ERIC V. SHUSTA • *Department of Chemical and Biological Engineering, University of Wisconsin-Madison, Madison, WI, USA*
ADAM J. STEVENS • *Department of Cellular and Molecular Pharmacology, University of California, San Francisco, San Francisco, CA, USA*
KRISTIAN STRØMGAARD • *Department of Drug Design and Pharmacology, Center for Biopharmaceuticals, University of Copenhagen, Copenhagen, Denmark*
JINGTAN SU • *Department of Pharmacology and Pharmaceutical Sciences, School of Pharmacy, University of Southern California, Los Angeles, CA, USA*
BENJAMIN J. UMLAUF • *Department of Chemical and Biological Engineering, University of Wisconsin-Madison, Madison, WI, USA*
ZHIPENG A. WANG • *Division of Genetics, Department of Medicine, Brigham and Women's Hospital, Harvard Medical School, Boston, MA, USA; Department of Biological Chemistry and Molecular Pharmacology, Harvard Medical School, Boston, MA, USA*
CHUNYU WANG • *Center for Biotechnology and Interdisciplinary Studies, Rensselaer Polytechnic Institute, Troy, NY, USA*
YU WANG • *School of Food and Biological Engineering, Hefei University of Technology, Hefei, China*
CAROLINE E. WELLER • *Department of Chemistry, University of Washington, Seattle, WA, USA*
JING ZHAO • *Center for Biotechnology and Interdisciplinary Studies, Rensselaer Polytechnic Institute, Troy, NY, USA*

Chapter 1

Methods and Applications of Expressed Protein Ligation

Zhipeng A. Wang and Philip A. Cole

Abstract

Expressed protein ligation is a method of protein semisynthesis and typically involves the reaction of recombinant protein C-terminal thioesters with N-cysteine containing synthetic peptides in a chemoselective ligation. The recombinant protein C-terminal thioesters are produced by exploiting the action of nature's inteins which are protein modules that catalyze protein splicing. This chapter discusses the basic principles of expressed protein ligation and recent advances and applications in this protein semisynthesis field. Comparative strengths and weaknesses of the method and future challenges are highlighted.

Key words Inteins, Protein splicing, Thioesters, Posttranslational modifications, Protein labeling

1 Introduction

1.1 Protein Semisynthesis and Expressed Protein Ligation (EPL)

The study of proteins has taken center stage in biomedical investigation over the past several decades [1]. Powerful genetic tools including recombinant protein expression and site-directed mutagenesis [2] have allowed for the investigation of particular domains and individual amino acid contributions to the structure and function of complex proteins. Over the past 20 years, unnatural amino acid mutagenesis using nonsense suppression and related strategies has expanded the scope of protein study by providing a wide range of new chemical functionality into increasingly large and diverse protein types, addressing many biological challenges [3]. Nevertheless, in some cases it becomes desirable in protein biochemistry to have even greater chemical precision than that provided by genetic methods. Examples of such cases where the power of chemistry is especially beneficial in the study of proteins include the analysis of proteins with multiple diverse posttranslational modifications, structural studies on segmentally isotopically labeled proteins, or the incorporation of two or more complex chemical probes at particular sites into proteins. One solution to these challenges is the use of chemical tools in the preparation of proteins [4]. Total synthesis of proteins typically involves generating peptides using

Miquel Vila-Perelló (ed.), *Expressed Protein Ligation: Methods and Protocols*, Methods in Molecular Biology, vol. 2133,
https://doi.org/10.1007/978-1-0716-0434-2_1, © Springer Science+Business Media, LLC, part of Springer Nature 2020

solid phase peptide synthesis (SPPS) and ligating such peptides together by chemoselective reactions including the powerful native chemical ligation reaction (NCL). The NCL reaction developed by Kent and colleagues [5] is based on the early work of Wieland [6] and exploits the chemoselective reaction of an N-terminal cysteine (Cys) and a C-terminal thioester of two unprotected peptides. Such chemoselectivity derives from an initial transthioesterification between the thiol of the N-terminal Cys and C-terminal thioester followed by intramolecular aminolysis to afford a stable peptide bond. Total chemical synthesis of proteins has been successfully used to generate small proteins, usually less than 200 aa, but becomes difficult to execute on larger proteins because of the need for five or more peptide ligation reactions.

A compromise between genetic and chemical approaches to the study of proteins is embodied by protein semisynthesis. Protein semisynthesis refers to the technological merger of recombinant protein expression and chemical modifications, which are often introduced via chemoselective ligation reactions [7]. One of the major modalities of protein semisynthesis is expressed protein ligation (EPL) [7]. The EPL method exploits nature's inteins for generating recombinant proteins with C-terminal thioesters, which will further undergo NCL with an external peptide with an N-terminal Cys. Functionally related to introns in mRNA, inteins are protein domains that exist in a number of prokaryotic and yeast proteins that undergo protein splicing [8]. Protein splicing is analogous to RNA splicing in which two lateral protein domains, the N-extein (Ext^N) and C-extein (Ext^C), fuse to form a new polypeptide at the expense of the internal intein which is extruded.

2 Mechanisms of Protein Splicing, PTS and EPL

2.1 Intein-Based Protein Splicing

Intein-mediated protein splicing proceeds through multiple steps [8]. To begin with, the amide bond between the intein and Ext^N is cleaved by the first residue of the intein, most commonly a cysteine (Cys), forming a thioester with the Cys side chain. Next, an acyl transfer reaction occurs in which the Ext^N is transferred to the first residue of the Ext^C, generally Cys, serine (Ser), or threonine (Thr). The intein appears to induce strain in the starting Ext^N-Cys peptide bond resulting in an equilibrium between the baseline amide and a thioester emanating from the intein Cys side chain. This initial equilibrium provides an energetic source for intramolecular transacylation with the Ext^C. Subsequently, an asparagine (Asn) residue cleaves the main polypeptide chain to form a succinimidyl intermediate, resulting in the release of a branched thioester/ester from the intein, and the thioester/ester reequilibrates to a normal backbone amide bond between the Ext^N and Ext^C (Fig. 1). Under some situations, the thioester/ester may undergo hydrolysis before the

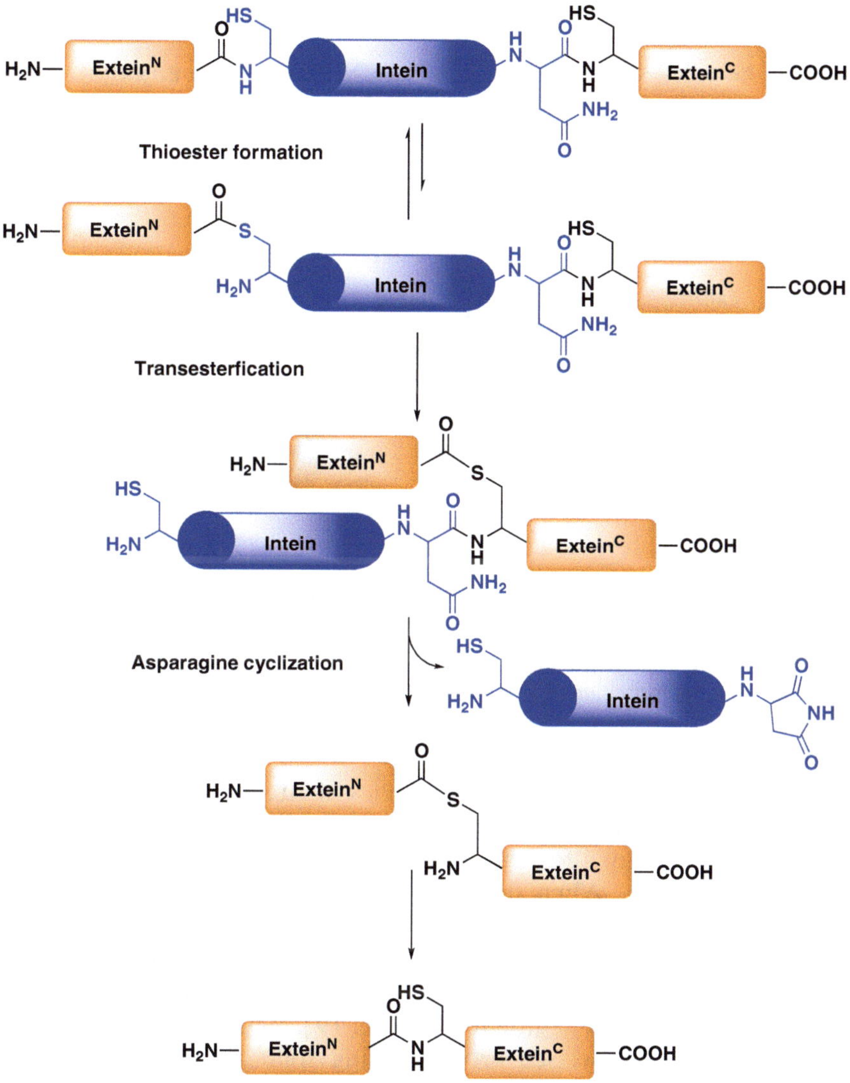

Fig. 1 Classical mechanism of protein splicing. Following the removal of the succinimidyl intein intermediate, Ext^N joins with Ext^C through a native amide bond

acyl transfer, leading to a side reaction commonly known as "N-terminal cleavage", which results in the release of free acid on the Ext^N. By judicious introduction of mutations in the intein–Ext^C junction, the initial amide–thioester equilibrium in the Ext^N-Cys is arrested but can be exploited by large concentrations of exogenous thiols to produced C-terminal thioesters. This paves the way to obtain stable thioesters for use in semisynthesis.

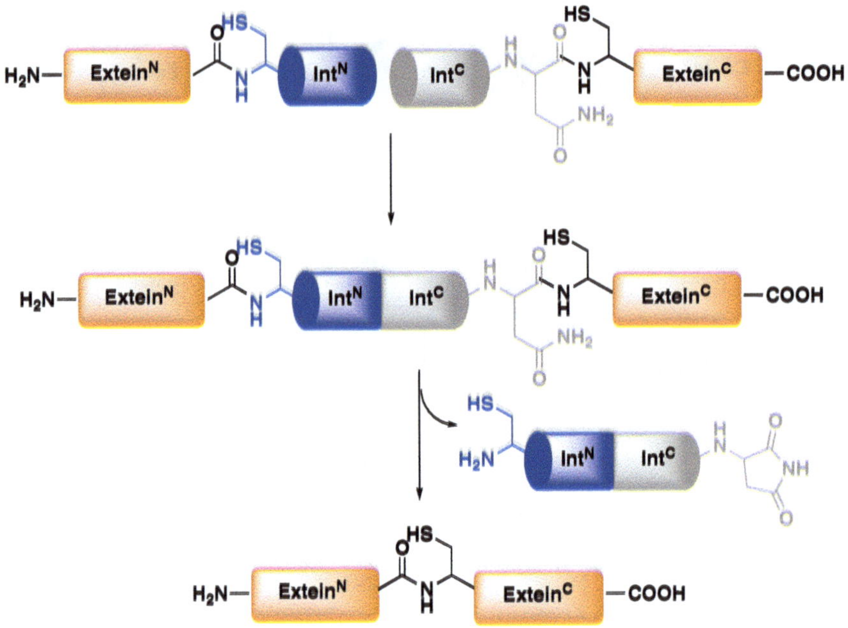

Fig. 2 Classical mechanism of protein trans-splicing (PTS). An intact intein in a functionally folded conformation is reconstituted after the formation of a heterodimeric complex of Int^N and Int^C, restoring normal splicing activity

2.2 Split Intein, Protein Trans-Splicing (PTS), and Conditional Protein Splicing (CPS)

There are different families of inteins. Some natural inteins are split into separate segments on two different polypeptide chains, known as Int^N for the N-terminal fragment and Int^C for the C-terminal one. When the Int^N and Int^C come together noncovalently and form a heterodimeric complex, the intact intein active fold is reconstituted and splicing activity is reestablished (Fig. 2). This mechanism has been termed "trans-splicing" because it links two separate peptide segments together [9]. Besides facilitating the purification of expressed proteins using split inteins as traceless affinity tags [10], like EPL, trans-splicing can also be used to generate site-specially modified proteins.

With the discovery of new intein families and the development of intein techniques, the speed [11] and efficiency [12] of split inteins (natural and artificial) have been largely improved [13]. To gain tighter control of the trans-splicing process with environmental stimuli such as light, temperature, pH, or small molecules [14], conditional protein splicing (CPS) has been developed [15]. Such split inteins can be used to generate protein thioesters for use in EPL.

2.3 The EPL Approach

In 1998, our group in collaboration with Tom Muir joined forces to develop EPL to generate site-specifically modified proteins. The fortuitous proximity of our two new groups at Rockefeller University helped to spark the development of this method. The Muir lab was expert in peptide synthesis and chemoselective ligations, and our group brought recombinant protein production and

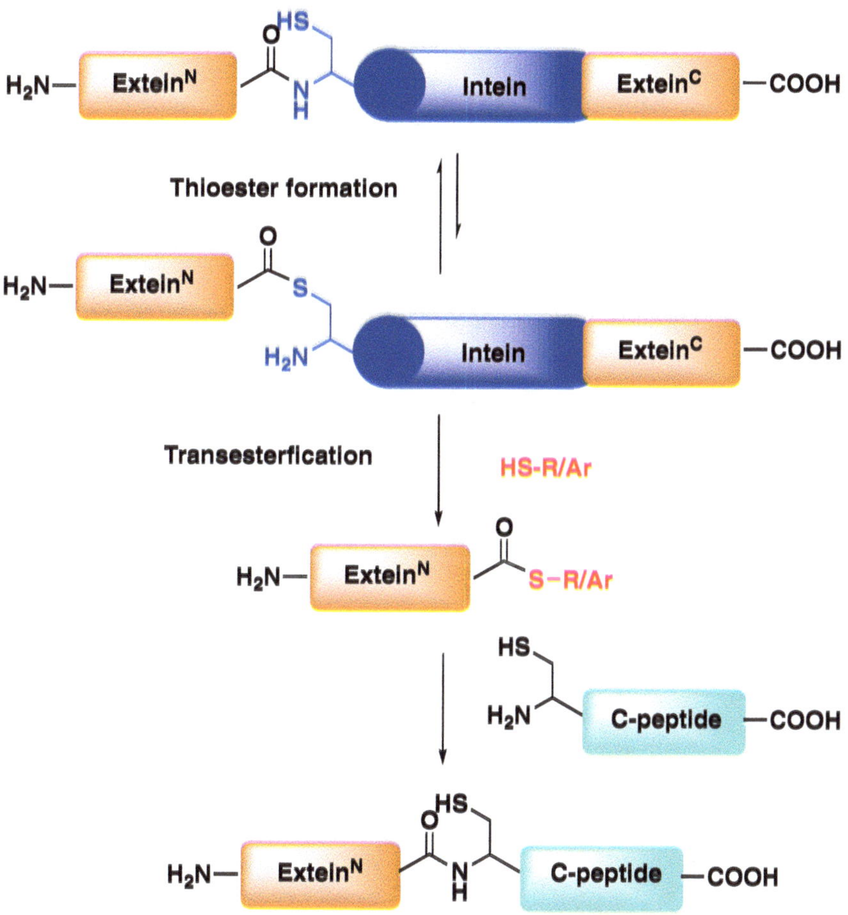

Fig. 3 The expressed protein ligation strategy. Intein cleavage generates the formation of a recombinant protein thioester fragment. This thioester fragment is then mixed with a synthetic N-terminal Cys-containing peptide affording a semisynthetic protein via the native chemical ligation reaction

enzymology skills. We both recognized that thioester peptides were difficult for most labs to access, especially at that time (although this situation has greatly improved [16, 17]), and that inteins offered a potentially attractive solution to their production. Furthermore, it was our thought that EPL would be important for applying native chemical ligation to much larger proteins than those accessible by total synthesis (Fig. 3).

In the first examples of this method, we generated a posttranslationally modified tyrosine kinase to explore the biochemistry of site-specific tyrosine phosphorylation at an unnatural location [7]. Shortly after our effort, Evans and colleagues [18] used EPL to generate a cytotoxic enzyme by producing it via two inert pieces that were ligated to form the full-length active enzyme. Over the years, EPL has been further refined and applied to address many biochemical challenges, particularly in the study of PTMs on proteins [19].

3 EPL for the Preparation of Nonnatural and Modified Proteins

A number of methods have been developed for the installation of modifications in proteins and they are summarized in Table 1. Several efforts have exploited Cys modification strategies to introduce PTM mimics into proteins [20, 25–27]. Overall, although

Table 1
Comparison of different chemical biology approaches for the preparation of posttranslationally modified proteins

Strategy	Example	Description	Advantage(s)	Disadvantage(s)	Reference
Amino acid (Cys) modification	H3K4me	PTM analogs are added to functional amino acid residue (mainly Cys)	Chemically simple, convenient	Bioorthogonal reactions needed; hard to separate side products with minor chemical differences	[20]
Enzyme-catalyzed ligation (e.g., Sortase)	H3K14ac	Enzyme catalyzed ligation of peptides	Rapid reactions, can avoid Cys	Specialized enzymes needed, yields can be variable, restricted to certain amino acid sequences	[21]
Noncanonical amino acid mutagenesis (ncAA)	H3K4succ	An orthogonal tRNA-synthetase pair to suppress the amber stop codon to incorporate a nonnatural amino acid	Similar to normal recombinant expression, accesses the middle of even larger proteins	Generally limited to one unnatural amino acid, tRNA synthetase mutant needed, can be difficult in eukaryotic expression systems	[22]
Protein total synthesis	H2B K34ub	Synthetic peptides are linked together via different ligation strategies	Complete chemical control	Generally for smaller proteins, refolding often required	[23]
EPL	H2B K120ub	A synthetic peptide is linked to a larger piece of expressed protein	Small peptide easier to synthesize; no size limitation for target protein, refolding can be avoided	Mainly focused on C-terminal or N-terminal modifications, reactions can be slow	[24]

3.1 EPL vs. Other Protein Chemical Modification Approaches in the Study of Modified Proteins

Cys modification can be convenient, it generally requires only one Cys to be present and is limited in chemical scope [28]. Noncanonical amino acid incorporation (ncAA) via amber codon suppression is an increasingly popular method to generating modified proteins purely genetically. By simply installing a TAG amber stop codon into the target protein DNA sequence, an orthogonal tRNA synthetase-tRNA pair can recognize it to incorporate the modified amino acids [29]. However, a specific tRNA synthetase-tRNA mutant pair is necessary for each single modification, which may not be suitable for all the nonnatural structures, such as large modifications like ubiquitination [22]. Although a combination of ncAA with bioorthogonal reactions can extend the accessibility [30], proteins with multiple modifications are still hard to generate via this strategy due to reduced production efficiency.

Another means of artificial protein creation is to catalyze peptide ligations between synthetic peptides and expressed protein fragments with specialized enzymes including sortase [21], butelase [31], subtiligase [32], and transglutaminase [33] for peptide ligation, as well as lipoic acid ligase (LplA) [34], Formylglycine generating enzyme (FGE) [35] for site-specific modification. These ligations can be rapid, but the yields are variable, and the enzymes can show particular sequence preferences that may be poorly compatible with particular applications [36].

3.2 Applications of EPL for the Synthesis of Nonnatural Proteins or Modified Proteins

As described above, EPL stands out as a robust and versatile strategy for the generation of proteins with side-specific PTMs. As an example, EPL has been often been used to investigate site-specific protein phosphorylation in cell signaling [37]. An early application [38] of this was in the analysis of SMAD2 and SMAD3 phosphorylation's role in oligomerization [39]. More recently, the influence of C-terminal phosphorylation of PTEN was analyzed [40]. It was found that such C-terminal phosphorylation induced a closed, inactive PTEN conformation with increased stability and resistance to HECT E3 ligase-mediated ubiquitination [41]. In this case, classical EPL which introduced an unnatural Cys residue in place of Tyr379 led to somewhat misleading results. That is, studies on Y379C phospho-PTEN underestimated the degree of conformational closure compared with the protein of native sequence (Fig. 4a) [42]. The use of subtiligase circumvented the Cys requirement and gave a more accurate assessment of phosphorylation impacts [43].

Recent EPL studies on the regulation of heavily investigated protein Ser/Thr kinase Akt helped clarify how it is modulated by C-terminal phosphorylation. Unexpectedly, two distinct mechanisms of activation were uncovered by Ser473 phosphorylation vs. Ser477/Thr479 dual phosphorylation (Fig. 4b) [44].

The tau protein is thought to play a central role in neurodegenerative diseases and is modified by Lys acetylation and other

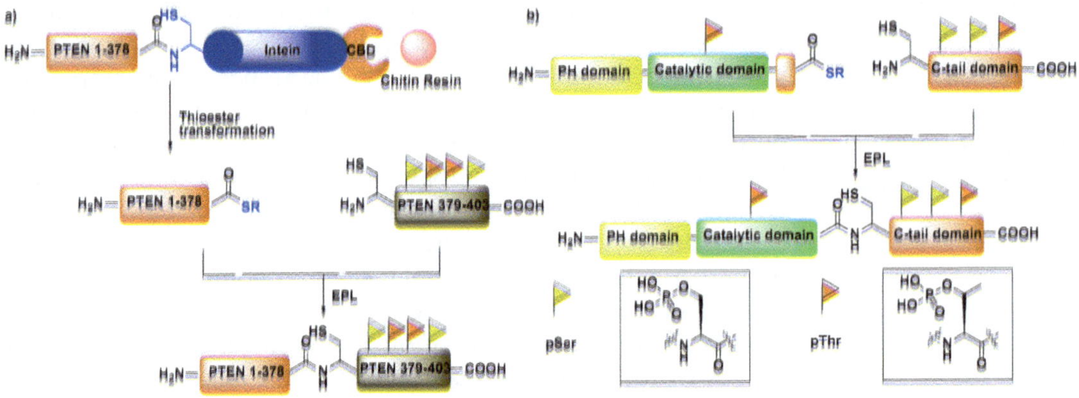

Fig. 4 Scheme for the semisynthesis of phosphorylated PTEN (**a**) or Akt (**b**). (**a**) The PTEN aa1-378 recombinant protein thioester fragment is generated with MESNA and then ligated to N-Cys synthetic peptide (PTEN aa379-403) possessing site-specific phosphoserines and/or phosphothreonines. (**b**) The recombinant Akt PH domain-catalytic domain thioester fragment (aa1-459) is ligated to the N-Cys synthetic peptide C-tail peptide (aa460-480) possessing site-specific phosphoserines or phosphothreonines

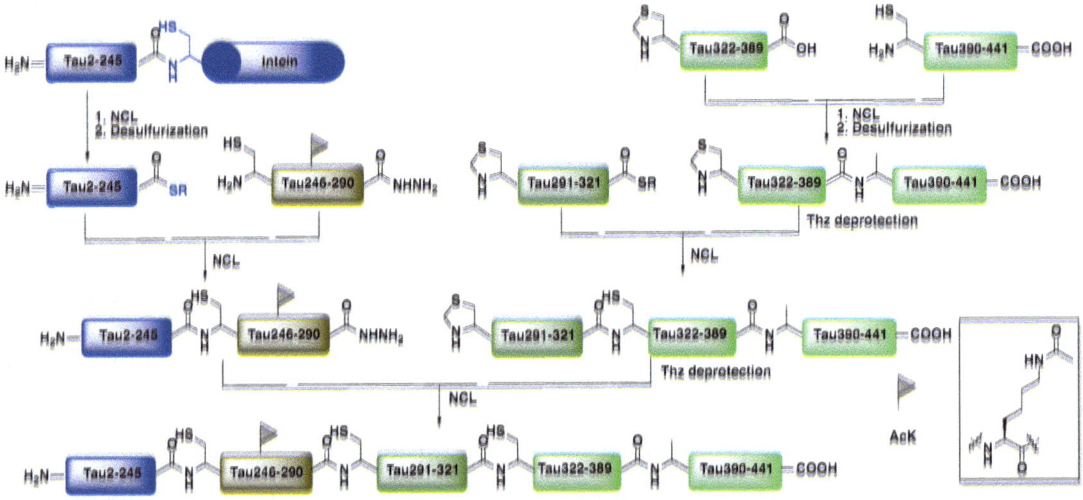

Fig. 5 The semisynthetic preparation of Tau protein with site-specific acetylations. A convergent strategy involving five-fragment stepwise ligations to produce full length Tau aa2-441 was achieved with site-specific acetylation. The longest N-terminal Tau aa2-245 recombinant protein thioester fragment was prepared using an intein

PTMs [45]. Elegant studies have recently been performed using EPL to elucidate the impact of a number of tau PTMs. As a protein with about 450 amino acid residues, a convergent ligation strategy involving five peptide segments was employed to assemble the full-length protein (Fig. 5). With these protein targets at hand, it was discovered that K280ac could strongly stimulate tau aggregation which might have implications in pathophysiology.

The functions of Ser *O*-β-*N*-acetylglucosamine (*O*-GlcNAc) modification on protein serine-threonine kinase casein kinase II (CK2) and on alpha-synuclein have been addressed using EPL [46, 47]. It was found that such glycosylation can affect protein stability and aggregation.

Ypt1 is a small GTPase that is monoprenylated and EPL was used to investigate this PTM [48]. The semisynthetic geranyl-Ypt1 formed a complex with RabGDI and a crystal structure was obtained to elucidate the structure of the complex.

Nucleosomes containing histone H2B with ubiquitination at the K120 site were prepared by EPL [24], and this helped reveal how H2B modification enhances methylation on histones H3 at Lys4 and Lys79 [49]. Very recently, a high resolution cryoEM structure of the Dot1L methyltransferase in complex with the semisynthetic ubiquitin-modified nucleosome (and a norleucine replacement at Lys79 in H3) has revealed how Dot1L recognizes the ubiquitin and nucleosome to achieve specific methylation at Lys79 [50].

3.3 Protein Labeling

One of other important applications of EPL is to label proteins of interest for biophysical studies [51]. Fluorescence labeling is commonly used in FRET and binding studies of proteins. With the help of protein semisynthesis, a large variety of fluorophores have been added to protein targets. An early and impactful example involved incorporating into an SH3 domain the fluorescent probe 7-azatryptophan, which has distinct noninvasive fluorescence features for biophysical experiments [52]. A recent variant of the EPL approach has involved exploiting commercial NHS esters that are pretreated with MESNA to form thioesters that can be selectively linked to the N-termini of proteins. This approach avoids peptide synthesis and can be used by labs that wish to minimize chemical manipulation. In the inaugural example, the NHS ester strategy was employed to label WWP2 E3 ubiquitin ligase for studies on binding to its substrate PTEN and to clarify the stoichiometry of autoubiquitination [53].

In addition to incorporating fluorescence tags, isotopic labeling is a widely used strategy for the biophysical analysis of proteins of interest. The site-specific installation of NMR-sensitive isotopes, such as ^{13}C and ^{15}N, into target protein allows the direct assignment of resonances for structural determination [54]. EPL can be used for the ligation of isotopically labeled polypeptide fragment to another fragment with natural abundance isotopes. Known as "segmental isotopic labeling" or "block-labeling," these methods can generate a sample for heteronuclear NMR analysis that possesses reduced spectral complexity relative to a fully labeled protein [55–57].

3.4 Other Applications of EPL

Besides the labeling of EPL for biophysical studies, EPL and semisynthesis have also been applied to protein microarrays in a high-density manner for proteomics experiments. Protein–small molecule conjugates with an eye toward therapeutic applications [58], such as antibody–small molecule conjugates [12], have also been enabled by EPL [59]. EPL has been used to facilitate the expression of some otherwise-hard-to-achieve peptides or proteins [60]. EPL has also been applied to the production of cyclic peptides, toxic proteins, and proteins with uniform N-termini [61].

4 Summary and Future Challenges

Protein semisynthesis by expressed protein ligation and related methods have been slowly transforming the study of protein science. Continued refinements in the methods including the use of trans splicing, enzymes for ligation, and alternative chemoselective reactions [62] allow for enhanced applications. Nevertheless, continuing challenges in accessing the middle of large proteins by semisynthesis and generating semisynthetic proteins in vivo motivate current and future efforts in the field. We remain optimistic that investigators in the molecular biology and chemical biology communities will rise to these challenges and produce innovations that will increase the scope and influence of this technique in the coming years.

Acknowledgments

We thank the NIH (GM62437 and CA74305) and the Leukemia and Lymphoma Society for financial support. We are grateful to Cole lab members, past and present, for their intellectual and experimental efforts that contributed to some of the work described here.

References

1. Kent SBH (2019) Novel protein science enabled by total chemical synthesis. Protein Sci 28:313–328. https://doi.org/10.1002/pro.3533
2. Fersht AR, Winter GP (1985) Redesigning enzymes by site-directed mutagenesis. Ciba Found Symp 111:204–218
3. Xiao H, Schultz PG (2016) At the Interface of chemical and biological synthesis : an expanded genetic code. Cold Spring Harb Perspect Biol 8:a023945
4. Dawson PE, Muir TW, Clark-Lewis I, Kent SBH (1994) Synthesis of proteins by native chemical ligation. Science 266:776
5. Dawson PE, Kent SBH (2000) Synthesis of native proteins by chemical ligation. Annu Rev Biochem 69:923–960
6. Wieland T, Bokelmann E, Bauer L et al (1953) Über Peptidsynthesen. 8. Mitteilung Bildung von S-haltigen Peptiden durch intramolekulare Wanderung von Aminoacylresten. Justus Liebigs Ann Chem 583:129–149

7. Muir TW, Sondhi D, Cole PA (1998) Expressed protein ligation: a general method for protein engineering. Proc Natl Acad Sci U S A 95:6705–6710. https://doi.org/10.1073/pnas.95.12.6705
8. Perler FB, Davis EO, Dean GE et al (1994) Protein splicing elements: inteins and exteins- a definition of terms and recommended nomenclature. Nucleic Acids Res 22:1125–1127
9. Mills KV, Lew BM, Jiang S -q, Paulus H (1998) Protein splicing in trans by purified N- and C-terminal fragments of the Mycobacterium tuberculosis RecA intein. Proc Natl Acad Sci U S A 95:3543–3548. https://doi.org/10.1073/pnas.95.7.3543
10. Vila-Perelló M, Liu Z, Shah NH et al (2013) Streamlined expressed protein ligation using split inteins. J Am Chem Soc 135:286–292. https://doi.org/10.1021/ja309126m
11. David Y, Vila-Perelló M, Verma S, Muir TW (2015) Chemical tagging and customizing of cellular chromatin states using ultrafast trans-splicing inteins. Nat Chem 7:394–402. https://doi.org/10.1038/nchem.2224
12. Stevens AJ, Brown ZZ, Shah NH et al (2016) Design of a split intein with exceptional protein splicing activity. J Am Chem Soc 138:2162–2165. https://doi.org/10.1021/jacs.5b13528
13. Stevens AJ, Sekar G, Mostafavi AZ et al (2017) A promiscuous split intein with expanded protein engineering applications. Proc Natl Acad Sci U S A 114:8538–8543. https://doi.org/10.1073/pnas.1701083114
14. Mootz HD, Muir TW (2002) Protein splicing triggered by a small molecule. J Am Chem Soc 124:9044–9045. https://doi.org/10.1021/ja026769o
15. Di Ventura B, Mootz HD (2018) Switchable inteins for conditional protein splicing. Biol Chem 400:467–475. https://doi.org/10.1515/hsz-2018-0309
16. Blanco-Canosa JB, Dawson PE (2008) An efficient Fmoc-SPPS approach for the generation of thioester peptide precursors for use in native chemical ligation. Angew Chem Int Ed Engl 47:6851–6855. https://doi.org/10.1002/anie.200705471
17. Zheng JS, Tang S, Huang YC, Liu L (2013) Development of new thioester equivalents for protein chemical synthesis. Acc Chem Res 46:2475–2484. https://doi.org/10.1021/ar400012w
18. Evans TC, Benner J, Xu MQ (1998) Semisynthesis of cytotoxic proteins using a modified protein splicing element. Protein Sci 7:2256–2264. https://doi.org/10.1002/pro.5560071103
19. Hackenberger CPR, Schwarzer D (2008) Chemoselective ligation and modification strategies for peptides and proteins. Angew Chem Int Ed Engl 47:10030–10074. https://doi.org/10.1002/anie.200801313
20. Simon MD, Chu F, Racki LR et al (2007) The site-specific installation of methyl-lysine analogs into recombinant histones. Cell 128 (5):12–1003. https://doi.org/10.1016/j.cell.2006.12.041
21. Wu M, Hayward D, Kalin JH et al (2018) Lysine-14 acetylation of histone H3 in chromatin confers resistance to the deacetylase and demethylase activities of an epigenetic silencing complex. Elife 7:e37231. https://doi.org/10.7554/eLife.37231
22. Wang ZA, Kurra Y, Wang X et al (2017) A versatile approach for site-specific lysine acylation in proteins. Angew Chem Int Ed Engl 56:1643–1647. https://doi.org/10.1002/anie.201611415
23. Siman P, Karthikeyan SV, Nikolov M et al (2013) Convergent chemical synthesis of histone H2B protein for the site-specific ubiquitination at Lys34. Angew Chemie Int Ed 52:8059–8063. https://doi.org/10.1002/anie.201303844
24. McGinty RK, Köhn M, Chatterjee C et al (2009) Structure-activity analysis of semisynthetic nucleosomes: mechanistic insights into the stimulation of Dot1L by ubiquitylated histone H2B. ACS Chem Biol 4:958–968. https://doi.org/10.1021/cb9002255
25. Huang R, Holbert MA, Tarrant MK et al (2010) Site-specific introduction of an acetyl-lysine mimic into peptides and proteins by cysteine alkylation. J Am Chem Soc 132:9986–9987. https://doi.org/10.1021/ja103954u
26. Spicer CD, Davis BG (2014) Selective chemical protein modification. Nat Commun 5:4740. https://doi.org/10.1038/ncomms5740
27. Wang ZA, Liu WR (2017) Proteins with site-specific lysine methylation. Chemistry 23 (49):11732–11737. https://doi.org/10.1002/chem.201701655
28. Bhat S, Hwang Y, Gibson MD et al (2018) Hydrazide mimics for protein lysine acylation to assess nucleosome dynamics and Deubiquitinase action. J Am Chem Soc 140:9478–9485. https://doi.org/10.1021/jacs.8b03572
29. Dai S, Tharp JM, Zeng Y et al (2016) A genetically encoded allysine for the synthesis of proteins with site-specific lysine dimethylation.

Angew Chemie Int Ed 56:212–216. https://doi.org/10.1002/anie.201609452

30. Chalker JM, Bernardes GJL, Davis BG (2011) A "tag-and-modify" approach to site-selective protein modification. Acc Chem Res 44 (9):730–741. https://doi.org/10.1021/ar200056q
31. Nguyen GKT, Qiu Y, Cao Y et al (2016) Butelase-mediated cyclization and ligation of peptides and proteins. Nat Protoc 11:1977–1988. https://doi.org/10.1038/nprot.2016.118
32. Henager SH, Chu N, Chen Z et al (2016) Enzyme-catalyzed expressed protein ligation. Nat Methods 13:925–927. https://doi.org/10.1038/nmeth.4004
33. Rachel NM, Toulouse JL, Pelletier JN (2017) Transglutaminase-catalyzed bioconjugation using one-pot metal-free bioorthogonal chemistry. Bioconjug Chem 28:2518–2523. https://doi.org/10.1021/acs.bioconjchem.7b00509
34. Uttamapinant C, White KA, Baruah H et al (2010) A fluorophore ligase for site-specific protein labeling inside living cells. Proc Natl Acad Sci U S A 107:10914–10919. https://doi.org/10.1073/pnas.0914067107
35. Rush JS, Bertozzi CR (2008) New aldehyde tag sequences identified by screening formylglycine generating enzymes in vitro and in vivo. J Am Chem Soc 130:12240–12241
36. Fottner M, Brunner A-D, Bittl V et al (2019) Site-specific ubiquitylation and SUMOylation using genetic-code expansion and sortase. Nat Chem Biol 15:276–284. https://doi.org/10.1038/s41589-019-0227-4
37. Chen Z, Cole PA (2015) Synthetic approaches to protein phosphorylation. Curr Opin Chem Biol 28:115–122. https://doi.org/10.1016/j.cbpa.2015.07.001
38. Wu J-W, Hu M, Chai J et al (2001) Crystal structure of a phosphorylated Smad2: recognition of Phosphoserine by the MH2 domain and insights on Smad function in TGF-signaling involved in a ligand-specific signaling pathway; and the inhibitory Smads (I-Smads), including Smad6 and-7, which ne. Mol Cell 8:1277–1289
39. Qin BY, Lam SS, Correia JJ, Lin K (2002) Smad3 allostery links TGF-β receptor kinase activation to transcriptional control. Genes Dev 16:1950–1963. https://doi.org/10.1101/gad.1002002
40. Bolduc D, Rahdar M, Tu-Sekine B et al (2013) Phosphorylation-mediated PTEN conformational closure and deactivation revealed with protein semisynthesis. Elife 2013:1–19. https://doi.org/10.7554/eLife.00691
41. Chen Z, Thomas SN, Bolduc DM et al (2016) Enzymatic analysis of PTEN Ubiquitylation by WWP2 and NEDD4-1 E3 ligases. Biochemistry 55:3658–3666. https://doi.org/10.1021/acs.biochem.6b00448
42. Chen Z, Dempsey DR, Thomas SN et al (2016) Molecular features of phosphatase and tensin homolog (PTEN) regulation by C-terminal phosphorylation. J Biol Chem 291:14160–14169. https://doi.org/10.1074/jbc.M116.728980
43. Dempsey DR, Cole PA (2018) Protein chemical approaches to understanding PTEN lipid phosphatase regulation. Methods Enzymol 607:405–422. https://doi.org/10.1016/bs.mie.2018.05.007
44. Chu N, Salguero AL, Liu AZ et al (2018) Akt kinase activation mechanisms revealed using protein Semisynthesis. Cell 174:897–907.e14. https://doi.org/10.1016/j.cell.2018.07.003
45. Haj-Yahya M, Lashuel HA (2018) Protein Semisynthesis provides access to tau disease-associated post-translational modifications (PTMs) and paves the way to deciphering the tau PTM code in health and diseased states. J Am Chem Soc 140:6611–6621. https://doi.org/10.1021/jacs.8b02668
46. Tarrant MK, Rho H-S, Xie Z et al (2012) Regulation of CK2 by phosphorylation and O-GlcNAcylation revealed by semisynthesis. Nat Chem Biol 8:262–269. https://doi.org/10.1038/nchembio.771
47. Levinea PM, Galesica A, Balanaa AT et al (2019) α-Synuclein O-GlcNAcylation alters aggregation and toxicity, revealing certain residues as potential inhibitors of Parkinson's disease. Proc Natl Acad Sci U S A 116:1511–1519. https://doi.org/10.1073/pnas.1808845116
48. Rak A, Pylypenko O, Durek T et al (2003) Structure of Rab GDP-dissociation inhibitor in complex with Prenylated YPT1 GTPase. Science 302:646–650
49. Kim J, Guermah M, McGinty RK et al (2009) RAD6-mediated transcription-coupled H2B Ubiquitylation directly stimulates H3K4 methylation in human cells. Cell 137:459–471. https://doi.org/10.1016/j.cell.2009.02.027
50. Worden EJ, Hoffmann NA, Hicks CW, Wolberger C (2019) Mechanism of cross-talk between H2B Ubiquitination and H3 methylation by Dot1L. Cell 176:1490–1501.e12. https://doi.org/10.1016/j.cell.2019.02.002

51. Muralidharan V, Muir TW (2006) Protein ligation: an enabling technology for the biophysical analysis of proteins. Nat Methods 3:429–438. https://doi.org/10.1038/nmeth886
52. Muralidharan V, Cho J, Trester-Zedlitz M et al (2004) Domain-specific incorporation of noninvasive optical probes into recombinant proteins. J Am Chem Soc 126:14004–14012. https://doi.org/10.1021/ja0466199
53. Dempsey DR, Jiang H, Kalin JH et al (2018) Site-specific protein labeling with N-Hydroxysuccinimide-esters and the analysis of ubiquitin ligase mechanisms. J Am Chem Soc 140:9374–9378. https://doi.org/10.1021/jacs.8b05098
54. Romanelli A, Shekhtman A, Cowburn D, Muir TW (2004) Semisynthesis of a segmental isotopically labeled protein splicing precursor: NMR evidence for an unusual peptide bond at the N-extein-intein junction. Proc Natl Acad Sci U S A 101:6397–6402. https://doi.org/10.1073/pnas.0306616101
55. Uegaki K, Oda N, Nakamura H et al (1998) Segmental isotope labeling for protein NMR using peptide splicing. J Am Chem Soc 120:5591–5592. https://doi.org/10.1021/ja980776o
56. Nabeshima Y, Mizuguchi M, Kajiyama A, Okazawa H (2014) Segmental isotope-labeling of the intrinsically disordered protein PQBP1. FEBS Lett 588:4583–4589. https://doi.org/10.1016/j.febslet.2014.10.028
57. Flynn JD, Jiang Z, Lee JC (2018) Segmental 13C-labeling and Raman microspectroscopy of α-Synuclein amyloid formation. Angew Chem Int Ed Engl 57:17069–17072. https://doi.org/10.1002/anie.201809865
58. Chiang MJ, Holbert MA, Kalin JH et al (2014) An Fc domain protein-small molecule conjugate as an enhanced immunomodulator. J Am Chem Soc 136:3370–3373. https://doi.org/10.1021/ja5006674
59. Flavell RR, Muir TW (2009) Expressed protein ligation (EPL) in the study of signal transduction, ion conduction, and chromatin biology. Acc Chem Res 42:107–116. https://doi.org/10.1002/chin.200918258
60. Leibly DJ, Nguyen TN, Kao LT et al (2012) Stabilizing additives added during cell Lysis aid in the solubilization of recombinant proteins. PLoS One 7:e52482. https://doi.org/10.1371/journal.pone.0052482
61. Topilina NI, Mills KV (2014) Recent advances in in vivo applications of intein-mediated protein splicing. Mob DNA 5:1–14. https://doi.org/10.1186/1759-8753-5-5
62. Bode JW (2017) Chemical protein synthesis with the α-Ketoacid-hydroxylamine ligation. Acc Chem Res 50:2104–2115. https://doi.org/10.1021/acs.accounts.7b00277

Chapter 2

Protein Splicing: From the Foundations to the Development of Biotechnological Applications

Alejandro Romero-Casañas, Verónica Gordo, Jessica Castro, and Marc Ribó

Abstract

Expressed protein ligation is a simple and powerful method in protein engineering to introduce sequences of unnatural amino acids, posttranslational modifications, and biophysical probes into proteins of any size. This methodology has been developed based on the knowledge obtained from protein splicing. Protein splicing is a multistep biochemical reaction that includes the concomitant cleavage and formation of peptide bonds carried out by self-processing domains named inteins. The natural substrates of protein splicing are essential proteins found in intein-containing organisms; inteins are also functional in nonnative frameworks and can be used to alter nearly any protein's primary amino acid sequence. Accordingly, different reactivity features of inteins have been largely exploited to manipulate proteins in countless methods encompassing fields from biochemical research to the development of biotechnological applications including the study of disease progression and validation of potential drug candidates. Here, we review almost three decades of research to uncover the chemical and biochemical enigmas of protein splicing and the development of inteins as potent protein engineering tools.

Key words Intein, Extein, Split intein, *Trans*-splicing, Expressed protein ligation, Conditional protein splicing

1 Introduction

Protein splicing is a posttranslational autoprocessing event that involves the precise removal of an internal protein sequence, named intein, from a precursor protein with the concurrent ligation through a peptide bond of the flanking polypeptide sequences, which are termed exteins. This rearrangement occurs posttranslationally, as intein genes are found embedded in frame within other protein-coding genes. Additionally, intein-mediated protein splicing is spontaneous; it requires no external factor or energy source, only the correct folding of the intein domain. The first intein was discovered in 1990, as a spacer within the *Saccharomyces cerevisiae* vacuolar ATPase gene [1, 2]. Since then, based on several highly

Miquel Vila-Perelló (ed.), *Expressed Protein Ligation: Methods and Protocols*, Methods in Molecular Biology, vol. 2133, https://doi.org/10.1007/978-1-0716-0434-2_2, © Springer Science+Business Media, LLC, part of Springer Nature 2020

conserved sequence motifs, more than 600 members of this protein domain family have been identified—many of them confirmed experimentally—from all three branches of life (archaea, bacteria, and eukaryota). Inteins have been described in unicellular organisms as well as in some viral genomes [3, 4]. Oddly, multicellular organisms have orthologous hedgehog-like proteins that contain autoprocessing domains that are structurally and mechanistically similar to intein splicing domains [5–7]. Inteins are typically embedded within essential proteins involved in DNA replication, transcription, and maintenance (e.g., DNA or RNA polymerase subunits, DNA helicases, DNA gyrases, and ribonucleotide reductase) or in other housekeeping genes including essential proteases and metabolic enzymes. Protein splicing is presumed to have an ancient evolutionary origin [8] and it has been suggested that inteins were once involved in enzyme evolution by peptide or domain shuffling [9, 10]. Beyond this possibility, protein splicing does not appear to be regulated in vivo, a finding that has so far frustrated attempts to find a defined biological role for the process in the host organism. Interestingly, several inteins contain an internal homing endonuclease domain (Fig. 1a) that, while not required for splicing activity, does confer genetic mobility to the gen [11]. This finding has led to suggest that inteins could be considered "selfish genes" and, moreover, that the number of inteins per organism is related to the ability to take up and integrate foreign DNA (either by natural competency or infection) [10, 12]. Thus far, there are at least 70 different intein alleles, differing not only by the sort of host gene in which the inteins are inserted, but also the integration point within that host gene. Likewise, some proteins have been found to contain as many as four inteins embedded at different integration points within their gene, and several organisms have more than one intein in their genome; standing out among them is *Methanocaldococcus jannaschii* with as many as 19 inteins.

1.1 Intein Domain Organization

According to their structural organization there are three broad categories of inteins: (1) *inteins* that contain a homing endonuclease domain inserted between the N- and C-terminal splicing regions in the intein primary sequence (Fig. 1a) [13, 14]; (2) *mini-inteins* comprising both the N- and C-terminal splicing regions but lacking a homing endonuclease domain and consequently smaller in size (Fig. 1b); and (3) *trans-splicing inteins* that lack a peptide linkage between the N- and C-terminal modules of the splicing domain (Fig. 1c).

In split inteins or *trans*-splicing inteins the intein domain is split into two fragments, the N-intein (IntN) and C-intein (IntC), encoded in two different genes, which have to associate to reconstitute the active intein to splice the surrounding extein fragments (N-extein and C-extein). The largest known class of split inteins is

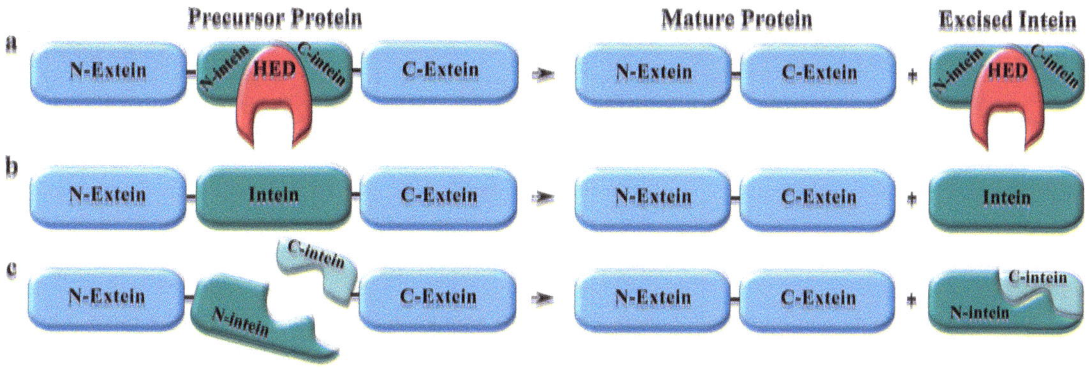

Fig. 1 Domain organization of inteins. The three different structural types of inteins are shown schematically in the context of the N- and C-exteins. (**a**) Inteins that contain a homing endonuclease domain (HED) inserted into the intein sequence. (**b**) Mini-inteins that comprise the N- and C-terminal splicing regions but lack a homing endonuclease domain. (**c**) *Trans*-splicing inteins are transcribed separately, each one coding for the N- and C-terminal modules of the splicing domain, respectively

the cyanobacterial DnaE family [15] constituted by 23 members registered in the In Base [3], which in the natural context, assemble the catalytic subunit of DNA polymerase III (DnaE). More recently, new *trans*-splicing family members have been identified from genomic databases and experimentally characterized, some of them showing the fastest splicing rates and efficiencies described so far [16–19].

1.2 Protein Splicing Mechanism

In spite of the low sequence homology between nonallelic inteins they share a similar biochemical mechanism for the protein splicing process, in which key residues are located in several conserved motifs, or Blocks, in the primary amino acid sequence of the polypeptide precursor (Fig. 2). Three variations of the multistep splicing process have been described so far which allow the mechanistic classification of inteins into three different classes [20].

1.3 Standard, or Class 1

Standard, or Class 1, mechanism of inteins was elucidated early on [21] and consists on a series of nucleophile-driven bond rearrangements and cleavage reactions (Fig. 3). The first step involves an N-to-O/S acyl shift in which the N-extein region is transferred to the side-chain sulfhydryl or hydroxyl group of a cysteine or serine residue in Block A (Cys1 or Ser1, where 1 stands for the first residue of the intein). Although threonine mutants were shown to splice, Thr1 has not been observed in nature [9]. This N to O/S acyl-shift is facilitated by a conserved threonine and histidine motif found in Block B (the TXXH motif), along with a commonly found aspartate in Block F, and results in the formation of a linear (thio)ester intermediate. In the next step, the entire N-extein region is transferred to a second conserved residue—cysteine, serine, or threonine—at the intein-C–extein boundary (+1 position) in a *trans*-

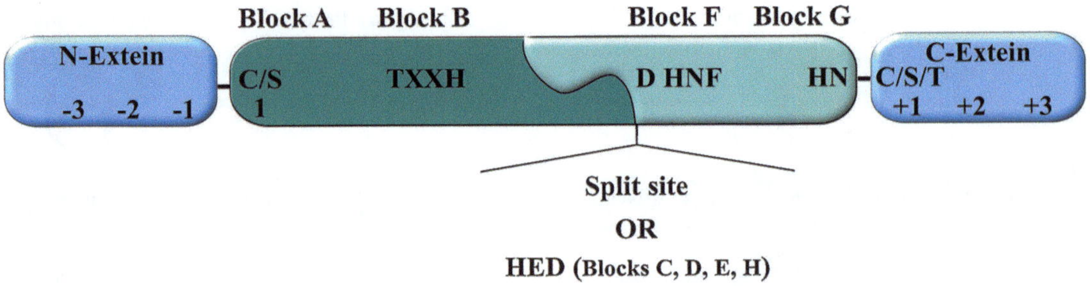

Fig. 2 Conserved motifs and intein numbering. Conserved sequence Blocks highlighting the most relevant residues for protein splicing. Blocks C, D, E, and H are found in inteins containing the homing endonucleases domain (HED)

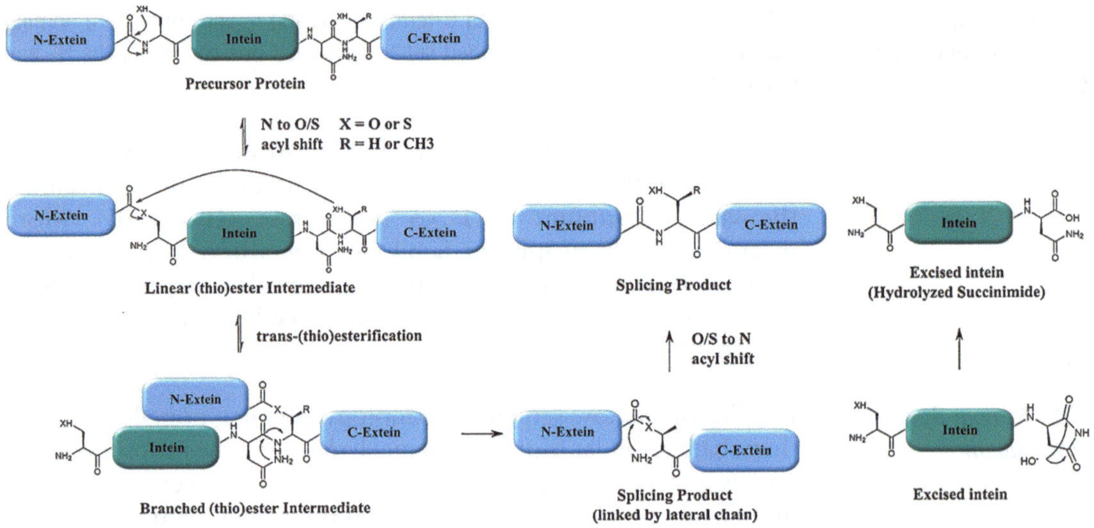

Fig. 3 Mechanism of protein splicing. The canonical mechanism of protein splicing of Class 1 inteins

(thio)esterification step. The resulting branched intermediate is then resolved through a cyclization reaction involving a conserved asparagine residue at the C-terminus of the intein. This process is facilitated by the Block F histidine and the penultimate Block G histidine, and may also involve the Block F aspartate [22, 23]. The intein is thus excised from the exteins as a C-terminal succinimide derivative. In the final step, an amide bond is formed between the two extein regions following an S/O to N acyl shift. The final step is known to be a thermodynamically favored chemical reaction and presumably does not require the structured intein [24].

Within inteins that use Class 1 mechanism of splicing some deviations on the canonical mechanism have been described, due to slight sequence differences in certain Blocks. A few inteins have been found in which the Block G asparagine has been substituted by a glutamine or aspartate residue. The main consequences of these changes affect the cyclization step. Whereas glutamine cyclization would proceed through a less stable six-membered, rather than the canonic five-membered cyclic intermediate, aspartate cyclization would render a succinic anhydride instead of a succinimide. Other observed variations are the lack of Block B or Block G histidines, which are crucial for the resolution of the branched intermediate. Different substitutions have been observed in these positions although, currently, it is not clear how these inteins bypass the need for the two histidines and if other proximal residues may act as alternate general acids or bases along splicing. Among split inteins, all belonging to Class 1, the most frequent 1/+1 catalytic residue (Fig. 2) combination is Cys1/Cys+1, followed by Cys1/Ser+1 found in several inteins identified in metagenomics databases [16, 17]. There are two known exceptions to these combinations among split inteins: the exceptional Ser1/Thr +1 found in the Neq pol intein, which employs only oxyester chemistry [25, 26], and the Ser1/Cys+1 from the GOS-TerL intein [19], which is the only intein described so far with this combination.

1.4 Class 2

Class 2 inteins [27] lack a nucleophilic Cys or Ser Block A residue at position 1. Class 2 inteins lacking a Block A nucleophile commonly have an alanine or proline at their N-terminal position and cannot initiate splicing through the formation of a linear (thio)ester intermediate. The attack of the amide −1 bond is prompted by the thiolate group of the Cys+1 of the C-extein in Block F, and consequently both, the N to O/S acyl shift and the *trans*-(thio)esterification steps, of the Class 1 mechanism, are circumvented.

1.5 Class 3

Class 3 inteins also lack the nucleophilic 1 residue in Block A found in Class 1 inteins but possess an internal tryptophan–cysteine–threonine triad, of which the cysteine residue plays the role of attacking the N-terminal splice junction [20].

Two recurrent side reactions have been observed during experimental analysis of protein splicing reactions [13, 28]. The first, N-extein cleavage or N-terminal cleavage can occur when an external nucleophile, typically water or a thiol, attacks the linear or branched (thio)ester intermediates to yield a free N-extein (Fig. 4a). The second, C-extein cleavage or C-terminal cleavage can occur from the precursor protein (Fig. 4b) or the N-extein cleaved protein, when the C-terminal asparagine cyclizes in the absence of the branched intermediate structure, freeing the C-extein. Both these reactions can be promoted by mutation of critical intein residues and have been exploited for several technological purposes, among which it should be highlighted the expressed protein ligation. This methodology benefits from these side-reactions to generate the recombinant component to generate semisynthetic proteins [29, 30].

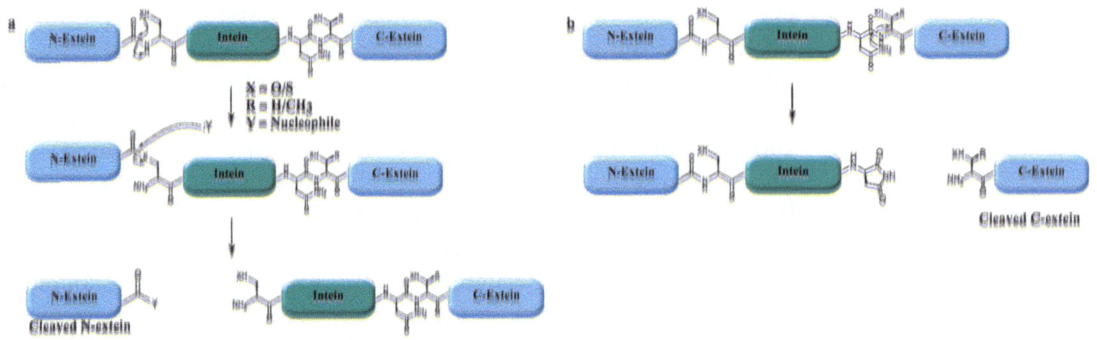

Fig. 4 Side reactions of protein splicing. (**a**) N-terminal cleavage from the linear intermediate. (**b**) C-terminal cleavage from the precursor protein

2 Applications of Protein Splicing

In their native context, inteins assist in both the cleavage and formation of peptide bonds between endogenous exteins. However, early studies on inteins demonstrated that they could also carry out protein splicing in exogenous contexts, often between polypeptides unrelated to the endogenous host protein [21]. Indeed, intein chemistry is currently exploited in a variety of powerful technological applications centered on the cleavage and/or formation of peptide bonds using nonnative exteins.

2.1 Protein Purification

The introduction of specific point mutations in the conserved splicing motifs can promote some of the side reactions of protein splicing, N- and C-extein cleavage and therefore facilitate the recovery of exogenous exteins recombinantly fused to the intein domain. For instance, mutation of the Block A nucleophile from cysteine/serine to alanine impedes the formation of the linear and branched (thio)ester intermediates. Alternatively, mutation of the Block G asparagine to alanine precludes the irreversible branch resolution step, resulting in a trapped equilibrium between the precursor amide and the two (thio)ester intermediates. These mutant inteins can be used to obtain recombinant proteins without the need to include protease cleavage sites between the protein of interest and purification tags typically used for recombinant protein expression. Precisely, a protein of interest can be fused to the N- or C-terminus of an intein-affinity purification tag pair bearing the appropriate mutations. After affinity enrichment on a solid support, chemical reagents or raising the pH can be used to induce N- or C-terminal cleavage, resulting in the release of an untagged protein (Fig. 5).

2.2 Conditional Protein Splicing

Controlling the function of proteins in vivo is of chief importance in drug development and small molecules are commonly used to directly inhibit a target protein or indirectly affect protein's function by modulating its expression. Inteins offer a unique

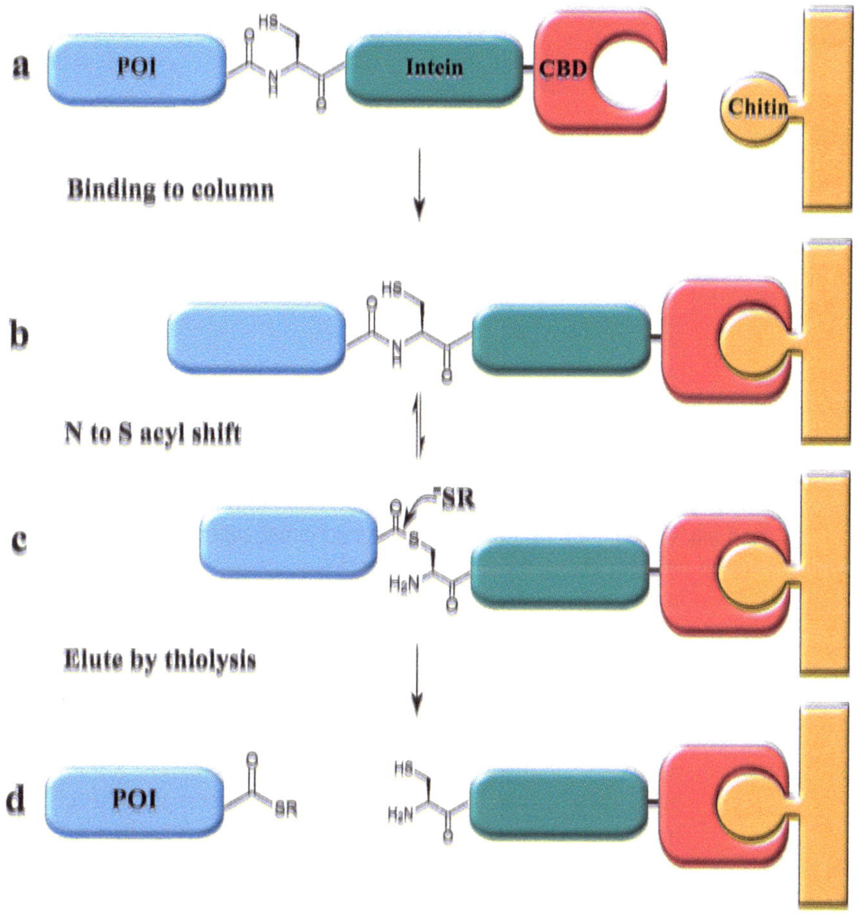

Fig. 5 Protein purification using an intein. (**a**) A protein of interest (POI) is fused to an intein and an affinity tag such as a chitin binding domain (CBD). (**b**) After the expression, the fusion protein is loaded onto an affinity column (**c**) where the N-to-S acyl shift occurs. (**d**) Protein is eluted by thiolysis

opportunity to develop strategies to control the function of a protein by controlling their primary sequence. A series of applications have been developed to accomplish this goal, which are based on controlling intein function by an external stimulus, called conditional protein splicing (CPS) (Fig. 6). In contrast to direct inhibition by small molecules, which requires a specific inhibitor for each target protein, conditional protein splicing is a wide-ranging solution that can be extended to several systems using the same conditional intein. CPS provides a way to control, in a tunable manner, the activation of any target protein in real-time. The examples of conditional protein splicing systems, either using natural or artificial split trans-inteins, are very diverse, as are the factors used to activate splicing: temperature, proteases, chemicals, redox, or light. Needless to mention that a successful design for controlling protein function requires engineering tightly controlled inteins [31, 32].

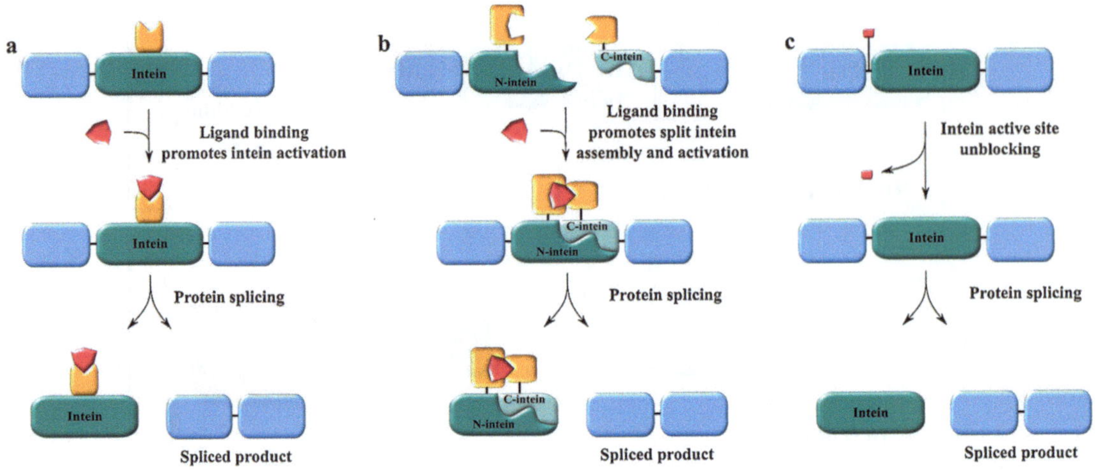

Fig. 6 Conditional protein splicing. (**a**) Intein activation through conformational change induced by ligand binding to a fused ligand binding domain. (**b**) Chemically induced dimerization and activation of a split intein. (**c**) Intein activation through deprotection of an active site residue

2.3 Segmental Isotopic Labeling

One of the applications of *trans*-splicing is the segmental isotopic labeling of proteins for Nuclear Magnetic Resonance (NMR) spectroscopic studies, to characterize the structure and dynamics of proteins (Fig. 7). Structural characterization of proteins by NMR is greatly facilitated by the stable enrichment of NMR-active ^{15}N and/or ^{13}C nuclei in the protein of interest. These heavy isotopes have very low abundance in natural sources but are the ones that provide the signals measured in NMR experiments. This limitation can be easily overcome by growing bacterial cells in an isotopically enriched medium. Nevertheless, for large proteins, significant peak overlap in uniformly labeled samples makes unambiguous interpretation of the spectra challenging. Protein *trans*-splicing allows for protein fragments with different isotope labeling arrangements to be ligated and generate segmentally labeled full-length protein samples providing simplified NMR spectra. Using protein *trans*-splicing for segmental protein labeling is advantageous as the reaction can be performed in vitro or in vivo, through the sequential expression of each fragment in the same cell culture with an intermediate exchange of the medium to include or exclude stable isotope sources. Furthermore, the use of a pair of orthogonal split inteins allows the segmentally labeling of internal segments of domains that can be structurally characterized by NMR techniques in the context of the entire protein [33].

2.4 Expressed Protein Ligation

N-extein cleavage from a mutant intein can be induced not only by hydrolysis but also through attack by nucleophiles such as small-molecule alkyl or aryl thiols rendering a protein of interest bearing a C-terminal α-thioester, rather than a free carboxylic acid. In native chemical ligation (NCL) reaction, a polypeptide with a C-terminal

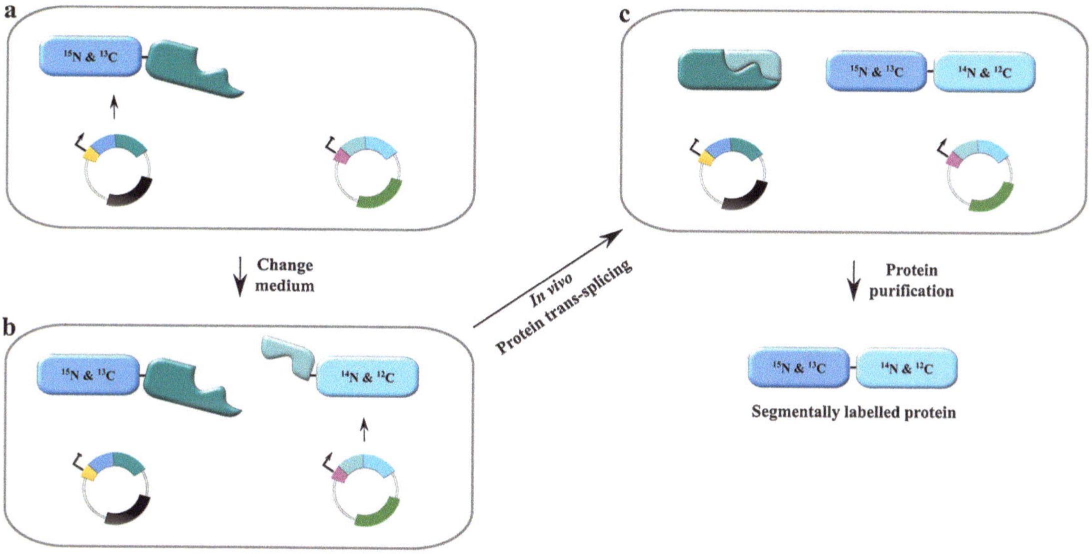

Fig. 7 In vivo isotopic segmental labeling of proteins using split inteins. (**a**) Induction of the first precursor expression from the yellow promoter in the labeled medium by inducer 1. (**b**) Replacement and washing of the cell culture with unlabeled medium, which turn off the expression of the first precursor and induction of the second precursor expression from promoter in pink with inducer 2 in unlabeled medium. (**c**) After the expression of the second precursor protein *trans*-splicing occurs in vivo. Both vectors have different origins of replication and antibiotic selection markers

α-thioester is condensed with a polypeptide containing an N-terminal cysteine to form a native peptide bond [34]. While NCL is appropriate to chemically synthesize small proteins through the ligation of several synthetic peptides, the use of inteins to generate protein thioesters recombinantly, allows to apply this reaction to significantly bigger macromolecules [30]. This semisynthetic approach of NCL, referred to as expressed protein ligation (EPL) (Fig. 8), has been used to site-specifically incorporate a wide array of chemical moieties into proteins that are typically inaccessible through purely recombinant methods. These include homogeneous post-translational modifications, unnatural amino acids, backbone modifications, photochemical cross-linkers, biophysical probes, or imaging probes [32]. Finally, it is remarkable that while intein-mediated EPL traditionally involves the production of the thioester fragment recombinantly, this fragment can also be generated synthetically, and the N-terminal cysteine-containing protein can be made recombinantly, exposing the cysteine either using proteolysis or an intein tag capable of C-extein cleavage. NCL and EPL have improved to a significant degree the ability to generate proteins with a semisynthetic backbone, in which the composition of the synthetic fragment is, in principle, only limited by chemical synthesis. However, NCL and EPL also face some limitations, some of which relate to the preparation of the reagents and

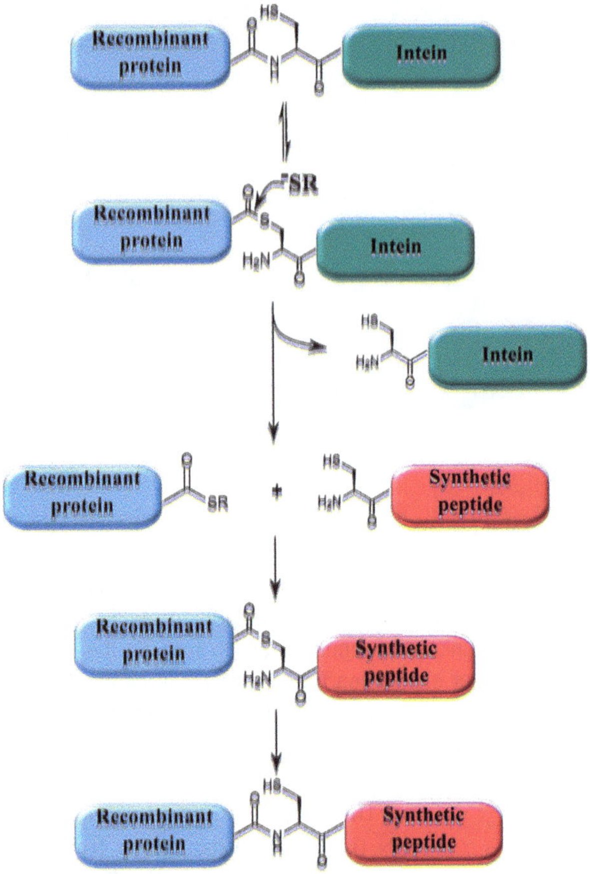

Fig. 8 In vitro protein semisynthesis. C-terminal thioester is generated by cleavage from the intein followed by condensation with an N-terminal cysteine-containing synthetic peptide

others to the chemical nature of the reaction. For instance, the synthesis of thioester peptides, both synthetic and recombinant, is not trivial. On the other side, NCL/EPL, as a bimolecular reaction, requires high concentrations of the peptides/proteins (ideally millimolar or high micromolar), which can be difficult to achieve for certain proteins, or require the need of carrying the reaction under denaturing conditions. Furthermore, because of the latter issue and the instability of the thioester moiety, NCL and EPL are more intricate when attempting chemical manipulation of proteins in living cells, although a few examples have been reported [35].

2.5 Protein Labeling via Split Inteins

In protein *trans*-splicing the intein domain is split into two parts, the IntN and the IntC fragments, each fused to the respective extein sequence. Thus, the intein subunits first have to associate with each other and fold into the active conformation before splicing takes place. Thus, while in EPL the intein is used to generate one of the functional groups for the ligation reaction, protein

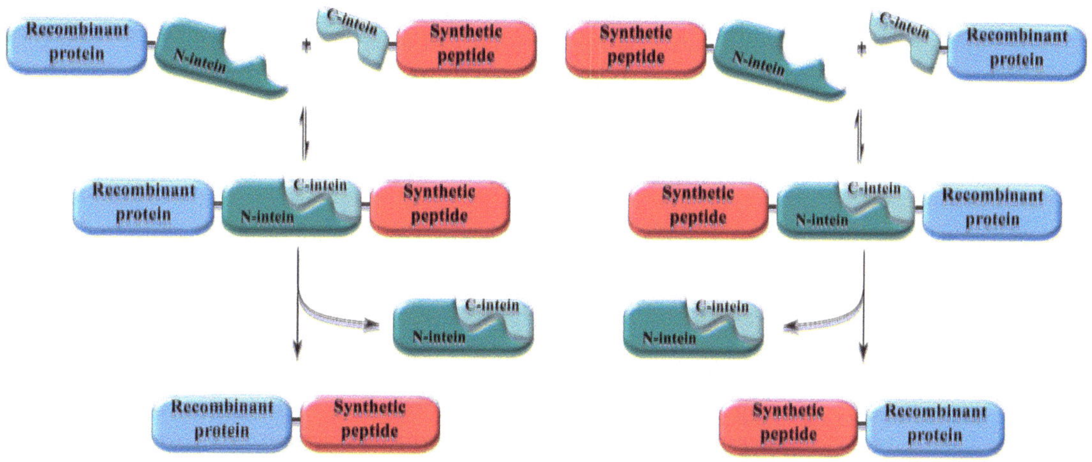

Fig. 9 In vitro protein semisynthesis using split inteins. Semisynthesis by protein *trans*-splicing using a synthetically accessible C-intein or N-intein

trans-splicing allows the direct formation of the peptide bond between the two polypeptides by the split intein [29, 36, 37] (Fig. 9). This provides several potential advantages over a chemical reaction. For example, no synthesis of a thioester or other functional groups is required and the reaction can be performed without high concentrations of thiol. Furthermore, the affinity of the split intein fragments for each other allows efficient performance of the reaction at a lower concentration, say, in the low micromolar to nanomolar range. Finally, as a result of their specific recognition, the reaction also proceeds efficiently in the presence of other proteins as it occurs in a cell extract or a living cell. As a consequence of what has just been exposed protein *trans*-splicing has been exploited for the semisynthesis of proteins. Since the protein of interest are assembled from two precursor parts (fused as extein sequences to the intein fragments), each of these fusion constructs can be prepared individually before the protein *trans*-splicing reaction. Importantly, the smaller C-intein fragment of naturally split inteins is naturally between 30 and 40 amino acids, making it synthetically accessible through solid-phase peptide synthesis. However the addition of any payload for protein labeling puts this fragment nearly out of synthetic reach. To address this issue, several groups have engineered artificially split inteins with noncanonical split sites as close as 6 residues from the intein C-terminus [38] or 11 residues from the N-terminus [39]. While these constructs often have lower splicing efficiencies and binding affinities than naturally split inteins, at least one fragment is synthetically accessible, providing a means to modify the N or C terminus of a protein using protein *trans*-splicing. Another key challenge for using protein *trans*-splicing as a tool to construct semisynthetic proteins is the residue requirements around the split site, which may also affect the yield of the reaction. Alternatively, both fusion proteins can be recombinant but are treated individually with bioconjugation chemicals to regioselectively introduce a synthetic label. This makes split inteins attractive tools for protein semisynthesis of challenging substrates where sample concentration, solubility, and the ability to refold the ligation product are limiting factors.

2.6 Cyclic Peptides

Another application developed based on expressed protein ligation and protein *trans*-splicing is the cyclization of polypeptide chains to increase thermodynamic and in vivo stability, which in turn can improve their biological function, making this an exciting approach for improving the efficacy of protein-based therapeutics. When using expressed protein ligation, an N-terminus containing an accessible cysteine and a C-terminus activated through an intein to yield an α-thioester are needed. The C-terminal α-thioester is then ready to be attacked by the N-terminal cysteine to yield a cyclic thioester, which will rearrange into the more stable peptide bond.

Whereas the efficiency of protein cyclization by expressed protein ligation is largely dependent on a close vicinity of the N- and C-termini, split inteins through their association endorse the approximation of the both termini (Fig. 10). Accordingly, cyclization is accomplished by the inversion of the natural order of the intein fragments at both sides of the target protein, which after splicing are excised. In addition, the use of split inteins provides an approach to generate large libraries of genetically encoded cyclic peptides which can be screened by high-throughput techniques.

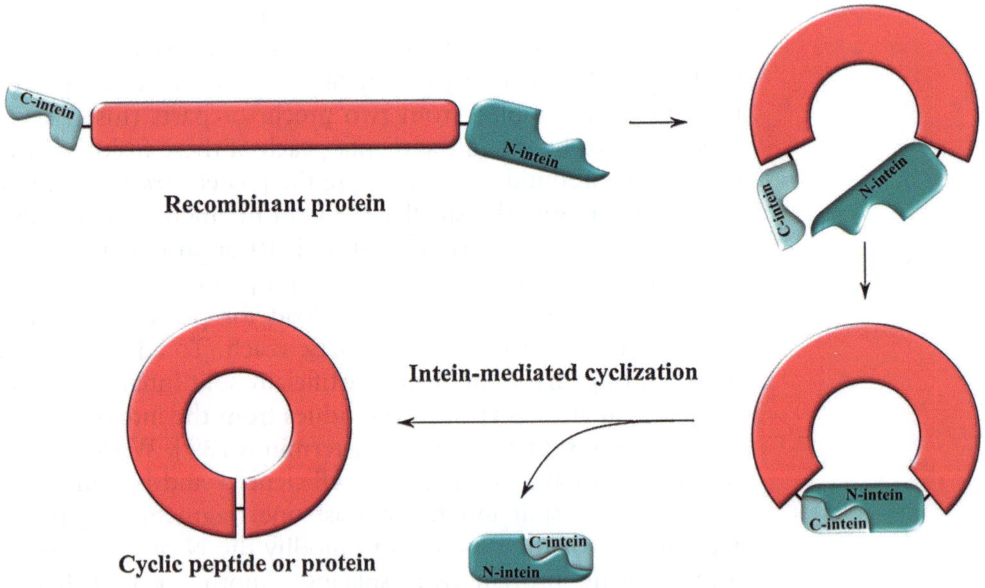

Fig. 10 Protein cyclization using split inteins. Association of intein subunits forces the approximation of the N- and C-terminal ends of a peptide or protein and forms the native peptide bond between them

3 Conclusion

In conclusion, the detailed understanding of the protein splicing mechanism has proven to be extraordinarily useful in the area of protein biotechnology, allowing for the development of an increasing number of in vitro and in vivo applications in the fields of biotechnology, protein and cellular chemistry, as well as molecular and structural biology and biomedicine. Among them, is remarkable the expressed protein ligation. Expressed protein ligation is a flexible platform for the chemical engineering of proteins. It provides researchers with a versatile tool for the study of protein function by allowing the preparation of proteins containing both natural and artificial modifications. The technology is certainly powerful for the biochemical and biophysical analysis of individual proteins. EPL may also be a valuable tool for more global analyses of protein function, that is, proteomic studies, and it might facilitate important advances in biomedicine through the generation of novel protein therapeutic drugs and diagnostic tools [32, 33, 36, 40–42].

Acknowledgments

This work has been supported by grants BIO2013-43517, from MINECO (Spain) and SING12/0 from UdG. V.G. acknowledges her fellowship from the Universitat de Girona.

References

1. Kane PM, Yamashiro CT, Wolczyk DF, Neff N, Goebl M, Stevens TH (1990) Protein splicing converts the yeast TFP1 gene product to the 69-kD subunit of the vacuolar H(+)-adenosine triphosphatase. Science 250(4981):651–657
2. Hirata R, Ohsumk Y, Nakano A, Kawasaki H, Suzuki K, Anraku Y (1990) Molecular structure of a gene, VMA1, encoding the catalytic subunit of H(+)-translocating adenosine triphosphatase from vacuolar membranes of Saccharomyces cerevisiae. J Biol Chem 265 (12):6726–6733
3. Perler FB (2002) InBase: the Intein Database. Nucleic Acids Res 30(1):383–384
4. Shah NH, Muir TW (2014) Inteins: nature's gift to protein chemists. Chem Sci 5 (1):446–461. https://doi.org/10.1039/C3SC52951G
5. Porter JA, Ekker SC, Park WJ, von Kessler DP, Young KE, Chen CH, Ma Y, Woods AS, Cotter RJ, Koonin EV, Beachy PA (1996) Hedgehog patterning activity: role of a lipophilic modification mediated by the carboxy-terminal autoprocessing domain. Cell 86(1):21–34
6. Hall TM, Porter JA, Young KE, Koonin EV, Beachy PA, Leahy DJ (1997) Crystal structure of a Hedgehog autoprocessing domain: homology between Hedgehog and self-splicing proteins. Cell 91(1):85–97
7. Varjosalo M, Taipale J (2008) Hedgehog: functions and mechanisms. Genes Dev 22 (18):2454–2472. https://doi.org/10.1101/gad.1693608
8. Perler FB (1998) Protein splicing of inteins and hedgehog autoproteolysis: structure, function, and evolution. Cell 92(1):1–4
9. Pietrokovski S (1998) Modular organization of inteins and C-terminal autocatalytic domains. Protein Sci 7(1):64–71. https://doi.org/10.1002/pro.5560070106
10. Perler FB (1999) A natural example of protein trans-splicing. Trends Biochem Sci 24 (6):209–211

11. Chevalier BS, Stoddard BL (2001) Homing endonucleases: structural and functional insight into the catalysts of intron/intein mobility. Nucleic Acids Res 29 (18):3757–3774
12. Giriat I, Muir TW, Perler FB (2001) Protein splicing and its applications. Genet Eng (NY) 23:171–199
13. Chong S, Williams KS, Wotkowicz C, Xu MQ (1998) Modulation of protein splicing of the Saccharomyces cerevisiae vacuolar membrane ATPase intein. J Biol Chem 273 (17):10567–10577
14. Chong S, Xu MQ (1997) Protein splicing of the Saccharomyces cerevisiae VMA intein without the endonuclease motifs. J Biol Chem 272 (25):15587–15590
15. Caspi J, Amitai G, Belenkiy O, Pietrokovski S (2003) Distribution of split DnaE inteins in cyanobacteria. Mol Microbiol 50 (5):1569–1577
16. Dassa B, London N, Stoddard BL, Schueler-Furman O, Pietrokovski S (2009) Fractured genes: a novel genomic arrangement involving new split inteins and a new homing endonuclease family. Nucleic Acids Res 37 (8):2560–2573. https://doi.org/10.1093/nar/gkp095
17. Carvajal-Vallejos P, Pallisse R, Mootz HD, Schmidt SR (2012) Unprecedented rates and efficiencies revealed for new natural split inteins from metagenomic sources. J Biol Chem 287 (34):28686–28696. https://doi.org/10.1074/jbc.M112.372680
18. Thiel IV, Volkmann G, Pietrokovski S, Mootz HD (2014) An atypical naturally split intein engineered for highly efficient protein labeling. Angew Chem Int Ed Engl 53(5):1306–1310. https://doi.org/10.1002/anie.201307969
19. Bachmann AL, Mootz HD (2015) An unprecedented combination of serine and cysteine nucleophiles in a split intein with an atypical split site. J Biol Chem 290(48):28792–28804. https://doi.org/10.1074/jbc.M115.677237
20. Tori K, Dassa B, Johnson MA, Southworth MW, Brace LE, Ishino Y, Pietrokovski S, Perler FB (2010) Splicing of the mycobacteriophage Bethlehem DnaB intein: identification of a new mechanistic class of inteins that contain an obligate block F nucleophile. J Biol Chem 285 (4):2515–2526. https://doi.org/10.1074/jbc.M109.069567
21. Xu MQ, Perler FB (1996) The mechanism of protein splicing and its modulation by mutation. EMBO J 15(19):5146–5153
22. Sun P, Ye S, Ferrandon S, Evans TC, Xu MQ, Rao Z (2005) Crystal structures of an intein from the split dnaE gene of Synechocystis sp. PCC6803 reveal the catalytic model without the penultimate histidine and the mechanism of zinc ion inhibition of protein splicing. J Mol Biol 353(5):1093–1105. https://doi.org/10.1016/j.jmb.2005.09.039
23. Du Z, Zheng Y, Patterson M, Liu Y, Wang C (2011) pK(a) coupling at the intein active site: implications for the coordination mechanism of protein splicing with a conserved aspartate. J Am Chem Soc 133(26):10275–10282. https://doi.org/10.1021/ja203209f
24. Frutos S, Goger M, Giovani B, Cowburn D, Muir TW (2010) Branched intermediate formation stimulates peptide bond cleavage in protein splicing. Nat Chem Biol 6 (7):527–533. https://doi.org/10.1038/nchembio.371
25. Waters E, Hohn MJ, Ahel I, Graham DE, Adams MD, Barnstead M, Beeson KY, Bibbs L, Bolanos R, Keller M, Kretz K, Lin X, Mathur E, Ni J, Podar M, Richardson T, Sutton GG, Simon M, Soll D, Stetter KO, Short JM, Noordewier M (2003) The genome of Nanoarchaeum equitans: insights into early archaeal evolution and derived parasitism. Proc Natl Acad Sci U S A 100 (22):12984–12988. https://doi.org/10.1073/pnas.1735403100
26. Choi JJ, Nam KH, Min B, Kim SJ, Soll D, Kwon ST (2006) Protein trans-splicing and characterization of a split family B-type DNA polymerase from the hyperthermophilic archaeal parasite Nanoarchaeum equitans. J Mol Biol 356(5):1093–1106. https://doi.org/10.1016/j.jmb.2005.12.036
27. Southworth MW, Benner J, Perler FB (2000) An alternative protein splicing mechanism for inteins lacking an N-terminal nucleophile. EMBO J 19(18):5019–5026. https://doi.org/10.1093/emboj/19.18.5019
28. Chong S, Shao Y, Paulus H, Benner J, Perler FB, Xu MQ (1996) Protein splicing involving the Saccharomyces cerevisiae VMA intein. The steps in the splicing pathway, side reactions leading to protein cleavage, and establishment of an in vitro splicing system. J Biol Chem 271 (36):22159–22168
29. Paulus H (2000) Protein splicing and related forms of protein autoprocessing. Annu Rev Biochem 69:447–496. https://doi.org/10.1146/annurev.biochem.69.1.447
30. Muir TW (2003) Semisynthesis of proteins by expressed protein ligation. Annu Rev Biochem 72:249–289. https://doi.org/10.1146/annurev.biochem.72.121801.161900

31. Mootz HD, Muir TW (2002) Protein splicing triggered by a small molecule. J Am Chem Soc 124(31):9044–9045
32. Vila-Perello M, Muir TW (2010) Biological applications of protein splicing. Cell 143 (2):191–200. https://doi.org/10.1016/j.cell.2010.09.031
33. Volkmann G, Iwai H (2010) Protein trans-splicing and its use in structural biology: opportunities and limitations. Mol Biosyst 6 (11):2110–2121. https://doi.org/10.1039/c0mb00034e
34. Muir TW (1995) A chemical approach to the construction of multimeric protein assemblies. Structure 3(7):649–652
35. Giriat I, Muir TW (2003) Protein semi-synthesis in living cells. J Am Chem Soc 125 (24):7180–7181. https://doi.org/10.1021/ja034736i
36. Cheriyan M, Perler FB (2009) Protein splicing: a versatile tool for drug discovery. Adv Drug Deliv Rev 61(11):899–907. https://doi.org/10.1016/j.addr.2009.04.021
37. Noren CJ, Wang J, Perler FB (2000) Dissecting the chemistry of protein splicing and its applications. Angew Chem Int Ed Engl 39 (3):450–466
38. Oeemig JS, Aranko AS, Djupsjobacka J, Heinamaki K, Iwai H (2009) Solution structure of DnaE intein from Nostoc punctiforme: structural basis for the design of a new split intein suitable for site-specific chemical modification. FEBS Lett 583(9):1451–1456. https://doi.org/10.1016/j.febslet.2009.03.058
39. Ludwig C, Schwarzer D, Zettler J, Garbe D, Janning P, Czeslik C, Mootz HD (2009) Semisynthesis of proteins using split inteins. Methods Enzymol 462:77–96. https://doi.org/10.1016/S0076-6879(09)62004-8
40. Mootz HD (2009) Split inteins as versatile tools for protein semisynthesis. Chembiochem 10(16):2579–2589. https://doi.org/10.1002/cbic.200900370
41. Volkmann G, Mootz HD (2013) Recent progress in intein research: from mechanism to directed evolution and applications. Cell Mol Life Sci 70(7):1185–1206. https://doi.org/10.1007/s00018-012-1120-4
42. Elleuche S, Poggeler S (2010) Inteins, valuable genetic elements in molecular biology and biotechnology. Appl Microbiol Biotechnol 87 (2):479–489. https://doi.org/10.1007/s00253-010-2628-x

Chapter 3

Identification, Characterization, and Optimization of Split Inteins

Neel H. Shah and Adam J. Stevens

Abstract

In recent years, split inteins have seen widespread use as molecular platforms for the design of a variety of peptide and protein chemistry technologies, most notably protein ligation. The development of these approaches is dependent on the identification and/or design of split inteins with robust activity, stability, and solubility. Here, we describe two approaches to characterize and compare the activities of newly identified or engineered split inteins. The first assay employs an *E. coli*-based selection system to rapidly screen the activities of many inteins and can be repurposed for directed evolution. The second assay utilizes reverse-phase high-performance liquid chromatography (RP-HPLC) to provide insights into individual chemical steps in the protein splicing reaction, information that can guide further engineering efforts. These techniques provide useful alternatives to common assays that utilize SDS-PAGE to analyze splicing reaction progress.

Key words Split intein, Protein *trans*-splicing, Protein engineering, Bacterial selection, RP-HPLC, Kinetic analysis, Chemical biology

1 Introduction

Protein splicing is a naturally occurring protein processing event in which a polypeptide chain undergoes autocatalytic excision of a conserved internal domain, known as an intein, concomitant with the ligation by the intein of the flanking polypeptide regions, called exteins (Fig. 1) [1]. Inteins were first identified in baker's yeast three decades ago [2], and they have remained a biochemical curiosity with no apparent biological function until recently. Evidence is now emerging that, when embedded in their natural host proteins, inteins can act as regulatory domains to control the assembly and activity of those host proteins, often in a stimulus-responsive manner [3–7]. While the biology of inteins is only now coming to light, chemical biologists have long exploited intein chemistry for a broad range of biotechnological applications [8]. In particular, methods for protein ligation and semisynthesis

Miquel Vila-Perelló (ed.), *Expressed Protein Ligation: Methods and Protocols*, Methods in Molecular Biology, vol. 2133,
https://doi.org/10.1007/978-1-0716-0434-2_3,

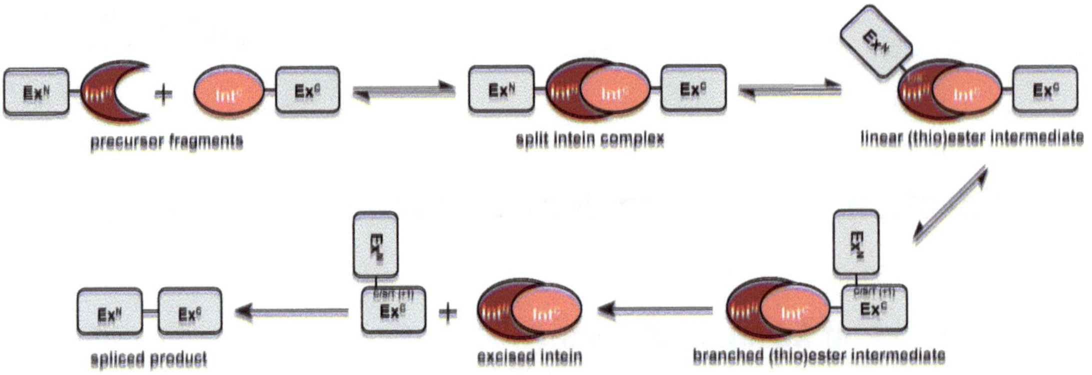

Fig. 1 Schematic diagram of the protein *trans*-splicing reaction. N- and C-intein fragments (Int^N and Int^C) are depicted in dark and light red, respectively. N- and C-exteins (Ex^N and Ex^C) are shown in gray. The linear and branched intermediates are shown attached to the first residue of the N-intein (cysteine or serine) and the first residue of the C-extein (the +1 residue, cysteine, serine, or threonine), respectively

have been substantially empowered by intein chemistry [8, 9]. Inteins are also useful tools for protein purification [10, 11], head-to-tail cyclization of polypeptides [12–14], the construction of biosensors [15–17], and the chemical manipulation of proteins in living systems [18–21].

Protein splicing comes in two flavors [1]. Most inteins exist as one contiguous polypeptide chain that mediates protein splicing *in cis.* A subset of inteins, however, is naturally split into two fragments that can spontaneously associate to form the canonical intein fold, and these proteins catalyze splicing *in trans* (Fig. 1). Notably, several artificially split inteins have been engineered from single-chain inteins [22–24], and naturally split intein fragments can be fused to generate *cis*-splicing systems [13, 25]. Most of the aforementioned technological applications of inteins benefit from the split intein architecture [26], as protein *trans*-splicing reactions and side reactions cannot occur prior to mixing of the naturally split intein protomers [27]. Once mixed, the fragments associate rapidly with high affinity and specificity, allowing for efficient peptide bond formation at low concentrations [28–30].

The broad utility of split inteins has motivated many efforts to identify or design new split inteins with enhanced biochemical and biophysical properties. Several of the first identified naturally split inteins were found through targeted genomic sequencing in organisms predicted to have inteins based on phylogenetic relationships to other organisms and genes containing known inteins [31–34]. Inteins have also been discovered through bioinformatic analyses of metagenomic datasets [35]. Now, new inteins sequences can be found readily through sequence homology searches against the vast gene and protein sequence databases [36], and the biochemical characterization of inteins from this apparently endless pool of sequences has been particularly fruitful. As a prime example, characterization of the naturally split *Npu*DnaE intein revealed that some inteins can catalyze protein *trans*-splicing in seconds or minutes, rather than hours [37, 38]. This observation has inspired numerous successful efforts to find naturally occurring split inteins that splice with similar or even faster splicing kinetics, with few side reactions [25, 39]. These "ultrafast" split inteins have substantially enhanced and enabled technologies that utilize protein *trans*-splicing [20, 21, 27, 40–42].

Another sought-after property in split inteins is tolerance to a broad range of extein sequences proximal to the splice junction. Naturally occurring split inteins often strongly prefer to splice sequences that resemble those of their endogenous host proteins [25, 37, 43–45]. These extein sequence constraints remain one of the largest barriers to the widespread use of split inteins by limiting the number of protein systems on which they can be deployed in a traceless manner. The practical use of split inteins is also constrained by the solubility of their fragments, their expression levels, and their tolerance to a broad range of reaction conditions (e.g., temperature, pH, ionic strength, and chaotropic agents).

Efforts to identify chemically and physically robust split inteins with broad sequence tolerance directly from nature's toolbox have been only modestly successful. One such example is the *Ava*DnaE intein, a close homolog of *Npu*DnaE that has similar splicing kinetics and extein sequence preferences, but augmented expression levels in *E. coli* and mammalian cells [25, 27]. Although a naturally occurring split "super-intein" has yet to be found, detailed studies on split inteins found in nature have generated a wealth of biochemical and structural data that have revealed deep mechanistic details about protein *trans*-splicing [1]. This information has, in turn, guided protein engineering strategies, using directed evolution [46–48], consensus sequence design [30, 49], and rational design [50, 51], to successfully generate inteins with each of the desirable properties described above. This emerging suite of engineered inteins holds great potential to yield new tools for protein science.

Essential to the identification or design of improved split inteins are assays that reliably report on intein activity. The quintessential *trans*-splicing assay entails fusion of split intein fragments to model extein domains (ubiquitin, SUMO, maltose binding protein, GB1, etc.), isolation of these fusion proteins, and monitoring of reaction progress by SDS-PAGE upon mixing of the fragments. This experiment can be conducted readily in virtually any biochemistry laboratory, and versions of it have been described extensively in the literature (e.g., *see* Refs. 25, 39). Notably, this assay provides a platform for determining the overall rate of *trans*-splicing, assessing the expression levels and stability of intein fragments, and identifying optimal reaction conditions. Despite its broad utility, however, the canonical gel-based splicing assay can be cumbersome when comparing large numbers of inteins, and it is often difficult to extract detailed kinetic information about individual steps in the splicing reaction using this approach.

Here, we describe two methods to characterize split intein activity that provide access to information not readily obtained using other common approaches. The first method relies on the coupling of antibiotic resistance in *E. coli* to intein activity, thereby providing a selection scheme to infer intein activity from bacterial viability [29, 46]. This platform allows for the rapid comparative screening of split intein splicing activity without the need for protein purification. Moreover, this assay is highly predictive of relative in vitro splicing rates under common laboratory conditions (i.e., temperatures between 25 °C and 37 °C, and buffers with moderate salt concentrations at neutral pH). The second method utilizes reverse-phase high performance liquid chromatography (RP-HPLC) to analyze protein splicing kinetics by monitoring the abundance not only of starting materials and products, but also a critical reaction intermediate [45]. Although this approach is time- and resource-intensive, it provides highly quantitative kinetic data that can be used to elucidate details of the *trans*-splicing mechanism. The two methods described here have been invaluable for characterizing previously uncharacterized inteins, identifying key features of inteins through comparative analyses, and guiding both the rational design and structure-guided directed evolution of inteins with new properties [25, 46, 49, 51].

2 Materials

2.1 Antibiotic Resistance Assay

2.1.1 Cloning of Constructs for Kanamycin Selection Assays

1. A gene, codon-optimized for expression in *E. coli,* encoding the intein fragments of interest. If a synthetic gene is purchased, consider including, in between the coding sequences for the intein fragments, a spacer containing an intervening ribosomal binding site (*see* **step 3** in Subheading 3.1.1).
2. A plasmid for constitutive expression of the aminoglycoside-3′-phosphotransferase gene (KanR) in *E. coli.* As previously reported, this plasmid is derived from the pBlueScript backbone and also encodes an ampicillin resistance marker [25, 29, 46, 49, 51].
3. DNA oligonucleotide primers for the insertion of an extein-coding sequence into the KanR construct by site-directed mutagenesis (*see* **Note 1**).
4. DNA oligonucleotide primers for the amplification and insertion of intein genes into the KanR construct.
5. A kit for plasmid DNA extraction and purification.
6. A kit for DNA purification by extraction from an agarose gel.

7. Chemically competent DH5α cells.
8. Ampicillin stock solution: 100 mg/mL (w/v) ampicillin in water.
9. Autoclaved lysogeny broth (LB).
10. LB/agar plates with 100 μg/mL of ampicillin.
11. A high-fidelity DNA polymerase kit.
12. A restriction free cloning kit (e.g., a kit for Gibson Assembly).

2.1.2 Kanamycin Growth Assays

1. Kanamycin stock solution: 50 mg/mL (w/v) kanamycin in water.
2. Ampicillin stock solution: 100 mg/mL (w/v) ampicillin in water.
3. Autoclaved LB.
4. Chemically competent DH5α cells.
5. Round-bottom polypropylene culture tubes (with a capacity of ~15 mL and a cap that can be snapped loosely for aerobic cultures).
6. A temperature-controlled incubator/shaker.
7. A 96-well plate reader with temperature control, plate shaking capability, and a light source and monochromator compatible with measuring absorbance at 650 nm.
8. Clear round-bottom 96-well plates with clear lids.

2.2 RP-HPLC Splicing Assays

2.2.1 Cloning of Recombinant Expression Plasmids for RP-HPLC Assay Constructs

1. Genes, codon-optimized for expression in *E. coli*, encoding the intein fragments of interest.
2. A plasmid for IPTG-inducible overexpression of proteins with an N-terminal His_6-SUMO tag from a T7 promoter, as described previously [45].
3. A modified pTXB1 plasmid bearing the *Mxe*GyrA intein with a C-terminal His_6-tag instead of a chitin binding domain, as described previously [45].
4. DNA oligonucleotide primers for the amplification and insertion of intein genes into the bacterial overexpression vectors.
5. Kanamycin stock solution: 50 mg/mL (w/v) kanamycin in water.
6. Ampicillin stock solution: 100 mg/mL (w/v) ampicillin in water.
7. LB media and LB/agar plates, with either 100 μg/mL of ampicillin or 50 μg/mL of kanamycin.
8. A high-fidelity DNA polymerase kit.
9. A restriction free cloning kit (e.g., a kit for Gibson Assembly).
10. A kit for plasmid DNA extraction and purification.

11. A kit for DNA purification by extraction from an agarose gel.
12. Chemically competent DH5α cells.

2.2.2 Preparation of Protein Samples

1. A temperature-controlled incubator/shaker.
2. Flasks for the growth of 1 L *E. coli* cultures (total capacity should be at least 2.5 L).
3. 1 L centrifuge bottles.
4. A centrifuge compatible with 1 L centrifuge bottles.
5. A centrifuge that can spin cell lysate at >30,000 rcf.
6. Centrifuge tubes with a 40–50 mL capacity that are compatible with high-speed centrifugation (at least 30,000 rcf).
7. LB media with either 100 μg/mL of ampicillin or 50 μg/mL of kanamycin.
8. Chemically competent BL21(DE3) cells for protein overexpression.
9. 1 M IPTG stock: 1 M Isopropyl β-D-1-thiogalactopyranoside (IPTG) in water.
10. A sonicator probe for lysing cells (e.g., a Branson digital sonifier).
11. Protease inhibitors (*see* **Note 2**).
12. β-mercaptoethanol (BME),
13. Ni column buffer A: 50 mM sodium phosphates, pH 8.0, 300 mM NaCl, 5 mM imidazole.
14. Ni column buffer B: 50 mM sodium phosphates, pH 8.0, 300 mM NaCl, 250 mM imidazole, with or without 2 mM BME.
15. Ni-NTA resin (*see* **Note 3**).
16. A fritted column with a stopcock and a capacity of at least 20 mL.
17. Dialysis tubing or a desalting column for buffer exchange.
18. Ulp1 cleavage buffer: 50 mM sodium phosphates, pH 8.0, 300 mM NaCl, 5 mM imidazole, 1 mM DL-dithiothreitol (DTT).
19. Ulp1, SUMO-specific protease (can be purchased or readily made using standard biochemical techniques).
20. Size exclusion buffer: 100 mM sodium phosphates, pH 7.2, 150 mM NaCl, 1 mM EDTA, with either 1 mM tris(2-carboxyethyl)phosphine (TCEP) or 1 mM DTT (*see* **Note 4**).
21. A fast protein liquid chromatography (FPLC) system.
22. A Superdex 75 column.

23. A solid-phase peptide synthesis vessel (fritted with a stopcock and a cap).
24. Resin for solid-phase peptide synthesis (e.g., Rink amide-MBHA resin).
25. Nitrogen gas under a regulator that can control pressure under 10 psi.
26. A vacuum line.
27. *N*,*N*-dimethylformamide (DMF).
28. Dichloromethane.
29. Methanol.
30. Standard Fmoc-protected amino acids.
31. 1-Hydroxybenzotriazole (HOBt).
32. 2-(1*H*-benzotriazol-1-yl)-1,1,3,3-tetramethyluronium hexafluorophosphate (HBTU).
33. HOBt/HBTU solution: 0.45 M HOBt, 0.45 M HBTU in DMF. Prepare fresh each day for solid-phase peptide synthesis.
34. *N*,*N*-Diisopropylethylamine (DIPEA).
35. Piperidine (*see* **Note 5**), diluted to 20% (v/v) in DMF.
36. Trifluoroacetic acid (TFA).
37. Triisopropylsilane (TIS).
38. Diethyl ether.
39. 1 M MESNa stock solution: 1 M sodium 2-mercaptoethane sulfonate (MESNa) in H_2O.
40. An HPLC system compatible with reverse-phase separations and equipped with a UV detector.
41. A C_{18} column capable of preparative or semipreparative-scale purifications (e.g., a C_{18} column with a 9.4 mm inner diameter, 250 mm length, and 5 μm particle size).
42. HPLC Solvent A: water with 0.1% TFA.
43. HPLC Solvent B: 90% acetonitrile with 0.1% TFA.
44. A lyophilizer to freeze-dry purified peptides and ligated proteins.

2.2.3 RP-HPLC Splicing Assays

1. An HPLC system compatible with reverse-phase separations and equipped with a UV detector.
2. A C_{18} column capable of analytical-scale purifications (e.g., a C_{18} column with a 4.6 mm inner diameter, 150 mm length, and 5 μm particle size).
3. HPLC Solvent A: water with 0.1% TFA.
4. HPLC Solvent B: 90% acetonitrile with 0.1% TFA.

5. 1 M TCEP stock solution: 1 M TCEP at neutral pH in water.
6. Reaction buffer: 100 mM sodium phosphate, pH 7.2, 150 mM NaCl, 1 mM EDTA, with or without 1 mM DTT (*see* **Note 6**).
7. Quenching solution: 8 M guanidine hydrochloride and 4% TFA in water.

3 Methods

3.1 Comparing Split Intein Activities Using a Kanamycin Resistance Assay in E. coli

This assay enables rapid comparison of the activities of split inteins without the need for protein overexpression and purification [25]. Instead, this assay uses a selection system where split intein activity in *E. coli* drives the assembly of a functional aminoglycoside-3′-phosphotransferase protein, which in turn confers resistance to the antibiotic kanamycin. The subsections below describe how to clone the appropriate plasmid constructs, introduce those plasmids into *E. coli* cells, execute the selection assay, and analyze data resulting from the experiment. For each intein analyzed, these experiments yield a dose–response curve for growth in the presence of kanamycin, where growth in the presence of higher kanamycin concentrations indicates a more active split intein.

This technique can also be adapted to engineer new inteins and to conduct sequence–activity relationship studies on individual inteins. For example, targeted mutagenesis within an intein has been applied in conjunction with this assay to rapidly identify residues that influence splicing activity [25]. This method has also been applied to screen extein dependency through mutation of the extein residues flanking the intein genes in the KanR constructs [25, 44, 46]. The assay has also been modified to serve as a platform for the directed evolution of inteins with enhanced activity (e.g., increased promiscuity, thermostability, or splicing rate), by generating mutant libraries using targeted saturation mutagenesis or error-prone PCR and selecting mutants with increased kanamycin resistance [46, 47, 51].

3.1.1 Cloning of Split Intein Constructs for Screening by Kanamycin Resistance

The general cloning strategy follows three steps. First, introduce an extein-coding sequence at an appropriate insertion site in the KanR gene (Fig. 2a). Next, insert the split intein fragment genes within that extein-coding sequence, with an intervening ribosomal binding site (Fig. 2b). Finally, introduce any necessary mutations to the intein, either to generate control constructs, or to assay the functional impact of specific mutations.

1. Using a PCR-based method for site-direct mutagenesis with the appropriate primers, introduce the proximal N- and C-extein (Ex^N and Ex^C) residues of the intein (generally the flanking three to five amino acids) within the insertion site of

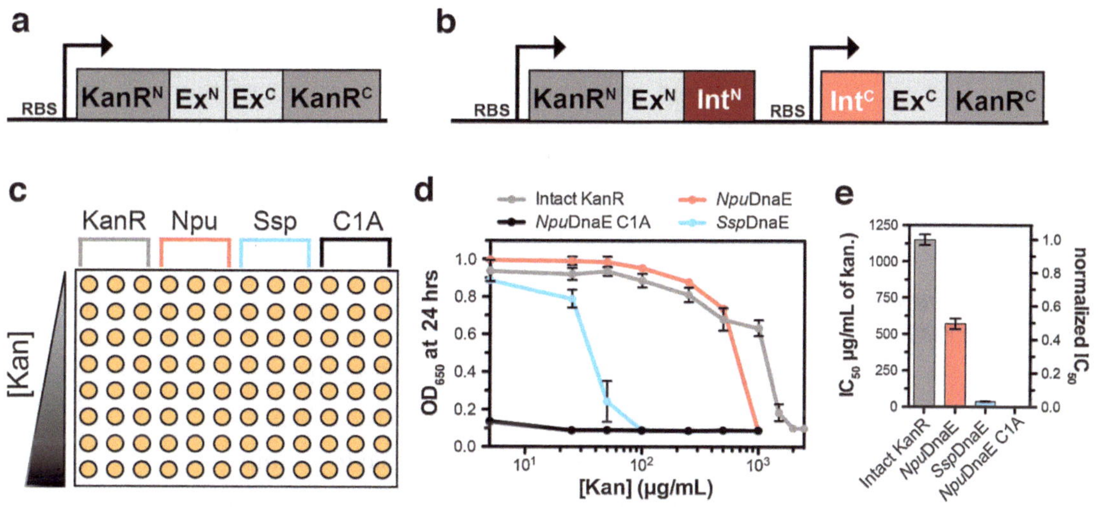

Fig. 2 A protein *trans*-splicing assay based on intein-activity-dependent kanamycin resistance in *E. coli.* (**a**) A schematic diagram of an aminoglycoside-3′-phosphotransferase (KanR) gene, with the insertion of a short coding sequence for N- and C-extein residues that do not perturb KanR function. (**b**) A schematic diagram of a split KanR gene, where the fragments are fused to the coding sequences of split intein protomers. The introduction of an intervening ribosomal binding site (RBS) allows the two fusion proteins to be translated separately from one transcript. (**c**) Setup of a 96-well plate for the comparison of two split inteins (Npu = *Npu*DnaE, Ssp = *Ssp*DnaE) and two control constructs (KanR = intact KanR protein, C1A = splicing deficient *Npu*DnaE C1A mutant) in parallel with three technical replicates per construct. (**d**) Representative data plotting endpoint optical densities from kanamycin resistance screens [25]. The graph shows kanamycin dose–response curves for two split inteins and two control constructs. Although technical replicates are often carried out, the error bars here represent the standard deviation from three or more biological replicates. (**e**) IC_{50} values for kanamycin resistance, derived from the four datasets shown in panel (**d**). The left vertical axis measures raw IC_{50} values, while the right vertical axis measures IC_{50} values normalized to that of the intact KanR gene depicted in panel (**a**). IC_{50} values correlate with the extent of KanR reconstitution, and thus with intein activity

the KanR gene (Fig. 2a). Several insertion sites within the KanR gene have been implemented previously for the analysis and evolution of inteins (*see* **Note 1**) [44, 46, 47, 52].

2. Amplify the split intein fragments by PCR using primers that allow for restriction-enzyme-free insertion within the extein sequences that have been embedded in the KanR gene.
3. Using restriction-free cloning, insert the split intein of interest into the KanR gene, such that each split intein fragment will be translated separately by including a stop codon and an additional ribosomal binding site (Fig. 2b). An example of a spacer sequence containing these features is the following: TAATGTACAGTAATACAAGGGGTGTTATG, where the underlined sequences correspond to the stop codon at the 3′ end of the Int^N gene, a ribosomal binding site, and the start codon at the 5′ end of the Int^C gene.

4. Using site-directed mutagenesis with a high-fidelity DNA polymerase and the appropriate oligonucleotide primers, generate a plasmid containing an inactive version of the split intein of interest by introducing an alanine residue in place of the nucleophilic residue at position 1 of the N-intein (*see* **Note 7**).
5. At each cloning step, isolate the pure plasmid from a clonal population, and confirm its identity by Sanger sequencing.

3.1.2 Transformation of E. coli with KanR Plasmids

1. Mix 20 μL aliquots of DH5α *E. coli* cells with 1 μL (~100 ng) of each plasmid.
2. Incubate the cells on ice for 15 min.
3. Heat-shock the cells at 42 °C for 1 min.
4. Allow the cells to recover on ice for 5 min.
5. Resuspend the cells in 200 μL of LB and incubate at 37 °C for 1 h while shaking.
6. Dilute the cells into 4 mL of LB containing 100 μg/mL ampicillin and grow for 18 h in culture tubes at 37 °C while shaking between 200 and 250 rpm.

3.1.3 Kanamycin Resistance Growth Assay

1. Prepare 8, 2.5 mL aliquots of LB containing 100 μg/mL ampicillin, added from the ampicillin stock solution, and a gradated concentration of kanamycin, added from the kanamycin stock solution (e.g., 1, 25, 50, 100, 250, 500, 1000, 2000 μg/mL, *see* **Note 8**).
2. Pipet 180 μL of each LB aliquot into the rows of the 96-well plate. Each row should contain a constant kanamycin concentration (Fig. 2c).
3. Dilute the *E. coli* cultures produced in **step 6** of the previous section 25-fold into LB containing 100 μg/mL ampicillin.
4. To achieve a final well volume of 200 μL on the plate, add 20 μL of the diluted cells into each well in a column (*see* **Note 9**). Cells from the overnight cultures will be diluted 250-fold, in total, onto the plate.
5. Incubate the cells at 30 °C for 24 h in the plate reader. The plate reader should measure the optical density of the wells at 650 nm (OD_{650}) every 5 min and agitate the plate for 1 min prior to this reading (*see* **Note 10**).

3.1.4 Analysis of Kanamycin Resistance Growth Assays

1. Inspect the raw data. For each well where bacteria have grown, a plot of the OD_{650} as a function of time should resemble a typical growth curve with a lag phase, an exponential growth phase, and a stationary phase. Choose an endpoint time such that the bacteria grown in the absence of kanamycin are in the stationary phase (typically 24 h when growing at 30 °C). If

there is a clear death phase visible after the stationary phase, an earlier endpoint should be chosen.

2. Using the plate reader, obtain the OD_{650} from the chosen endpoint of the growth assay in order to calculate IC_{50}.
3. Plot the OD_{650} ($t = 24$ h) as a function of kanamycin concentration using a standard graphic and curve-fitting software (Fig. 2d).
4. Fit the plot to the following variable-slope dose–response equation to determine the IC_{50} value:

$$OD_{obs} = OD_{min} + \frac{(OD_{max} - OD_{min})}{1 + 10^{[(\log IC_{50} - \log[Kan]) \times Hillslope]}}$$

where OD_{max} is the value of OD_{650} in the absence of kanamycin and OD_{min} is the background absorbance at 650 nm in the absence of growth. The IC_{50} and Hill slope should be allowed to vary during the fit.

5. Determine the IC_{50} ratio between each intein-containing construct and the contiguous KanR gene with inserted extein residues (Fig. 2e).

3.2 Analysis of Protein Trans-Splicing Kinetics by RP-HPLC

The technique described in this section uses reverse-phase high-performance liquid chromatography to monitor the rates of protein *trans*-splicing reactions in a highly quantitative manner, while also providing kinetic information about intermediate steps in the reaction pathway [45]. This additional kinetic information provides deep insights into the mechanism of protein *trans*-splicing and can guide efforts to engineer enhanced inteins by augmenting the interpretation of sequence–activity relationships [45, 51]. The sub-sections below describe the design and production of the required protein samples for this method, characterization of the splicing reaction by RP-HPLC, and analysis of the numerical data extracted from these experiments

3.2.1 Cloning of Protein Expression Plasmids

To ensure sufficient resolution of starting materials and products by RP-HPLC, this assay requires split intein fragments bearing short extein sequences (typically three amino acid residues in length). The N-terminal fragment is generated as a SUMO fusion, where the SUMO protein can be cleaved off to liberate an N-intein with a short N-extein (Fig. 3a). The C-terminal fragment can be produced in two ways: (1) the C-intein, lacking any extein residues, is produced as a fusion to the GyrA contiguous intein, liberated from GyrA by thiolysis, and ligated to a short C-extein-containing peptide (Fig. 3b), or (2) a C-intein–C-extein sequence is produced as a GyrA fusion and liberated by hydrolysis.

1. Design a set of primers that allow for the amplification of an N-intein gene with three 5′ extein codons and flanking regions

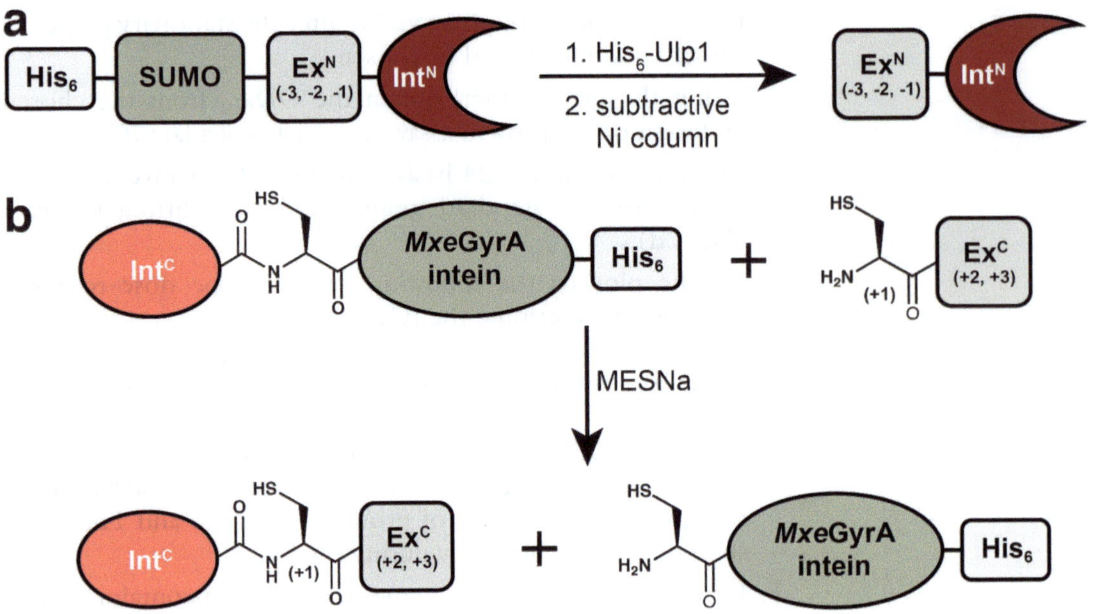

Fig. 3 Preparation of protein samples for the RP-HPLC splicing assay. (**a**) Isolation of an N-intein construct with a short N-extein by cleavage from a SUMO fusion protein using the SUMO-specific protease Ulp1. (**b**) Semisynthesis of a C-intein construct with a short C-extein by thiolysis from an *Mxe*GyrA fusion protein and ligation to a C-extein peptide

for insertion in-frame with SUMO in a bacterial expression vector. Frequently, we utilize the endogenous N-extein sequence.

2. Amplify the N-intein (Int^N) coding sequence from a template plasmid using standard PCR techniques with the appropriate oligonucleotides and a high-fidelity DNA polymerase.

3. Using a restriction-free cloning method, insert the amplicon obtained from **step 2** of this section into the expression vector that encodes for an N-terminal His_6-SUMO tag so that the Ex^N-Int^N gene is in-frame with the SUMO gene, with the first codon of the N-extein immediately downstream of the codons for the C-terminal SUMO Gly-Gly motif (Fig. 3a).

4. Transform 50 μL of DH5α cells as described in Subheading 3.1.2, grow the cells under selective pressure, isolate the pure plasmid from a clonal population, and confirm its identity by Sanger sequencing.

5. For the C-intein expression plasmid, follow a similar protocol to that described for the N-intein, with two additional considerations: (1) Insert the C-intein coding gene into the modified pTXB1 plasmid so that the C-terminal intein residue (typically an asparagine) is immediately upstream of the N-terminal cysteine of the *Mxe*GyrA intein gene that is native

to the vector (Fig. 3b). (2) If a peptide ligation strategy will not be used to append a C-extein, as described previously [45], the C-extein residues can be encoded directly in between the C-intein gene and the *Mxe*GyrA gene (*see* **Note 11**).

3.2.2 Recombinant Production of the N-Terminal Fragment

1. Transform BL21(DE3) cells, as described in Subheading 3.1.2, with the His_6-SUMO-Ex^N-Int^N fusion plasmid, and use the transformation reaction to inoculate 15 mL of LB with the appropriate antibiotic. Grow the culture overnight at 37 °C until dense.
2. Use 5 mL of overnight culture to inoculate 1 L of LB with antibiotics. Grow the cells at 37 °C with vigorous shaking (200–250 rpm). Typically, overexpression is carried out in two, 1 L cultures of LB.
3. Grow the cells until the optical density at 600 nm reaches between 0.4 and 0.6, then cool the cells and the incubator down to 18 °C.
4. Induce protein overexpression by adding 0.5 mL of the 1 M IPTG stock solution to each culture flask, then continue to incubate and shake the cells at 18 °C for 16 h.
5. Transfer the cells to 1 L centrifuge bottles and pellet the cells at 10,000 rcf for 30 min.
6. Decant the supernatant and resuspend the pellets from each 1 L bottle in 15 mL of Ni column buffer A supplemented with 2 mM BME and protease inhibitors at the concentration recommended in the manufacturer's protocol.
7. Lyse the cells by sonication while on ice. A typical sonication protocol might have 10 pulses, each 20 s long at 35–50% amplitude, with intervening rest periods of 30 s each.
8. After sonication, clarify the lysate by centrifugation for 30 min at 30,000 rcf.
9. Mix the lysate supernatant (30–40 mL of liquid) with 4 mL of NiNTA resin that has been washed and equilibrated with Ni column buffer A.
10. Rock this slurry at 4 °C for 30 min, then load it onto a fritted column and collect the flow through.
11. Wash the column with 5 column volumes (CV) of Ni column buffer A, 5 CV of a 19:1 mixture of buffers A and B, and then 3 CV of a 4:1 mixture of buffers A and B, collecting each wash in a separate tube.
12. Finally, elute the protein in 5, 1.5 CV fractions of Ni column buffer B.
13. Analyze the fractions by SDS-PAGE and pool the purest fractions.

14. Next, add TCEP to the samples from the 1 M TCEP stock solution to a final concentration of 10 mM, and carry out a buffer exchange into Ulp1 buffer, either by dialysis or using a desalting column.
15. Add 10 μg of purified Ulp1 to the protein mixture and keep it overnight at room temperature (Fig. 3a, *see* **Note 12**).
16. The next day, mix the cleavage mixture with 4 mL of NiNTA resin for 30 min, collect the flow through, and collect two 1.5 CV washes with Ni column wash buffer. The cleaved N-intein construct should elute in these low-imidazole fractions. The column can be washed with Ni column buffer B to determine what remains bound to the resin.
17. Finally, purify the protein by size exclusion chromatography over a Superdex 75 column in size exclusion buffer. Fractions from this final purification step can be assessed by SDS-PAGE to locate the desired product and assess purity.
18. Measure the absorbance of the purified protein at 280 nm and estimate the concentration using a calculated extinction coefficient (e.g., from the ExPASY ProtParam web tool). Protein concentrations greater than 5–10 μM are sufficient.
19. The protein can be used immediately or flash-frozen in liquid nitrogen and stored at −80 °C (*see* **Note 13**).

3.2.3 Solid-Phase Synthesis of a C-Extein Peptide

1. Weigh out each Fmoc-protected amino acid, and transfer it to a glass vial (5 eq. of amino acid relative to the resin, so 1.25 mmol for a 0.25 mmol scale synthesis).
2. Prepare the HOBt/HBTU solution. 2.5 mL of solution are required per coupling reaction for a 0.25 mmol scale synthesis.
3. Prepare the 20% piperidine solution for Fmoc deprotection steps. It is most convenient to have this solution in a squirt bottle and to also have a squirt bottle with neat DMF for washes.
4. Weigh out enough dry Rink amide-MBHA resin for a 0.25 mmol scale synthesis (e.g., if the loading level is 0.5 mmol/g, use 500 mg of resin), and transfer the resin to a fritted synthesis vessel.
5. Wash the resin with methanol by applying sufficient solvent to submerge the resin completely, and then removing the solvent by filtration under vacuum. Do this several times, then repeat the process with dichloromethane.
6. To swell the resin, add DMF to roughly 2–3 times the volume of the resin, and agitate the resin by gently bubbling nitrogen gas through the suspension for 30 min. This mixing strategy of bubbling nitrogen gas through the vessel will be used for all deprotection and coupling steps.

7. Remove the first Fmoc protecting group by quickly flow-washing the resin (*see* **Note 14**) with 20% piperidine in DMF for 1 min and then incubating the resin in 20% piperidine in DMF for 20 min.
8. Wash the resin thoroughly by a combination of batch and flow washes with neat DMF, then drain the excess solvent.
9. For each coupling reaction, dissolve the preweighed protected amino acid in 2.5 mL of the HOBt/HBTU solution so that the amino acid is at a concentration of 0.5 M. Add 10 eq. of DIPEA (413 μL, for a 0.25 mmol scale synthesis) to the solution, mix it rapidly, and add it to the washed resin.
10. Let the coupling reaction proceed for at least 30 min while agitating with a stream of nitrogen gas.
11. The extent of each coupling and deprotection reaction can be monitored by a ninhydrin test [53].
12. After the coupling reaction, wash the resin as in **step 8**, then repeat **steps 7–11** to couple and deprotect the next amino acid residue.
13. Once all rounds of coupling, deprotection, and washing are complete, wash the resin thoroughly with dichloromethane, then methanol. Dry the resin completely under vacuum.
14. To fully deprotect the peptide and cleave it off of the resin, suspend the peptidyl-resin in 4 mL of a cleavage cocktail that contains 95% TFA, 2.5% TIS, and 2.5% water, and vigorously shake the suspension for 2 h (*see* **Note 15**).
15. After cleavage and deprotection, precipitate the peptide by adding 45 mL of cold diethyl ether. This mixture can be kept on ice to facilitate precipitation.
16. Centrifuge the sample for 10 min at 4 °C and 4000 rcf and decant the supernatant (*see* **Note 16**).
17. The precipitated peptide can be dissolved in water or a mixture of water and acetonitrile, then filtered to remove the resin. Excess TFA and ether can be removed by lyophilization if desired.
18. Finally, purify the peptide by RP-HPLC, lyophilize the final product, and store it as a dry powder or as a frozen stock dissolved in water to a concentration greater than 20 mM.

3.2.4 Production of the C-Terminal Fragment by Protein Semisynthesis

1. The Int^C-GyrA-His_6 fusion protein can be expressed from the modified pTXB1 vector and enriched from cell lysates using the same procedure described in Subheading 3.2.2 for the His_6-SUMO-Ex^N-Int^N construct, with two additional considerations: (1) The induction time and temperature during overexpression may be altered, depending on the intein. For

example, for the *Npu*C and *Ssp*C intein fragments, 3 h of induction at 37 °C is sufficient. (2) No thiols (e.g., BME and DTT) should be present in any of the purification buffers.

2. After elution from a nickel column, treat the eluate with 10 mM TCEP, then use dialysis or a desalting column to exchange the buffer to thiol-free Ni column buffer A.
3. Initiate the thiolysis/ligation reaction (Fig. 3b) by adding 100 mM MESNa (from the 1 M MESNa stock solution) and 2 mM of the desired C-extein peptide (from the stock solution prepared at the end of Subheading 3.2.3) to 10–50 μM of the Int^C-GyrA-His_6 fusion protein, and ensure that the pH of the solution remains between 7 and 8 (*see* **Note 17**).
4. Allow the ligation to proceed at room temperature overnight. Reaction progress can be monitored by analyzing 10–20 μg of total protein by analytical RP-HPLC at various time points.
5. Once the ligation is complete, quench the reaction with the addition of neat TFA to a final concentration of 0.5% (v/v). Some of the GyrA protein will precipitate during this quenching step.
6. Centrifuge the sample at 4000 rcf for 10 min to remove the precipitated GyrA protein, then filter the supernatant.
7. Purify the Int^C-Ex^C fusion by RP-HPLC and confirm the identity of the product by mass spectrometry.
8. Pool and lyophilize the fractions that contain the pure peptide, redissolve the peptide in water or reaction buffer, and then determine its concentration by measuring absorbance at 280 nm.
9. *See* **Note 18** for an alternative strategy to produce the Int^C-Ex^C construct.

3.2.5 Monitoring Protein Trans-Splicing by Reverse-Phase HPLC

1. Prepare a 650 μL solution of the Ex^N-Int^N protein at 15 μM in reaction buffer supplemented with 1 mM TCEP.
2. Prepare a 650 μL solution of the Int^C-Ex^C protein at 10 μM in reaction buffer supplemented with 1 mM TCEP.
3. Prepare and label 12 time point quenching tubes containing 33 μL of quenching buffer.
4. Generate an artificial time zero sample by mixing 50 μL of each of the intein fragments sequentially into one time point tube.
5. Preincubate the intein fragments at the desired temperature, and then initiate the reaction by mixing the remainder of the two solutions together.
6. At the desired time points, transfer 100 μL from the reaction tube to a quenching tube and mix thoroughly. Time points

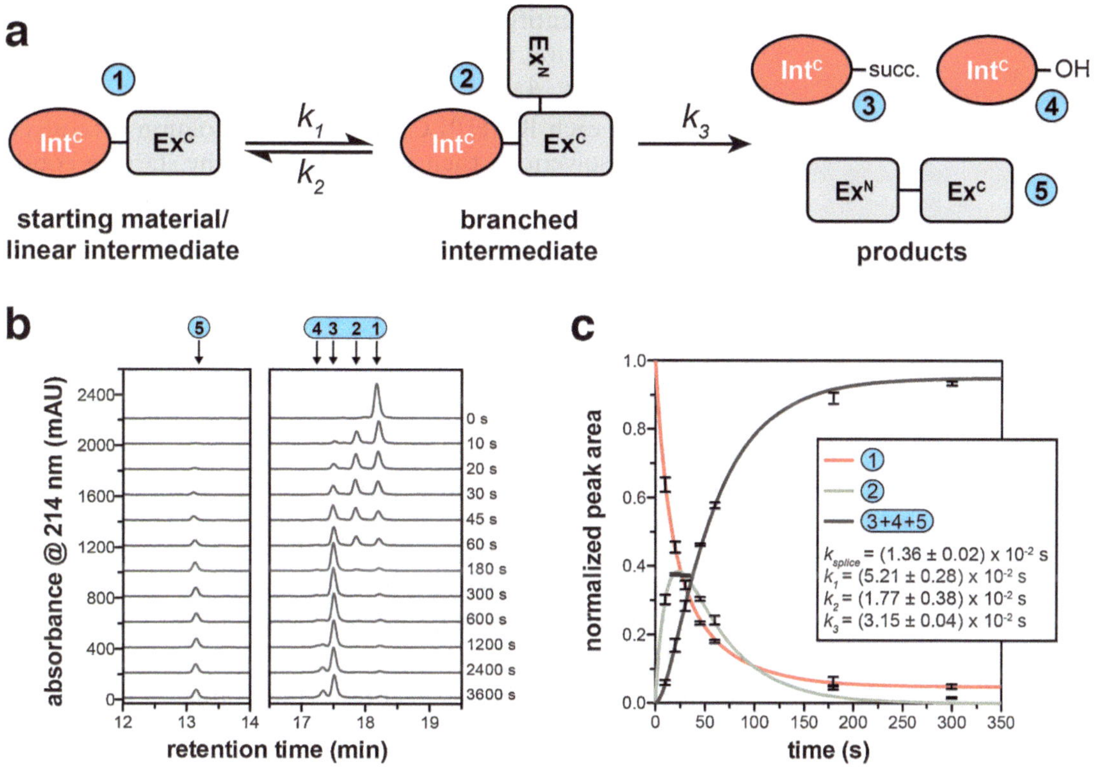

Fig. 4 An RP-HPLC assay to monitor the kinetics of protein *trans*-splicing. (**a**) The five molecules in this assay that can be readily resolved by RP-HPLC and identified by mass spectrometry. The label "-succ." in molecule **3** refers to a C-terminal succinimide, whereas "-OH" in molecule **4** indicates a C-terminal carboxylic acid. (**b**) Representative RP-HPLC chromatograms from a reaction of the *Npu*DnaE intein, showing changes in the abundance of each species in panel (**a**) over the course of a splicing reaction. In this reaction, the Ex^N sequence is Ala-Glu-Tyr and the Ex^C sequence is Cys-Phe-Asn. (**c**) Reaction progress curves showing the normalized abundance of the starting material, the branched intermediate, and the products, for the reaction with the *Npu*DnaE intein shown in panel (**b**). Error bars represent the standard deviation from three replicates. Rate constants were derived using the fitting procedures described in Subheading 3.2.6. (The figures in panels (**b**) and (**c**) are derived from data reported previously [45])

should be chosen to be close together early in the reaction and more spread out as the reaction goes to completion (Fig. 4b).

7. Run 100 μL of each quenched sample on an analytical C_{18} column. As described previously for *Npu*-derived inteins [45], the following RP-HPLC parameters are a good starting point: at a flow rate of 1 mL/min, first run a 2 min isocratic segment with 100% solvent A, then separate the polypeptides over a 30 min linear gradient from 0% to 73% solvent B in solvent A (*see* **Note 19**). Monitor absorbance at 214 nm.
8. The first time a new intein-extein combination is analyzed, the gradient may need to be optimized, and peaks should be

collected to identify species by mass spectrometry. For DnaE-family split inteins, typically, Int^C-derived molecules are readily identifiable (Fig. 4a, b).

9. In between each RP-HPLC run, wash the column extensively with 100% solvent B, then reequilibrate in the starting conditions of the gradient before injecting the next sample.
10. Reactions should be carried out in triplicate.

3.2.6 Quantitative Analysis of Reaction Progress from RP-HPLC Data

An extensive description of the kinetic analysis of data from this type of RP-HPLC splicing assay was reported in the supplementary information of the original study where this technique was first used [45]. A summary of the important steps is given below.

1. Most HPLC instruments have a data analysis software that allows for the integration of peaks. Use this software to determine the area under every peak that corresponds to a species in Fig. 4a: the starting material Int^C-Ex^C (1), the branched intermediate $Int^C(Ex^N)$-Int^C (2), the excised Int^C in its succinimide (3) and free acid forms (4), and the spliced product Ex^N-Ex^C (5).
2. Sum the areas of the two excised Int^C species and the spliced product (3 + 4 + 5) to represent the full product of the splicing reaction.
3. For simplicity of analysis and curve fitting, express the area of each peak as a fraction of the total area of all peaks in that time point (Fig. 4c).
4. To determine the overall rate of splicing (k_{splice}), fit the normalized data for the formation of the product species as a function of time to a first order rate equation:

$$[P](t) = [P]_{max}\left(1 - e^{-k_{splice} \times t}\right)$$

where t is time $[P]$ is the normalized peak area for the sum of the products, $[P]_{max}$ is the extrapolated maximum abundance of product at the end of the reaction, and k_{splice} is the rate of product formation. $[P]_{max}$ and k_{splice} should be allowed to vary during the fit.

5. To obtain the kinetic parameters describing the formation and resolution of the branched intermediate, the data can be fit to a three-state kinetic model that describes the relative concentrations of the starting material, branched intermediate, and combined products as a function of time (Fig. 4a).
6. To fit the normalized data to this kinetic model, use the following collection of equations, which are the analytical solution to the differential rate equations describing that model.

$$p = k_1 + k_2 + k_3$$

$$q = \sqrt{p^2 - 4(k_1 k_3)}$$

$$a = \frac{1}{2}(p + q)$$

$$b = \frac{1}{2}(p - q)$$

$$[\mathrm{S/L}](t) = [\mathrm{S/L}]_0 \left[\left(\frac{k_1(a - k_3)}{a(a - b)} \right) \mathrm{e}^{-at} + \left(\frac{k_1(k_3 - b)}{b(a - b)} \right) \mathrm{e}^{-bt} \right]$$

$$[\mathrm{B}](t) = [\mathrm{S/L}]_0 \left[\left(\frac{-k_1 a}{a(a - b)} \right) \mathrm{e}^{-at} + \left(\frac{k_1 b}{b(a - b)} \right) \mathrm{e}^{-bt} \right]$$

$$[\mathrm{P}](t) = [\mathrm{S/L}]_0 \left[\left(\frac{k_1 k_3}{ab} \right) + \left(\frac{k_1 k_3}{a(a - b)} \right) \mathrm{e}^{-at} - \left(\frac{k_1 k_3}{b(a - b)} \right) \mathrm{e}^{-bt} \right]$$

where p, q, a, and b are algebraic combinations of rate constants k_1, k_2, and k_3 that simplify expression of the rate equations. [S/L] is the normalized intensity of the **S**tarting Material/**L**inear intermediate (**1**), [B] is the normalized intensity of the **B**ranched Intermediate (**2**), and [P] is the normalized intensity of the **P**roducts (**3** + **4** + **5**). The variable t is the reaction time, and $[\mathrm{S/L}]_0$ can be considered a normalization factor that represents the extent of the reaction (the fraction of starting material that gets converted to product).

7. For all experiments, fit the data for individual replicates separately and average the kinetic parameters to determine the rates and extrapolated yields of reactions for a given intein-extein combination.

4 Notes

1. We frequently split the KanR gene between the codons for F198 and S199, and codons 199-201 can be mutated to accommodate a + 1 Cys residue and other C-extein residues, if needed, without compromising KanR activity [25, 29, 46, 51]. Residue numbering here is based on NCBI Reference Sequence WP_000018329.1.

2. Although this protocol has been optimized using the Roche cOmplete EDTA-free protease inhibitor cocktail, any cocktail of protease inhibitors that can target the most common types of proteases in *E. coli* is sufficient. Common reagents include leupeptin, benzamidine, phenylmethylsulfonyl fluoride, and 4-(2-aminoethyl)benzenesulfonyl fluoride. The inhibitors should be tested to ensure that they do not inhibit intein chemistry.

3. This protocol was optimized using Thermo Scientific HisPure resin. Other NiNTA resins should also be applicable, and adjustments should be made according to the manufacturer's instructions.
4. The buffer can be degassed under vacuum with vigorous stirring to extend the lifetime of the reducing agents.
5. Piperidine is on a list of chemicals, maintained by the US Drug Enforcement Administration, that can be used for the production of controlled substances. As such, it can be difficult to purchase; 4-methylpiperidine is a suitable alternative that circumvents this issue.
6. EDTA is added to the reaction buffer to scavenge divalent cations such as Zn^{2+}, which have been shown to inhibit protein splicing [54].
7. The difference in IC_{50} values between the inactive intein plasmid and the contiguous KanR control plasmid represents the dynamic range of this assay for a given intein and KanR split site. If a poor dynamic range is observed, additional split sites in the KanR gene should be explored [44, 46, 47, 52].
8. Use a serial dilution of the LB/Kan aliquots to ensure consistency in the assay. Since kanamycin will be diluted 10% when adding cells, kanamycin concentrations in these stock solutions can be prepared at a 1.1-fold higher concentration than the target concentration.
9. It is advisable to have technical replicates for each construct on the same plate (ideally three).
10. Cells can also be grown at 25 °C and 37 °C, with longer or shorter incubation periods, respectively. During these long growth periods on the 96-well plate, ensure that minimal evaporation of the LB occurs. This may be achieved by slightly sealing the lid to the plate with parafilm or using plates that are made for low evaporation rates.
11. The ligation strategy for producing C-inteins with small C-exteins is useful when introducing functional groups into the C-extein that are not genetically encodable. This approach also provides a useful modular platform for rapidly altering C-extein composition [45, 49, 51]. By contrast, the alternative strategy, in which the C-extein is encoded in the plasmid, may be necessary for inteins that require a + 1 serine or threonine, where a ligation reaction to append the C-extein may not be efficient (*see* **Note 18**).
12. The amount of Ulp1 and the exact temperature that is best for cleavage without compromising the protein sample by aggregation, precipitation, or misfolding may need to be optimized, by doing a series of small-scale reaction tests.

13. Ideal cryogenic conditions should be assessed on a case-by-case basis. Often, storage in the purification buffer, supplemented with 10% glycerol, is sufficient for most proteins.
14. To flow-wash resin, first prepare a squirt bottle with the desired solvent. Submerge the resin in this solvent, then, without completely drying the resin, constantly evacuate solvent and replenish it with fresh solvent from the squirt bottle. Do this for roughly 30 s. Alternatively, resuspend the resin in solvent and evacuate that solvent rapidly and repeatedly for 30 s.
15. Depending on the peptide sequence, alternative cleavage cocktails may be more appropriate and should be explored empirically. The cleavage conditions given in this protocol are sufficient for many small peptides synthesized using standard Fmoc-protected amino acids.
16. Some small hydrophobic peptides may not precipitate completely in cold diethyl ether, and so caution should be taken in doing excessive ether precipitation and washing. If yields are low, the supernatant from the ether precipitation or washes should be checked. For larger peptides, it may be favorable to wash the pellet one or more times with cold ether.
17. Since the Int^C-GyrA-His_6 protein has only been partially purified by Ni affinity chromatography, it may be desirable to add protease inhibitors to the ligation reaction mixture to curb any undesired proteolysis.
18. In some cases, it may be more practical to generate the Int^C-Ex^C polypeptide by encoding the whole construct genetically, rather than producing it semisynthetically. We have observed, for the *Npu*DnaE intein, that a His_6-SUMO-Int^C-Ex^C fusion is proteolyzed substantially during overexpression in *E. coli*, although other fusion systems may work. One alternative is to produce an Int^C-Ex^C-GyrA-His_6 fusion and then thiolyze the Int^C-Ex^C polypeptide from the GyrA intein with the addition of β-mercaptoethanol (200 mM, pH 7.5, 22 °C), which will ultimately yield the hydrolyzed product [55].
19. In cases where the species of interest do not resolve, the HPLC method can be refined to use a longer and/or shallower gradient from solvents A to B. For some intein-extein combinations, the species of interest may not be resolvable by HPLC. In these cases, the quenched sample can be desalted using a zip-tip or hydrophobic extraction cartridge and analyzed by electrospray ionization mass spectrometry without an initial reverse-phase chromatography step. As shown previously, since all of the Int^C species are similar in size and composition, the relative abundance by mass spectrometry correlates well with abundances measured by RP-HPLC [45].

Acknowledgments

N.H.S and A.J.S. are both supported by grants from the Damon Runyon Cancer Research Foundation.

References

1. Shah NH, Muir TW (2014) Inteins: nature's gift to protein chemists. Chem Sci 5:446–461
2. Kane PM, Yamashiro CT, Wolczyk DF, Neff N, Goebl M, Stevens TH (1990) Protein splicing converts the yeast TFP1 gene product to the 69-kD subunit of the vacuolar H(+)-adenosine triphosphatase. Science 250:651–657
3. Topilina NI, Novikova O, Stanger M, Banavali NK, Belfort M (2015) Post-translational environmental switch of RadA activity by extein-intein interactions in protein splicing. Nucleic Acids Res 43:6631–6648
4. Topilina NI, Green CM, Jayachandran P, Kelley DS, Stanger MJ, Piazza CL, Nayak S, Belfort M (2015) SufB intein of Mycobacterium tuberculosis as a sensor for oxidative and nitrosative stresses. Proc Natl Acad Sci U S A 112:10348–10353
5. Lennon CW, Stanger M, Belfort M (2016) Protein splicing of a recombinase intein induced by ssDNA and DNA damage. Genes Dev 30:2663–2668
6. Lennon CW, Stanger M, Banavali NK, Belfort M (2018) Conditional protein splicing switch in hyperthermophiles through an intein-extein partnership. MBio 9(1):e02304–e02317. https://doi.org/10.1128/mBio.02304-17
7. Kelley DS, Lennon CW, Li Z, Miller MR, Banavali NK, Li H, Belfort M (2018) Mycobacterial DnaB helicase intein as oxidative stress sensor. Nat Commun 9:4363
8. Vila-Perelló M, Muir TW (2010) Biological applications of protein splicing. Cell 143:191–200
9. Muir TW, Sondhi D, Cole PA (1998) Expressed protein ligation: a general method for protein engineering. Proc Natl Acad Sci U S A 95:6705–6710
10. Chong S, Mersha FB, Comb DG et al (1997) Single-column purification of free recombinant proteins using a self-cleavable affinity tag derived from a protein splicing element. Gene 192:271–281
11. Lahiry A, Fan Y, Stimple SD, Raith M, Wood DW (2017) Inteins as tools for tagless and traceless protein purification. J Chem Technol Biotechnol 93:1827–1835
12. Camarero JA, Muir TW (1999) Biosynthesis of a head-to-tail cyclized protein with improved biological activity. J Am Chem Soc 121:5597–5598
13. Evans TC, Martin D, Kolly R, Panne D, Sun L, Ghosh I, Chen L, Benner J, Liu XQ, Xu MQ (2000) Protein trans-splicing and cyclization by a naturally split intein from the dnaE gene of Synechocystis species PCC6803. J Biol Chem 275:9091–9094
14. Tavassoli A (2017) SICLOPPS cyclic peptide libraries in drug discovery. Curr Opin Chem Biol 38:30–35
15. Ozawa T, Nogami S, Sato M, Ohya Y, Umezawa Y (2000) A fluorescent indicator for detecting protein-protein interactions in vivo based on protein splicing. Anal Chem 72:5151–5157
16. Jeon H, Lee M, Jang W, Kwon Y (2016) Intein-mediated protein engineering for biosensor fabrication. Biochip J 10:277–287
17. Slomovic S, Collins JJ (2015) DNA sense-and-respond protein modules for mammalian cells. Nat Methods 12:1085–1090
18. Mootz HD, Blum ES, Tyszkiewicz AB, Muir TW (2003) Conditional protein splicing: a new tool to control protein structure and function in vitro and in vivo. J Am Chem Soc 125:10561–10569
19. Giriat I, Muir TW (2003) Protein semi-synthesis in living cells. J Am Chem Soc 125:7180–7181
20. Borra R, Dong D, Elnagar AY, Woldemariam GA, Camarero JA (2012) In-cell fluorescence activation and labeling of proteins mediated by FRET-quenched split inteins. J Am Chem Soc 134:6344–6353
21. David Y, Vila-Perelló M, Verma S, Muir TW (2015) Chemical tagging and customizing of cellular chromatin states using ultrafast trans-splicing inteins. Nat Chem 7:394–402
22. Southworth MW, Adam E, Panne D, Byer R, Kautz R, Perler FB (1998) Control of protein splicing by intein fragment reassembly. EMBO J 17:918–926
23. Mills KV, Lew BM, Jiang S, Paulus H (1998) Protein splicing in trans by purified N- and C-terminal fragments of the Mycobacterium

tuberculosis RecA intein. Proc Natl Acad Sci U S A 95:3543–3548

24. Brenzel S, Kurpiers T, Mootz HD (2006) Engineering artificially split inteins for applications in protein chemistry: biochemical characterization of the split Ssp DnaB intein and comparison to the split Sce VMA intein. Biochemistry 45:1571–1578
25. Shah NH, Dann GP, Vila-Perelló M, Liu Z, Muir TW (2012) Ultrafast protein splicing is common among cyanobacterial split inteins: implications for protein engineering. J Am Chem Soc 134:11338–11341
26. Shah NH, Muir TW (2011) Split inteins: nature's protein ligases. Isr J Chem 51:854–861
27. Vila-Perelló M, Liu Z, Shah NH, Willis JA, Idoyaga J, Muir TW (2013) Streamlined expressed protein ligation using split inteins. J Am Chem Soc 135:286–292
28. Shi J, Muir TW (2005) Development of a tandem protein trans-splicing system based on native and engineered split inteins. J Am Chem Soc 127:6198–6206
29. Shah NH, Vila-Perelló M, Muir TW (2011) Kinetic control of one-pot trans-splicing reactions by using a wild-type and designed split intein. Angew Chem Int Ed Engl 50:6511–6515
30. Stevens AJ, Sekar G, Gramespacher JA, Cowburn D, Muir TW (2018) An atypical mechanism of split intein molecular recognition and folding. J Am Chem Soc 140:11791–11799
31. Wu H, Hu Z, Liu XQ (1998) Protein trans-splicing by a split intein encoded in a split DnaE gene of Synechocystis sp. PCC6803. Proc Natl Acad Sci U S A 95:9226–9231
32. Perler FB (2002) Inbase: the intein database. Nucleic Acids Res 30:383–384
33. Waters E, Hohn MJ, Ahel I et al (2003) The genome of Nanoarchaeum equitans: insights into early archaeal evolution and derived parasitism. Proc Natl Acad Sci U S A 100:12984–12988
34. Caspi J, Amitai G, Belenkiy O, Pietrokovski S (2003) Distribution of split DnaE inteins in cyanobacteria. Mol Microbiol 50:1569–1577
35. Dassa B, London N, Stoddard BL, Schueler-Furman O, Pietrokovski S (2009) Fractured genes: a novel genomic arrangement involving new split inteins and a new homing endonuclease family. Nucleic Acids Res 37:2560–2573
36. Novikova O, Topilina N, Belfort M (2014) Enigmatic distribution, evolution, and function of inteins. J Biol Chem 289:14490–14497
37. Iwai H, Züger S, Jin J, Tam P-H (2006) Highly efficient protein trans-splicing by a naturally split DnaE intein from Nostoc punctiforme. FEBS Lett 580:1853–1858
38. Zettler J, Schütz V, Mootz HD (2009) The naturally split Npu DnaE intein exhibits an extraordinarily high rate in the protein trans-splicing reaction. FEBS Lett 583:909–914
39. Carvajal-Vallejos P, Pallissé R, Mootz HD, Schmidt SR (2012) Unprecedented rates and efficiencies revealed for new natural split inteins from metagenomic sources. J Biol Chem 287:28686–28696
40. Gramespacher JA, Stevens AJ, Nguyen DP, Chin JW, Muir TW (2017) Intein zymogens: conditional assembly and splicing of split inteins via targeted proteolysis. J Am Chem Soc 139:8074–8077
41. Pan D, Xuan B, Sun Y, Huang S, Xie M, Bai Y, Xu W, Qian Z (2016) An intein-mediated modulation of protein stability system and its application to study human cytomegalovirus essential gene function. Sci Rep 6:26167
42. Wang T, Badran AH, Huang TP, Liu DR (2018) Continuous directed evolution of proteins with improved soluble expression. Nat Chem Biol 14:972–980
43. Amitai G, Callahan BP, Stanger MJ, Belfort G, Belfort M (2009) Modulation of intein activity by its neighboring extein substrates. Proc Natl Acad Sci U S A 106:11005–11010
44. Cheriyan M, Pedamallu CS, Tori K, Perler F (2013) Faster protein splicing with the Nostoc punctiforme DnaE intein using non-native extein residues. J Biol Chem 288:6202–6211
45. Shah NH, Eryilmaz E, Cowburn D, Muir TW (2013) Extein residues play an intimate role in the rate-limiting step of protein trans-splicing. J Am Chem Soc 135:5839–5847
46. Lockless SW, Muir TW (2009) Traceless protein splicing utilizing evolved split inteins. Proc Natl Acad Sci U S A 106:10999–11004
47. Appleby-Tagoe JH, Thiel IV, Wang Y, Wang Y, Mootz HD, Liu X-Q (2011) Highly efficient and more general cis- and trans-splicing inteins through sequential directed evolution. J Biol Chem 286:34440–34447
48. Thiel IV, Volkmann G, Pietrokovski S, Mootz HD (2014) An atypical naturally split intein engineered for highly efficient protein labeling. Angew Chem Int Ed Engl 53:1306–1310
49. Stevens AJ, Brown ZZ, Shah NH, Sekar G, Cowburn D, Muir TW (2016) Design of a split intein with exceptional protein splicing activity. J Am Chem Soc 138:2162–2165
50. Ramirez M, Valdes N, Guan D, Chen Z (2013) Engineering split intein DnaE from Nostoc punctiforme for rapid protein purification. Protein Eng Des Sel 26:215–223

51. Stevens AJ, Sekar G, Shah NH, Mostafavi AZ, Cowburn D, Muir TW (2017) A promiscuous split intein with expanded protein engineering applications. Proc Natl Acad Sci U S A 114:8538–8543
52. Daugelat S, Jacobs WR (1999) The Mycobacterium tuberculosis recA intein can be used in an ORFTRAP to select for open reading frames. Protein Sci 8:644–653
53. Sarin VK, Kent SBH, Tam JP, Merrifield RB (1981) Quantitative monitoring of solid-phase peptide synthesis by the ninhydrin reaction. Anal Biochem 117:147–157
54. Sun P, Ye S, Ferrandon S, Evans TC, Xu M-Q, Rao Z (2005) Crystal structures of an intein from the split dnaE gene of Synechocystis sp. PCC6803 reveal the catalytic model without the penultimate histidine and the mechanism of zinc ion inhibition of protein splicing. J Mol Biol 353:1093–1105
55. Batjargal S, Walters CR, Petersson EJ (2015) Inteins as traceless purification tags for unnatural amino acid proteins. J Am Chem Soc 137:1734–1737

Chapter 4

Methods to Study the Structure and Catalytic Activity of *cis*-Splicing Inteins

Jing Zhao, Zhenming Du, Chunyu Wang, and Kenneth V. Mills

Abstract

The autocatalytic process of protein splicing is facilitated by an intein, which interrupts flanking polypeptides called exteins. The mechanism of protein splicing has been studied by overexpression in *E. coli* of intein fusion proteins with nonnative exteins. Inteins can be used to generate reactive α-thioesters, as well as proteins with N-terminal Cys residues, to facilitate expressed protein ligation. As such, a more detailed understanding of the function of inteins can have significant impact for biotechnology applications. Here, we provide biochemical methods to study splicing activity and NMR methods to study intein structure and the catalytic mechanism.

Key words Intein, Protein splicing, NMR spectroscopy, SDS-PAGE, Protein overexpression

1 Introduction

Protein splicing is a posttranslational modification facilitated by an intein [1–3]. Inteins are intervening protein sequences that catalyze their own excision from the flanking polypeptides, or exteins, concomitant to extein ligation. Some inteins are also interrupted by an endonuclease domain that can initiate the homing of intein-containing genes into inteinless alleles. Other inteins are the so-called mini-inteins, in which the homing endonuclease domain is lost and replaced by a short linker. Yet other inteins are split in this region, with N- and C-terminal extein–intein segments separated in the genome, requiring reassociation of the inteins to facilitate splicing in *trans*. In this chapter we focus on methods to study the mechanisms of *cis*-splicing inteins.

Most studies of the mechanism of protein splicing in *cis* have been undertaken with inteins that are overexpressed in *E. coli*, interrupting nonnative exteins that serve dual roles as solubility enhancers and affinity tags. Usually, a number of native extein residues flanking the intein are retained, as these residues can influence the efficiency of splicing [4–9]. Given that most inteins

Miquel Vila-Perelló (ed.), *Expressed Protein Ligation: Methods and Protocols*, Methods in Molecular Biology, vol. 2133,
https://doi.org/10.1007/978-1-0716-0434-2_4,

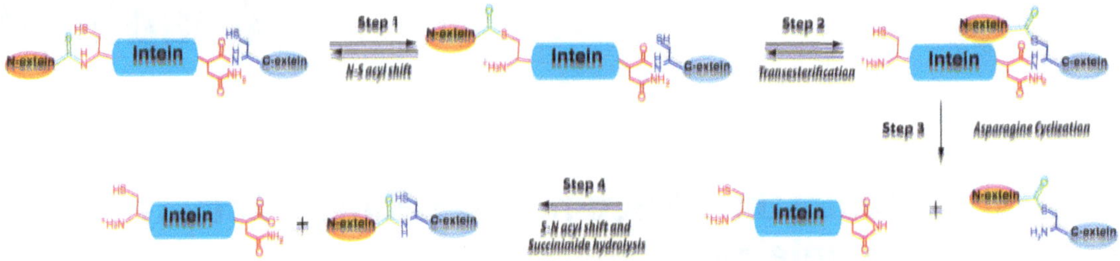

Fig. 1 Schematic representation of the four steps of the protein splicing mechanism

that are well expressed in *E. coli* splice efficiently on translation, in vitro study of protein splicing or of intein structures is usually performed either with inteins that have mutations at conserved catalytic residues or with inteins from extremophiles that can be purified as unspliced precursors and induced to splice by a physical or chemical cue, including change in temperature, pH, oxidation state, or salinity. Alternatively, some inteins have been induced to splice on a conditional basis by directed evolution or random mutagenesis [10–15].

The canonical protein splicing mechanism consists of four steps (Fig. 1): (1) an amide-ester (or thioester) rearrangement of the peptide bond that links the N-extein and intein, called an N–O (or N–S) acyl shift; (2) a transesterification that results in transfer of the N-extein from the side chain of the first residue of the intein (usually a Ser or Cys) to the side chain of the first residue of the C-extein (usually a Ser, Thr, or Cys); (3) cyclization of the C-terminal Asn residue of the intein that cleaves the peptide bond between the intein and C-extein; and (4) spontaneous conversion of the ester or thioester bond linking the excised exteins back to a peptide bond [1–3].

Inteins can be manipulated to create tools for protein biotechnology both in vitro and in vivo [16–18]. In the case of expressed protein ligation (EPL), the reactive α-thioester can be created if steps 2 or 3 of the splicing reaction are blocked by mutation. Induced C-terminal cleavage via Asn cyclization uncoupled from splicing can generate a C-extein with an N-terminal Cys residue, which can serve as the other reactant for EPL in place of a synthetic peptide [16, 19].

Our labs have studied the activity and structure of inteins using assays via SDS-PAGE [20, 21] and NMR spectroscopy [21, 22]. Below, we describe and discuss methods for intein fusion protein expression in *E. coli*, analysis of splicing activity via SDS-PAGE, and analysis of protein structure and pK_a determination of key catalytic residues by NMR spectroscopy.

2 Materials

All solutions for protein analysis should be prepared using ultra-pure water. All solutions for bacterial media should be sterilized before use.

1. LB Media: 20 g/L Tryptone, 10 g/L Yeast Extract, 10 g/L NaCl, pH 7.5. Sterilize by autoclave.
2. LB Agar: LB media supplemented with 15 g/L agar. Sterilize by autoclave.
3. *E. coli* NovaBlue and *E. coli* BL21(DE3).
4. 100 mg/mL carbenicillin or ampicillin. Sterilize by filtration.
5. 1.0 M Isopropyl β-D-1-thiogalactopyranoside (IPTG): 238 mg/mL IPTG in water.
6. 0.25 M Phenylmethylsulfonyl fluoride in isopropanol.
7. Benzonase nuclease.
8. EDTA-free protease inhibitor cocktail.
9. Cobalt- or Nickel-based metal affinity resin.
10. Centrifugal filter unit for protein concentration with appropriate molecular weight cut off.
11. 10× M9 Media: 30 g/L KH_2PO4, 68 g/L Na_2HPO_4 and 10 g/L NaCl. Sterilize by autoclave.
12. 1 M $MgSO_4$. Sterilize by autoclave.
13. 1 M $CaCl_2$. Sterilize by autoclave.
14. 1 g/mL $^{15}NH_4Cl$. Sterilize by filtration.
15. 1 g/mL glucose. Sterilize by filtration.
16. 1 g/mL $^{13}C_6$-D-glucose. Sterilize by filtration.
17. Protein splicing buffer solution: 20 mM HEPES buffer, pH 7.5, 500 mM NaCl.
18. Precast 4–20% gradient gels with stacking for SDS-PAGE.
19. 3× SDS loading buffer.
20. 1.2 M dithiothreitol (DTT).
21. 20 mM tris(2-carboxyethyl) phosphine (TCEP).
22. 50 mM EDTA in protein splicing buffer solution.
23. 6% aqueous acetic acid.
24. Coomassie Blue staining solution.
25. High-resolution flatbed scanner.
26. Protein buffer solution for NMR analysis: 20 mM sodium phosphate, pH 7.1, 5 mM TCEP.
27. Plasmids: pET21-b(+), pMal-c2X, pMal-c5X, pETM-44.

28. Bruker Advance II 800 MHz or Bruker Advance II 600 MHz spectrometer, equipped with a triple-resonance cryogenic probe.
29. A pH electrode for small volume NMR samples.
30. French Pressure Cell Press with appropriate cell.

3 Methods

3.1 Construction of Expression Vectors for Protein Expression in E. coli

We utilize one of two strategies for the design of *E. coli* expression vectors. In order to study protein splicing via SDS-PAGE, it is useful to have N- and C-terminal exteins that are soluble, can facilitate affinity chromatography, can be used to distinguish spliced products from precursor inteins via SDS-PAGE, and can be detected by commercially available antibodies. We use *E. coli* maltose binding protein (MBP) as the N-extein and a polypeptide with a C-terminal His-tag as the C-extein. Alternatively, for applications where we plan to study the structure or folding of the intein, we use expression vectors that encode for a very short segment of the N-extein in place of the MBP.

1. In order to generate an expression vector with N-terminal MBP, insert the intein gene in-frame following the gene for *E. coli* maltose binding protein (MBP) and a short linker sequence that also encodes for a protease-recognition site. (*See* **Note 1** for possible plasmid vectors to use.) Fuse the 3′-end of the intein gene in-frame to a sequence that codes for another affinity tag, such as a poly-His tag or the C-terminal domain of I-TevI.
2. To generate an expression vector without the MBP, either delete the MBP segment so that expression is still under control of the Tac promoter in the pMal plasmid, or transfer of the intein gene into an expression vector such as pET-21b(+), so that expression is under the control of the T7 promoter (*see* **Notes 2** and **3**).

3.2 Protein Overexpression and Purification of Unlabeled Intein Fusion Protein

Intein fusion proteins can be expressed in *E. coli*. Many inteins splice in *E. coli*, and their activity can be analyzed by comparing the extent of splicing or splicing side reactions of wild type inteins to those modified by mutation to key active site residues. Other inteins, such as those described here, can be purified from *E. coli* as unspliced precursors and induced to splice after purification.

1. Transform the expression plasmid into *E. coli* BL21(DE3) on an LB agar plate with appropriate antibiotic selection (100 μg/mL carbenicillin or ampicillin) and incubate overnight at 37 °C (*see* **Note 4**).

2. Pick a single colony and incubate overnight in 2 mL of LB media at 37 °C with shaking.
3. Inoculate 1 mL of overnight culture into 100 mL of LB media and grow with shaking at 37 °C until OD_{600} reaches about 0.6.
4. Induce expression with 1.0 mM Isopropyl β-D-1-thiogalactopyranoside (IPTG) and incubate with shaking at 20 °C for 16 h (*see* **Note 5**).
5. Harvest cells by centrifugation at 3000 × *g* for 10 min. Decant supernatant and either store cell pellets at −20 °C or process immediately.
6. Resuspend cells in 2.5 mL of protein splicing buffer solution supplemented with 0.1 mM phenylmethylsulfonyl fluoride (PMSF), 10 units of benzonase nuclease, and EDTA-free protease inhibitor cocktail.
7. Disrupt cells via passage through a French Pressure cell at approximately 18,000 psi (with 900 psi gauge pressure using a 3/8 inch diameter piston) (*see* **Note 6**).
8. Centrifuge clarified lysate at 12,000 × *g* for 30 min at 4 °C.
9. Purify protein via C-terminal His-tag using immobilized metal affinity chromatography. Add 0.7 mL of a 50% slurry of the resin to a 15 mL centrifuge tube, and centrifuge at 500 × *g* to settle resin. Decant liquid and resuspend resin with 5 mL cold protein splicing buffer solution, repeating twice. Add clarified cell lysate to resin and incubate with rotation for 30 min at 4 °C. Centrifuge at 500 × *g* for 5 min, decant liquid, resuspend resin in 5 mL of cold protein splicing buffer solution, and add to a small chromatography column at 4 °C. Wash one time with 10 column volumes (CV) of protein splicing buffer solution supplemented with 10 mM imidazole and 0.1% Tween-20, two times with 10 CV of protein splicing buffer solution supplemented only with 10 mM imidazole, and elute with three 500 μL fractions of protein splicing buffer solution supplemented with 200 mM imidazole (*see* **Notes 7** and **8**).
10. In order to remove the high concentration of imidazole from the protein solution, use a centrifugal filter with appropriate molecular weight cut off. First add the elution to the filter, centrifuge at 14,000 × *g* for 10 min, add protein splicing buffer solution to 500 μL, centrifuge and repeat six times. Recover the final buffer-exchanged protein by inverting the filter into a 1.5 mL Eppendorf tube and centrifuging at 14,000 × *g* for 2 min.
11. Determine concentration of protein in exchanged fraction using a Bradford assay. Using the reagent brought to room temperature, add 20 μL of protein sample to 980 μL of reagent, measure absorbance at 595 nm, and use a standard curve to estimate protein concentration.

3.3 Standard Method for Observing Splicing Activity

For intein fusion proteins that can be purified as unspliced precursors, we can induce splicing or protein splicing side reactions by incubation at conditions similar to those native to the host organism. For inteins with active site Cys residues, this includes incubation with a reducing agent such as DTT or TCEP.

1. Create a reaction cocktail of 16 μL total volume on ice. Prepare stock solutions of EDTA (50 mM) and TCEP (20 mM) in protein splicing buffer solution. Create a master mix such that 16 μL aliquots of the mix have 3 μg of protein, and are 5 mM EDTA and 2 mM TCEP, all in protein splicing buffer solution. Transfer 16 μL aliquots into thin-walled 0.5 mL PCR tubes, and incubate the reaction for the appropriate temperature or time for the intein under study (*see* **Notes 9–11**).
2. Analyze the extent of protein splicing via SDS-PAGE, using precast 4–20% gradient gels with stacking, which provide good separation in the molecular weight region of interest. For each sample, add 8 μL of 3× SDS loading buffer, 1 μL of 1.2 M DTT, heat for 3 min at 100 °C, and load. Include a sample lane that contains the same reaction cocktail, but stored at 4 °C as a negative control for splicing.
3. Stain the SDS-PAGE gel in Coomassie Blue overnight, and destain in 6% acetic acid (*see* **Note 12**) [23].
4. Scan the gel with a high resolution flat-bed scanner. Protein splicing converts the precursor protein (MBP-Intein-C-extein) into MBP-C-extein and Intein (*see* **Note 13**). *See* Fig. 2 for an example of a SDS-PAGE analysis of an intein from *Pyrococcus abyssi*. Densitometry of the intensity of the bands can be used as a measure of the extent of protein splicing. Using the free Java-based program ImageJ [24], estimate the relative intensity of each band, and correct for the M_r of each protein. Given that protein splicing results in the conversion of the unspliced precursor (MIH) to the spliced product (MH) and excised intein (I), one can estimate the percentage of splicing using the ratio of $\frac{(MH)}{(MH)+(MIH)}$.

3.4 NMR Analysis of Intein Structure and Dynamics

In order to gain structural and mechanistic insight into protein splicing, NMR spectroscopy has been used to study the structure and dynamics of a hyperthermophilic intein. The ionization states of active site residues make crucial contributions to the atom-level mechanism of the catalysis, with highly elevated or depressed pK_a values observed in many enzymes [25]. We used a pH titration coupled with NMR measurement to determine the pK_a of catalytic intein residues in protein splicing. The methods presented are based on our studies of *Mtu* RecA intein [22, 26], *Pab* PolII intein [21] and Hint domain of Hedgehog protein [27] and can be generally applicable to other inteins and other proteins.

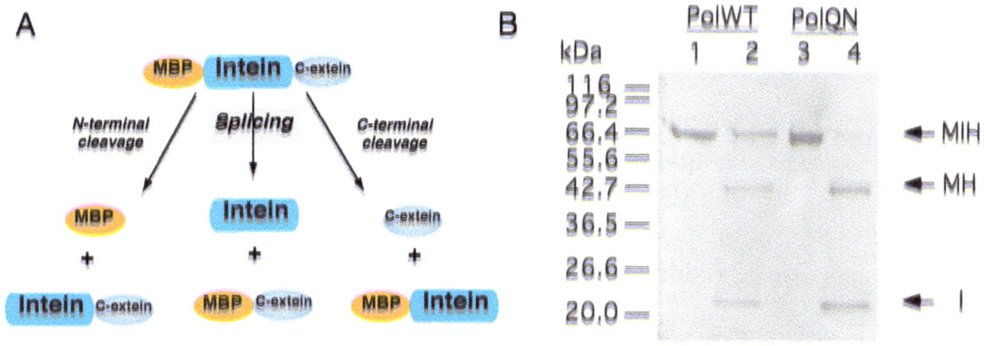

Fig. 2 In vitro assay of protein splicing activity. (**a**) Schematic representation of the process of protein splicing, in which the precursor protein (MIH in panel B) is converted to the spliced product and excised intein (MH and I in panel B, respectively.) Peptide bond cleavage uncoupled from splicing can result in the side reactions of N-terminal or C-terminal cleavage. For the *Pyrococcus abyssi* PolII intein, the fusion protein contains M [residues 27–392 of *E. coli* MBP, a 23 residue Asn-rich linker, and the 7 C-terminal residues of the N-extein (H-A-A-K-R-R-N)], I [the 185 residue intein], and H [the first six residues of the C-extein (C-D-G-D-E-D) and a His-6 tag] [56]. (**b**) SDS-PAGE analysis of the in vitro splicing of the *Pyrococcus abyssi* PolII intein. Lanes 1 and 3 contain purified, unspliced intein fusion protein, whereas lanes 2 and 4 are the result of incubation for 18 h at 70 °C. The intein has a native C-terminal Gln (PolWT); splicing efficiency was compared to the intein with conversion to the more highly conserved Asn (PolQN). (This research was originally published in *The Journal of Biological Chemistry*. Mills, K.V.; Manning, J.S.; Garcia, A.M., and Wuerdeman, L.A. Protein Splicing of a *Pyrococcus abyssi* Intein with a C-terminal Glutamine. *J. Biol. Chem.* **2004**, *279* (20), 20,685–20,691. © The American Society for Biochemistry and Molecular Biology)

3.4.1 Overexpression of Isotopically Labeled Proteins for NMR Analysis

For NMR analysis, clone the intein of interest into a pETM-44 vector, which expresses a fusion protein with an N-terminal His-tag fused to MBP and the intein. There is a linker sequence (TPGSLEVLKQGPM) between MBP and the intein. The sample data shown are from the *Pab* PolII intein (*Pyrococcus abyssi* DNA polymerase II intein).

1. Transform the pETM-44 plasmid with the intein gene into *E. coli* BL21(DE3) on an LB agar plate with appropriate antibiotic selection (100 μg/mL carbenicillin or ampicillin) and incubate overnight at 37 °C.
2. Pick single colony and inoculate overnight in LB media at 37 °C with shaking.
3. Prepare 10× M9 salt stock solution, 1 M $MgSO_4$, and 1 M $CaCl_2$ stock solutions. All stock solutions should be prepared in water and sterilized before use.
4. Dilute 10× M9 salt 1:10 with sterilized water in a 1 L flask. Add 2 mL of $MgSO_4$ stock solution and 0.1 mL of $CaCl_2$ stock solution.
5. Add ampicillin or carbenicillin to a final concentration of 100 μg/mL.
6. Add isotopes. For ^{15}N labeled samples, add 1 mL 1 g/mL of $^{15}NH_4Cl$ (final concentration is 1 g/L) and 4 mL 1 g/mL of glucose (final concentration is 4 g/L). For ^{15}N and ^{13}C doubly

labeled samples, add 1 mL 1 g/mL of $^{15}NH_4Cl$ (final concentration is 1 g/L) and 4 mL 1 g/mL of $^{13}C_6$-D-glucose (final concentration is 4 g/L) (*see* **Note 14**).

7. Swirl flask until all components are mixed well.
8. Resuspend the pellets obtained from an overnight culture into M9 medium to obtain a starting OD_{600} of 0.1.
9. Incubate the M9 cultures at 37 °C until OD_{600} reaches 0.3–0.4 and induce with 1 mM IPTG at 20 °C for an additional 16 h.

3.4.2 NMR Resonance Assignment and Structure Determination of Intein

To solve the solution structure of the intein, carry out the assignment of NMR resonance and structure determination of intein using a series of 2D and 3D NMR spectra (*see* **Notes 15** and **16**).

1. As described in Subheading 3.4.1, overexpress and purify ^{15}N labeled and $^{15}N,^{13}C$ doubly labeled intein samples.
2. Record a 2D ^{1}H–^{15}N HSQC spectrum using ^{15}N labeled intein.
3. Record a 2D ^{1}H–^{13}C HSQC spectrum using ^{13}C labeled intein.
4. To assign the backbone atoms (H_N, N_H, C_α, C_β, and C′), record 3D HN(CA)CO, 3D HNCO, 3D HNCACB, and 3D HN(CO)CACB spectra using $^{15}N,^{13}C$ doubly labeled intein.
5. To assign the sidechain atoms (H_α, H_β, H_γ, H_δ, C_γ, C_δ) of the intein, record a 3D ^{15}N-TOCSY ($\tau_m = 55$ ms) spectrum using ^{15}N labeled intein, and 3D (H)CC(CO)NH-TOCSY, 3D H(CC)(CO)NH-TOCSY, and 3D HCCH-TOCSY ($\tau_m = 15$ ms) spectra using $^{15}N,^{13}C$ doubly labeled intein.
6. To determine the structure of the intein, record a 3D ^{15}N-NOESY spectrum in 90% H2O, and two 3D ^{13}C-NOESY spectra in 100% D2O and 90% H2O, respectively.
7. Generate peak lists from 3D ^{15}N-NOESY and 3D ^{13}C-NOESY spectra. Use the peak lists, along with chemical shifts from the resonance assignment, as input for a program such as CYANA3.0 [28, 29], which can generate automatic NOE assignments and 3D structure through iterative assignment and structural calculation (*see* **Note 17**).
8. Derive angle restraints from TALOS+ [30, 31] prediction based on chemical shifts or from three-bond J-coupling using the Karplus relationship [32, 33] (*see* **Note 18**).
9. Subject the structure from CYANA with the lowest target function values to refinement with residual dipolar coupling in explicit water in Xplor-NIH [34].
10. Assess the quality of the final structures with PSVS (*see* **Note 19**) [35].

See Table 1 for the statistics of experimental NMR data for the structure calculation.

Table 1
The experimental NMR data for the structure calculation and the structural statistics of the 20 energy-minimized conformers of Pab PolII intein

(a) NMR constraints	*Value*		
Distance constraints	3231		
Intra-residue [$i = j$]	774		
Sequential [$\|i - j\|=1$]	843		
Medium range [$1<\|i - j\|<5$]	345		
Long range [$\|i - j\|\geq 5$]	1269		
Dihedral-angle constraints	223		
Φ	110		
Ψ	113		
Hydrogen bonds	51		
Residual dipolar couplings	90		
Total number of restricting constraints	3595		
(b) RMSD from ideal geometry			
Bond lengths, Å	0.016		
Bond angle, °	0.9		
(c) RMSD to mean coordinates, Å			
All backbone atoms excluding disordered loop residues 121–146	0.46 ± 0.10		
All heavy atoms excluding disordered loop residues 121–146	0.93 ± 0.12		
(d) Ramachandran plot statistics[a]			
Residues in most favored regions	82.0%		
Residues in additionally allowed regions	15.4%		
Residues in generally allowed regions	2.6%		
Residues in disallowed regions	0.0%		
(e) Structure quality factors	*Raw score*	*Z-score of Pab PolII intein structure*[b]	*Z-score of pdb_nmr*[c]
Verify3D	0.43	−0.48	−2.15
ProsaII	0.67	0.08	−0.62
Procheck (phi-psi)	−0.67	−0.32	−2.55
Procheck (all dihedrals)	−0.63	−3.73	−5.08
MolProbityClashscore	54.5	−7.82	−10.74

This research was originally published in *The Journal of Biological Chemistry.* Du, Z.; Liu, J.; Albracht, C. D.; Hsu, A.; Chen, W.; Marieni, M. D.; Colelli, K. M.; Williams, J. E.; Reitter, J. N.; Mills, K. V.; and Wang, C. Structural and Mutational Studies of a Hyperthermophilic Intein from DNA Polymerase II of Pyrococcus Abyssi. *J. Biol. Chem.* **2011**, *286* (44), 38,638–38,648. © The American Society for Biochemistry and Molecular Biology

[a]Residues for Ramachandran statistics are 5–90, 93–120, 147–158, 161–170, 174–183

[b]With respect to mean and standard deviation for a set of 252 X-ray structures <500 residues, of resolution ≦1.80 Å, *R*-factor ≦ 0.25 and *R*-free ≦ 0.28; a positive value indicates a "better" score

[c]183 NMR structures determined by nonstructural genomics consortia groups

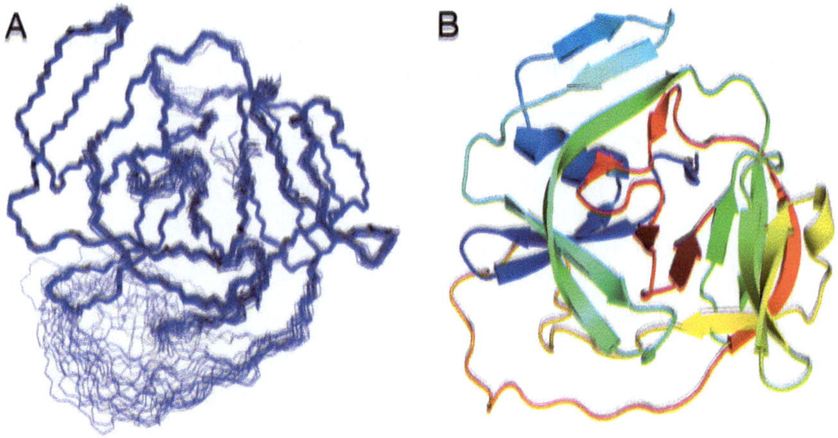

Fig. 3 Solution NMR structure of *Pab* PolII intein. (**a**) Ensemble of 20 conformers representing the *Pab* PolII intein structure. (**b**) Ribbon presentation of the structure closest to the average *Pab* PolII intein NMR structure. (This research was originally published in *The Journal of Biological Chemistry*. Du, Z.; Liu, J.; Albracht, C. D.; Hsu, A.; Chen, W.; Marieni, M. D.; Colelli, K. M.; Williams, J. E.; Reitter, J. N.; Mills, K. V.; and Wang, C. Structural and Mutational Studies of a Hyperthermophilic Intein from DNA Polymerase II of Pyrococcus Abyssi. *J. Biol. Chem.* **2011**, *286* (44), 38,638–38,648. © the American Society for Biochemistry and Molecular Biology)

See Fig. 3 for the ensemble of 20 energy-minimized conformers of Pab PolII intein and the ribbon presentation of the structure closest to average NMR structure.

3.4.3 Using NMR to Determine the pK_a of Key Catalytic Residues

The ionization states of amino acid residues are determined by their pK_a and the pH of the environment. Elevated or depressed pK_a has been found in many catalytic residues of enzymes [25]. One can use solution NMR to measure the pK_a of key catalytic residues in enzymes such as an intein.

Histidine pK_a Determination

1. As described in Subheading 3.4.1, overexpress and purify the protein sample labeled with ^{15}N.
2. Start with a sample in a buffer at pH 7, and split it in two: one for pH titration in the range from 7 to 10.4, and the other for pH titration from 7 to 3.5. Adjust pH values with 0.1 M HCl or 0.1 M NaOH, and measure with a pH meter equipped with an electrode for small volume NMR samples.
3. Record a long-range ^{1}H-^{15}N HMQC spectrum of the intein at each pH point (*see* **Note 20**).
4. Plot the observed ^{15}N chemical shifts of the histidine imidazole ring against pH (*see* **Note 21**). *See* Fig. 4 for the pK_a determination of histidine residues by long range ^{1}H-^{15}N HMQC NMR spectra.
5. Fit with nonlinear least-squares regression analysis according to the Henderson–Hasselbalch equation. The Henderson–

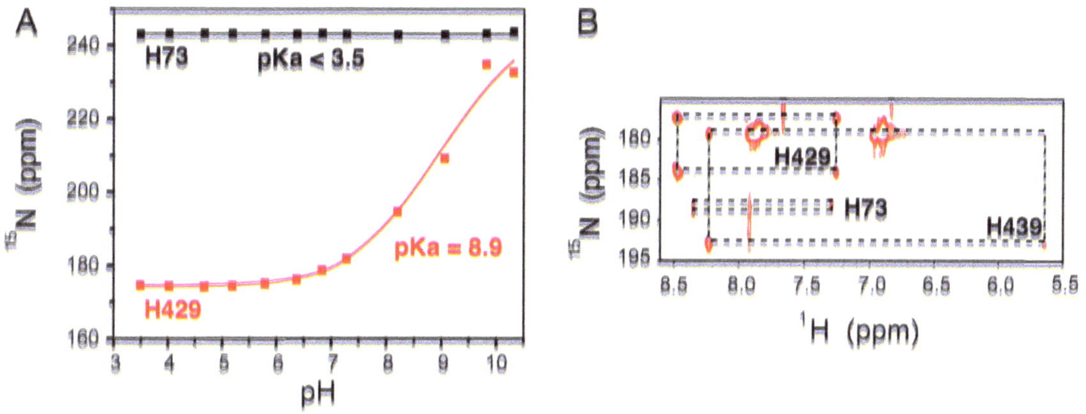

Fig. 4 pK_a determination of conserved histidine residues in excised intein. (**a**) Plot of the ring ^{15}N chemical shift ($N_{\delta 1}$) versus pH for conserved histidines, B-block H73 and F-block H429. (**b**) HMQC spectrum for intein precursor with an N-extein at pH 7.1. (Figure reprinted with permission from Du, Z.; Shemella, P. T.; Liu, Y.; McCallum, S. A.; Pereira, B.; Nayak, S. K.; Belfort, G.; Belfort, M.; Wang, C. Highly Conserved Histidine Plays a Dual Catalytic Role in Protein Splicing: A pK_a Shift Mechanism. *Journal of the American Chemical Society* **2009**, *131* (32), 11581–11589. Copyright 2009 American Chemical Society)

Hasselbalch equation using a Hill parameter (n_H) to account for nonideality is given by

$$\delta_{obs} = \frac{\left(\delta_{HA} + \delta_{A^-} \times 10^{n_H(pH - pK_a)}\right)}{\left(1 + 10^{n_H(pH - pK_a)}\right)}$$

where δ_{obs}, δ_{HA}, and δ_{A^-} are the chemical shifts for the observed, protonated, and deprotonated species, respectively [36].

Cysteine and Aspartate pK_a Determination

1. As described in Subheading 3.4.1, overexpress and purify the protein sample labeled with ^{13}C.
2. pH Titration. Same as in Subheading 3.4.3.1.
3. For cysteine, use a 2D ^{1}H-^{13}C HSQC spectrum to monitor the ^{13}C chemical shift change of the β-carbon ($^{13}C^{\beta}$) (*see* **Note 22**).
4. For aspartate, use a 2D ^{1}H-^{13}C HSQC or 2D HB(CB)CO spectrum to monitor the ^{13}C chemical shift change of the β-carbon ($^{13}C^{\beta}$) or side chain carbonyl ($^{13}C'$), respectively (*see* **Note 23**).
5. Plot the observed ^{13}C chemical shifts ($^{13}C^{\beta}$ or $^{13}C'$) as a function of sample pH value.
6. Fit with nonlinear least-squares regression analysis according to the Henderson–Hasselbalch equations as given above in Subheading 3.4.3.1.

3.4.4 ^{15}N Relaxation Rates and Analysis of Intein

Both conformational changes and protein dynamics are important in intein catalysis [37, 38]. Dynamic properties can be probed by ^{15}N spin relaxation rates, R_1, R_2, and heteronuclear steady-state NOE experiments (*see* **Note 24**). These relaxation parameters are sensitive to motions occurring at the time scale faster the time scale of overall protein tumbling, on the order of picosecond to nanosecond [39].

1. As described in Subheading 3.4.1, overexpress and purify the protein sample labeled with ^{15}N.
2. In order to measure the ^{15}N R_1 parameter, set approximately 10 relaxation time points with at least two duplicate time points. The longest delay should cause about a 30% decrease in signal intensity. For example, for the *Pab* PolII intein, we used the following time points: 10, 100, 200, 300, 400 (×2), 500, 600, 700, 800, and 900 ms.
3. In order to measure the ^{15}N R_2 parameter, again set about 10 relaxation time points with at least two duplicate time points. The longest delay should cause about a 30% decrease in signal intensity. For example, for the *Pab* PolII intein, we used the following time points: 2, 16 (×2), 30, 44, 58, 72, 86 (×2), and 100 ms.
4. For ^{1}H-^{15}N steady-state heteronuclear NOEs, interleave the proton saturation experiment and no proton saturation experiment at each t1 point (*see* **Note 25**).
5. For each R_1 and R_2 experiment, pick peaks in the spectrum with the shortest relaxation time (highest intensities) with Sparky. Use the program CURVEFIT (*see* **Note 26**) to fit the NMR data to obtain relaxation rates.
6. For heteronuclear NOE, calculate the ratio of the peak heights for saturated and unsaturated spectra.
7. Analyze NMR relaxation data using a r_{NH} of 1.02 Å as the mean amide nitrogen-hydrogen bond length, and $\Delta\sigma = -172$ ppm, which is the chemical shift anisotropy of backbone ^{15}N.
8. Determine the amplitudes and time scales of the internal motions of the protein according to the model-free formalism [40, 41] using the program Fast-Modelfree [42] (*see* **Note 27**).

See Fig. 5 for the ^{15}N backbone dynamics of intein detected by R_1, R_2, and heteronuclear steady-state NOE experiments, demonstrating the unusual rigidity at the termini of the intein, which also is the location of the active site.

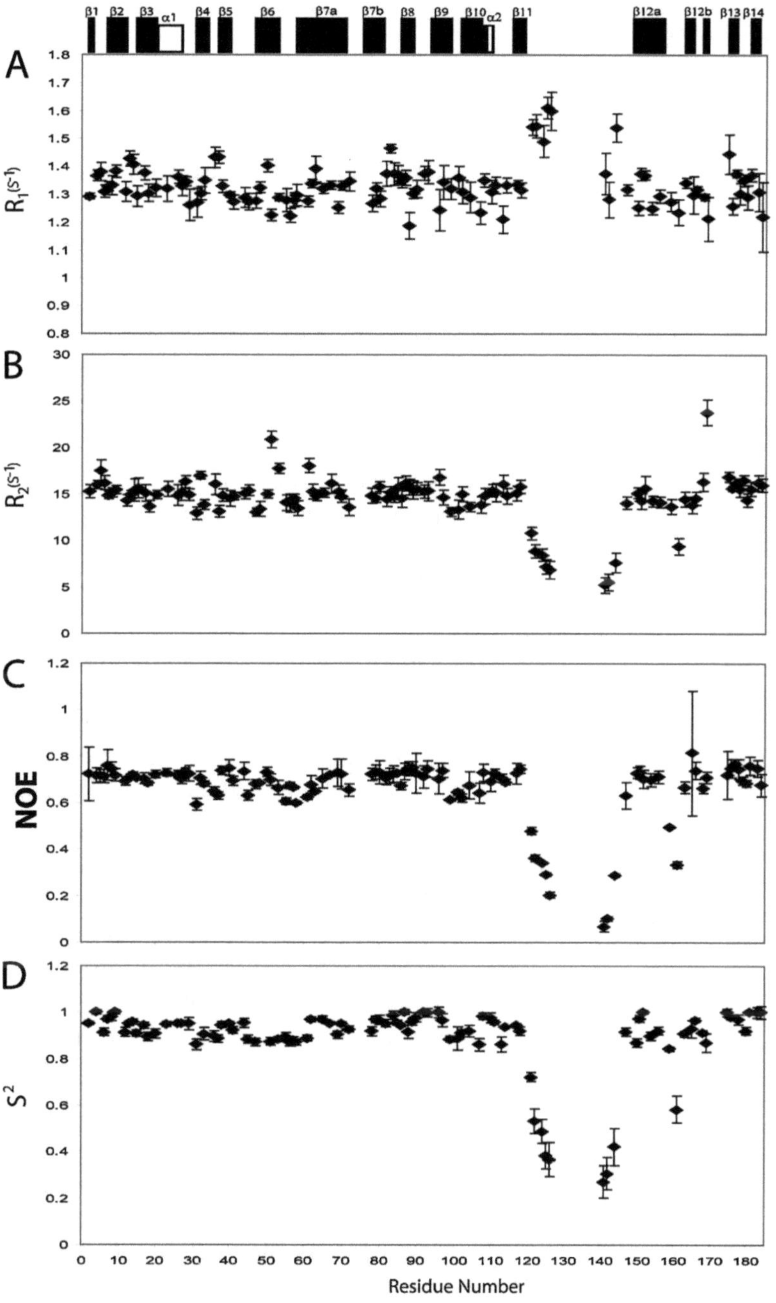

Fig. 5 ^{15}N backbone dynamics for the *Pab* PolII intein. (**a**) Longitudinal relaxation rates R_1 (s^{-1}) with regular secondary structure indicated on the top of the panel. (**b**) Transverse relaxation rates R_2 (s^{-1}). (**c**) ^{1}H-^{15}N steady-state heteronuclear NOEs. (**d**) Generalized order parameters. (This research was originally published in *The Journal of Biological Chemistry*. Du, Z.; Liu, J.; Albracht, C. D.; Hsu, A.; Chen, W.; Marieni, M. D.; Colelli, K. M.; Williams, J. E.; Reitter, J. N.; Mills, K. V.; and Wang, C. Structural and Mutational Studies of a Hyperthermophilic Intein from DNA Polymerase II of Pyrococcus Abyssi. *J. Biol. Chem.* **2011**, *286* (44), 38,638–38,648. © The American Society for Biochemistry and Molecular Biology)

4 Notes

1. Although other vectors are available, we use expression vectors based on the pMal-c2X or pMal-c5X vectors from New England Biolabs because they already include the gene for MBP. Users should consider vectors with antibiotic resistance compatible with their *E. coli* expression strain, and may wish to consider whether to use a T7 or Tac promoter (*see* **Note 2**). In the case of the *P. abyssi* PolII intein, we included seven and 6 residues of both the native N- and C-exteins, respectively; extein residues flanking the intein can have a significant influence on the folding of the active site and on splicing activity [4, 6, 8].
2. We have found that if the intein fusion protein expressed poorly, which can be an issue particularly for inteins with active homing endonuclease domains or those from extreme halophiles, we can achieve better levels of expression from pET expression vectors.
3. We have found that some inteins with active homing endonuclease domains express either poorly or into inclusion bodies, and that the soluble expression of these inteins can be enhanced with mutation of key conserved homing endonuclease residues to Ala, which is unlikely to affect splicing activity [43, 44].
4. For expression of inteins whose splicing is conditional on reduction of a native disulfide bond, we have used *E. coli* BL21(DE3) Origami cells [45] to prevent or reduce in vivo splicing. Either carbenicillin or ampicillin is appropriate; carbenicillin may reduce the number of satellite colonies but also is usually more expensive.
5. We generally overexpress at 20 °C in order to promote intein solubility as well as to prevent splicing of intein precursors. We have found that the fraction of unspliced to spliced precursor can be higher at shorter expression times.
6. Alternatively, we use Novagen's Bugbuster protocol, although we have found that some intein fusion proteins fractionate into inclusion bodies using the detergent protocol and that the French Press technique yields more soluble protein. For the detergent lysis protocol, we add 250 μL of Novagen Bugbuster (10×) to Subheading **3.2**, **step 6**, incubate on ice for 30 min, and skip the French Press step.
7. We find that Cobalt-based affinity resin does not result in co-purification of nonspecific *E. coli* proteins of similar size to *E. coli* MBP and is useful for this application. (We utilize Talon metal-affinity resin from Takara Bio and base our protocol, in part, on their instructions.) For purification of His-tag

containing proteins lacking *E. coli* MBP, we often use instead the nickel-based HisLink resin, which has fewer nonspecific binding proteins near the size of the isolated intein. Our lab uses a similar method for either resin. We have also used an amylose resin to purify fusion proteins via their N-terminal MBP.

8. It is important that the pH of stock imidazole solutions is adjusted to pH 7.5, and that no solutions containing EDTA or reducing agents are used in the purification.
9. To study inteins from halophiles, we also prepare salt solutions at high ionic strength in 20 mM HEPES, pH 7.5, to vary the salt concentration as appropriate, usually in a range from 0.5 M to 2.5 M. We usually incubate inteins from halophiles at the growth temperature of their host. For inteins from thermophiles, we initially survey a range of temperatures from 30 °C to 70 °C, as their activities can vary and higher temperatures can result in protein precipitation. We initially screen for activity at an extended time (16 h), and then use increasingly shorter incubations to determine the optimal time to study activity.
10. EDTA serves to chelate any residual divalent cation, as these have been shown to inhibit splicing [46–49]. TCEP is a non-nucleophilic reducing agent, and is important for inteins that have disulfide bonds involving catalytic Cys residues.
11. We find that the reactivity of some intein fusion proteins declines rapidly on storage at 4 °C, indicating that these inteins may be susceptible to oxidative or other damage. When this is the case, we purify the protein and run the reaction on the same day.
12. We have observed two conditions that result in the aberrant migration of bands on SDS-PAGE. Some inteins have disulfide bonds that are particularly resistant to reduction, and a disulfide bond between Cys residues near the termini of the intein results in aberrant migration [13, 50–52]. We also have observed that inteins from halophiles that have particularly low pI values migrate aberrantly on SDS-PAGE [20], which is common for proteins with a highly negative surface charge [53].
13. We also verify the identity of the protein bands on SDS-PAGE via Western blot. When using either commercial antibodies against poly-His tags or direct reagents such as KPL's HisDetector Nickel-AP, we have found when analyzing inteins from halophiles that the performance of the reagents varies and a trial-and-error technique is required to determine which antibody is appropriate.

14. Isotopic labeling was accomplished by growing cultures in M9 minimal medium containing either 1 g/L of $^{15}NH_4Cl$ for uniformly ^{15}N-labeled samples or 1 g/L of $^{15}NH_4Cl$ and 1 g/L of $^{13}C_6$-D-glucose for uniformly ^{15}N- and ^{13}C-labeled samples.
15. All 2D/3D spectra were acquired at an appropriate temperature on a Bruker Advance II 800 MHz or Bruker Advance II 600 MHz spectrometer, each equipped with a triple-resonance cryogenic probe. For the *Pab* PolII intein, temperature up to 47 °C can be utilized to take advantage of the enhanced NMR sensitivity at high temperature due to the extreme thermostability of proteins from hyperthermophiles.
16. The ^{1}H chemical shifts were referenced by 4,4-dimethyl-4-silapentane-1-sulfonic acid and the ^{15}N and ^{13}C chemical shifts were referenced using frequency ratios between ^{15}N, ^{13}C, and ^{1}H ($^{15}N/^{1}H = 0.101329118$, $^{13}C/^{1}H = 0.251449530$). All chemical shifts were deposited in the BioMagResBank under accession number 17418 [21].
17. CYANA carries out automatic NOE assignment and uses distance and angular constraints from NMR data to calculate the structure of a protein, using torsion angle dynamics [28, 29].
18. TALOS+ uses chemical shifts of N_H, C', C^{α}, C^{β}, H^{α}, and H^{N} for a given residue to predict protein backbone torsion angles [30, 31].
19. PSVS analyzes standard NMR constraint and NMR structures, and provides structure quality scores [35].
20. In our experiment [22], long range ^{1}H-^{15}N HMQC spectra [54] were acquired with 2048 × 288 complex data points, spectral widths of 100 ppm in ^{15}N and 16 ppm in ^{1}H and 128 scans. The spectrum center was set as 200.0 ppm for ^{15}N and 4.76 ppm for ^{1}H.
21. The assignments of histidine residues are based on the (HB)CB (CGCD)HD spectrum, which connects backbone C_{β} and H_{δ}.
22. The assignments of cysteine side chains in our experiment [27] were based on H_{β} chemical shifts from ^{15}N-TOCSY, C_{β} chemical shifts from HNCACB and ^{1}H-^{13}C aliphatic HSQC.
23. ^{15}N-TOCSY, ^{1}H-^{13}C HSQC, and HBCBCO spectra were recorded for the assignment of aspartate side chains.
24. All spectra were recorded at an appropriate temperature on a Bruker SB 600 MHz Spectrometer equipped with a $^{1}H/^{2}H/^{13}C/^{15}N$ cryoprobe with *z*-axis gradients. For Pab PolII intein, temperature up to 47 °C can be utilized to take advantage of the enhanced NMR sensitivity at high temperature due to the extreme thermostability of proteins from hyperthermophiles.

25. The NOE experiment was performed with a recycle delay of 7.5 s to ensure that ^{15}N magnetization reaches equilibration before each scan starts. A 120° proton pulse at 5 ms delay was applied to achieve proton saturation.
26. CURVEFIT is a program that computes a nonlinear least square fit with an arbitrary number of parameters, using a gradient-expansion algorithm.
27. Fast-Modelfree [42] is a graphic interface for Modelfree [55], a program for rapid automated analysis of NMR relaxation data. The generalized order parameters (S^2) can describe the amplitude of the internal motion for individual amide bonds at the picosecond to nanosecond time scale.

Acknowledgement

This work was supported by NSF grant MCB-1517138 (KVM), a Henry Dreyfus Teacher-Scholar Award (KVM), and by NIH Grants 1R15GM132817-01 (KVM) and 1R01CA206592 (CW).

References

1. Novikova O, Topilina N, Belfort M (2014) Enigmatic distribution, evolution, and function of Inteins. J Biol Chem 289:14490–14497. https://doi.org/10.1074/jbc.R114.548255
2. Mills KV, Johnson MA, Perler FB (2014) Protein splicing: how Inteins escape from precursor proteins. J Biol Chem 289:14498–14505. https://doi.org/10.1074/jbc.R113.540310
3. Eryilma E, Shah NH, Muir TW, Cowburn D (2014) Structural and dynamical features of Inteins and implications on protein splicing. J Biol Chem 289:14506–14511. https://doi.org/10.1074/jbc.R113.540302
4. Friedel K, Popp MA, Matern JCJ et al (2019) A functional interplay between intein and extein sequences in protein splicing compensates for the essential block B histidine. Chem Sci 10:239–251. https://doi.org/10.1039/C8SC01074A
5. Chong S, Williams KS, Wotkowicz C, Xu MQ (1998) Modulation of protein splicing of the Saccharomyces cerevisiae vacuolar membrane ATPase intein. J Biol Chem 273:10567–10577. https://doi.org/10.1074/jbc.273.17.10567
6. Amitai G, Callahan BP, Stanger MJ et al (2009) Modulation of intein activity by its neighboring extein substrates. Proc Natl Acad Sci U S A 106:11005–11010. https://doi.org/10.1073/pnas.0904366106
7. Nogami S, Satow Y, Ohya Y, Anraku Y (1997) Probing novel elements for protein splicing in the yeast Vmal protozyme: a study of replacement mutagenesis and intragenic suppression. Genetics 147:73–85
8. Shah NH, Eryilmaz E, Cowburn D, Muir TW (2013) Extein residues play an intimate role in the rate-limiting step of protein trans-splicing. J Am Chem Soc 135:5839–5847. https://doi.org/10.1021/ja401015p
9. Pearl EJ, Bokor AAM, Butler MI et al (2007) Preceding hydrophobic and β-branched amino acids attenuate splicing by the CnePRP8 intein. Biochim Biophys Acta 1774:995–1001. https://doi.org/10.1016/j.bbapap.2007.05.015
10. Di Ventura B, Mootz HD (2019) Switchable inteins for conditional protein splicing. Biol Chem 400:467–475
11. Buskirk AR, Ong Y-C, Gartner ZJ, Liu DR (2004) Directed evolution of ligand dependence: small-molecule-activated protein splicing. Proc Natl Acad Sci U S A 101:10505–10510. https://doi.org/10.1073/pnas.0402762101
12. Callahan BP, Stanger M, Belfort M (2013) A redox trap to augment the intein toolbox.

Biotechnol Bioeng 110:1565–1573. https://doi.org/10.1002/bit.24821

13. Callahan BP, Topilina NI, Stanger MJ et al (2011) Structure of catalytically competent intein caught in a redox trap with functional and evolutionary implications. Nat Struct Mol Biol 18:630–633. https://doi.org/10.1038/nsmb.2041
14. Skretas G (2005) Regulation of protein activity with small-molecule-controlled inteins. Protein Sci 14:523–532. https://doi.org/10.1110/ps.04996905
15. Zeidler MP, Tan C, Bellaiche Y et al (2004) Temperature-sensitive control of protein activity by conditionally splicing inteins. Nat Biotechnol 22:871–876. https://doi.org/10.1038/nbt979
16. Wood DW, Camarero JA (2014) Intein applications: from protein purification and labeling to metabolic control methods. J Biol Chem 289:14512–14519
17. Topilina NI, Mills KV (2014) Recent advances in in vivo applications of intein-mediated protein splicing. Mob DNA 5:5
18. Elleuche S, Pöggeler S (2010) Inteins, valuable genetic elements in molecular biology and biotechnology. Appl Microbiol Biotechnol 87:479–489
19. Berrade L, Camarero JA (2009) Expressed protein ligation: a resourceful tool to study protein structure and function. Cell Mol Life Sci 66:3909–3922
20. Reitter JN, Cousin CE, Nicastri MC et al (2016) Salt-dependent conditional protein splicing of an Intein from Halobacterium salinarum. Biochemistry 55:1279–1282. https://doi.org/10.1021/acs.biochem.6b00128
21. Du Z, Liu J, Albracht CD et al (2011) Structural and mutational studies of a hyperthermophilic intein from DNA polymerase II of Pyrococcus abyssi. J Biol Chem 286:38638–38648. https://doi.org/10.1074/jbc.M111.290569
22. Du Z, Shemella PT, Liu Y et al (2009) Highly conserved histidine plays a dual catalytic role in protein splicing: a pK_a shift mechanism. J Am Chem Soc 131:11581–11589. https://doi.org/10.1021/ja904318w
23. Laemmli UK (1970) Cleavage of structural proteins during the assembly of the head of bacteriophage T4. Nature 227:680–685. https://doi.org/10.1038/227680a0
24. Abràmofff MD, Magalhães PJ, Ram SJ (2005) Image processing with ImageJ part II. Biophoton Int 11:36–43
25. Fersht A (1999) Structure and mechanism in protein science. A guide to enzyme catalysis and protein folding. World Scientific, Singapore
26. Du Z, Zheng Y, Patterson M et al (2011) pK_a coupling at the intein active site: implications for the coordination mechanism of protein splicing with a conserved aspartate. J Am Chem Soc 133:10275–10282. https://doi.org/10.1021/ja203209f
27. Xie J, Owen T, Xia K et al (2016) A single aspartate coordinates two catalytic steps in hedgehog autoprocessing. J Am Chem Soc 138:10806–10809. https://doi.org/10.1021/jacs.6b06928
28. Güntert P, Mumenthaler C, Wüthrich K (1997) Torsion angle dynamics for NMR structure calculation with the new program DYANA. J Mol Biol 273:283–298. https://doi.org/10.1006/jmbi.1997.1284
29. Herrmann T, Güntert P, Wüthrich K (2002) Protein NMR structure determination with automated NOE assignment using the new software CANDID and the torsion angle dynamics algorithm DYANA. J Mol Biol 319:209–227. https://doi.org/10.1016/S0022-2836(02)00241-3
30. Cornilescu G, Delaglio F, Bax A (1999) Protein backbone angle restraints from searching a database for chemical shift and sequence homology. J Biomol NMR 13:289–302. https://doi.org/10.1023/A:1008392405740
31. Shen Y, Delaglio F, Cornilescu G, Bax A (2009) TALOS+: a hybrid method for predicting protein backbone torsion angles from NMR chemical shifts. J Biomol NMR 44:213–223. https://doi.org/10.1007/s10858-009-9333-z
32. Karplus M (1959) Contact electron-spin coupling of nuclear magnetic moments. J Chem Phys 30:11–15. https://doi.org/10.1063/1.1729860
33. Karplus M (1963) Vicinal proton coupling in nuclear magnetic resonance. J Am Chem Soc 85:2870–2871
34. Schwieters CD, Kuszewski JJ, Marius Clore G (2006) Using Xplor-NIH for NMR molecular structure determination. Prog Nucl Magn Reson Spectrosc 48:47–62. https://doi.org/10.1016/j.pnmrs.2005.10.001
35. Bhattacharya A, Tejero R, Montelione GT (2007) Evaluating protein structures determined by structural genomics consortia. Proteins 795:75–82. https://doi.org/10.1002/prot
36. Webb H, Tynan-Connolly BM, Lee GM et al (2011) Remeasuring HEWL pK_a values by NMR spectroscopy: methods, analysis, accuracy, and implications for theoretical pK_a

calculations. Proteins 79:685–702. https://doi.org/10.1002/prot.22886

37. Frutos S, Goger M, Giovani B et al (2010) Branched intermediate formation stimulates peptide bond cleavage in protein splicing. Nat Chem Biol 6:527–533. https://doi.org/10.1038/nchembio.371
38. Sun P, Ye S, Ferrandon S et al (2005) Crystal structures of an intein from the split dnaE gene of Synechocystis sp. PCC6803 reveal the catalytic model without the penultimate histidine and the mechanism of zinc ion inhibition of protein splicing. J Mol Biol 353:1093–1105. https://doi.org/10.1016/j.jmb.2005.09.039
39. Kalbitzer HR (2011) Protein NMR spectroscopy. Principles and practice. Z Phys Chem. https://doi.org/10.1524/zpch.1997.202.part_1_2.297
40. Lipari G, Szabo A (1982) Model-free approach to the interpretation of nuclear magnetic resonance relaxation in macromolecules. 1. Theory and range of validity. J Am Chem Soc 104:4546–4559. https://doi.org/10.1021/ja00381a009
41. Clore GM, Szabo A, Bax A et al (1990) Deviations from the simple two-parameter model-free approach to the interpretation of Nitrogen-15 nuclear magnetic relaxation of proteins. J Am Chem Soc 112:4989–4991. https://doi.org/10.1021/ja00168a070
42. Cole R, Loria JP (2003) FAST-Modelfree: a program for rapid automated analysis of solution NMR spin-relaxation data. J Biomol NMR 26:203–213. https://doi.org/10.1023/A:1023808801134
43. Kerrigan AM, Powers TL, Dorval DM et al (2009) Protein splicing of the three Pyrococcus abyssi ribonucleotide reductase inteins. Biochem Biophys Res Commun 387:153–157. https://doi.org/10.1016/j.bbrc.2009.06.145
44. Saves I (2002) Investigating the endonuclease activity of four Pyrococcus abyssi inteins. Nucleic Acids Res 30:4158–4165. https://doi.org/10.1093/nar/gkf556
45. Bessette PH, Aslund F, Beckwith J, Georgiou G (1999) Efficient folding of proteins with multiple disulfide bonds in the Escherichia coli cytoplasm. Proc Natl Acad Sci U S A 96 (24):13703–13708
46. Zhang L, Xiao N, Pan Y et al (2010) Binding and inhibition of copper ions to RecA inteins from mycobacterium tuberculosis. Chemistry 16:4297–4306. https://doi.org/10.1002/chem.200903584
47. Nichols NM, Benner JS, Martin DD, Evans TC (2003) Zinc ion effects on individual Ssp DNaE intein splicing steps: regulating pathway progression. Biochemistry 42:5301–5311. https://doi.org/10.1021/bi020679e
48. Ghosh I, Sun L, Xu MQ (2001) Zinc inhibition of protein trans-splicing and identification of regions essential for splicing and Association of a Split Intein. J Biol Chem 276:24051–24058. https://doi.org/10.1074/jbc.M011049200
49. Mills KV, Paulus H (2001) Reversible inhibition of protein splicing by zinc ion. J Biol Chem 276:10832–10838. https://doi.org/10.1074/jbc.M011149200
50. Chen W, Li L, Du Z et al (2012) Intramolecular disulfide bond between catalytic cysteines in an intein precursor. J Am Chem Soc 134:2500–2503. https://doi.org/10.1021/ja211010g
51. Nicastri MC, Xega K, Li L et al (2013) Internal disulfide bond acts as a switch for intein activity. Biochemistry 52:5920–5927. https://doi.org/10.1021/bi400736c
52. Xu Y, Zhang L, Ma B et al (2018) Intermolecular disulfide bonds between unpaired cysteines retard the C-terminal trans-cleavage of Npu DnaE. Enzym Microb Technol 118:6–12. https://doi.org/10.1016/j.enzmictec.2018.06.013
53. Iakoucheva LM, Kimzey AL, Masselon CD et al (2010) Aberrant mobility phenomena of the DNA repair protein XPA. Protein Sci 10:1353–1362. https://doi.org/10.1110/ps.ps.40101
54. Bax A, Summers MF (1986) ^{1}H and ^{13}C assignments from sensitivity-enhanced detection of Heteronuclear multiple-bond connectivity by 2D multiple quantum NMR. J Am Chem Soc 108:2093–2094. https://doi.org/10.1021/ja00268a061
55. Mandel AM, Akke M, Palmer AG (1995) Backbone dynamics of Escherichia coli ribonuclease HI: correlations with structure and function in an active enzyme. J Mol Biol 246:144–163. https://doi.org/10.1006/jmbi.1994.0073
56. Mills KV, Manning JS, Garcia AM, Wuerdeman LA (2004) Protein splicing of a pyrococcus abyssi intein with a C-terminal glutamine. J Biol Chem 279:20685–20691. https://doi.org/10.1074/jbc.M400887200

Chapter 5

Expressed Protein Ligation: General Experimental Protocols

Joshua Diaz and Jean-Philippe Pellois

Abstract

Expressed protein ligation allows for the attachment of a chemically labeled peptide to the N- or C-terminus of a recombinant protein. In this book chapter, the practical considerations involved in using this protein engineering technology are described. In particular, approaches used to design optimal ligation sites are discussed. In addition, several methods used to generate the reactive fragments required for EPL are highlighted in practical details. Finally, strategies that one can implement to achieve efficient ligation reactions are presented.

Key words Expressed protein ligation, N-terminal cysteine, C-terminal thioester, Solid-phase peptide synthesis, Ligation site, Protein engineering, Protein chemistry, Protein labeling

1 Introduction

Expressed protein ligation (EPL) typically involves the joining of a synthetic peptide to a recombinant protein. The ligation chemistry involves the reaction between the N-terminal cysteine residue and the C-terminal thioester of two participating fragments. The result of this reaction is the formation of a native peptide bond at the site of ligation. Because of its generality, high yields of product formation, and its ease of adoption, EPL is a popular option for protein engineering projects requiring the site-specific modification of proteins with non-genetically encodable modifications (NGEMs). These NGEMs include small molecule probes, post-translational modifications, and fluorescent dyes. In this book chapter, practical considerations for designing ligation sites optimal to achieve efficient EPL are discussed. An overview of techniques useful for the generation of proteins and peptides containing a C-terminal thioester or an N-terminal cysteine residue is also presented. Furthermore, general EPL protocols used to attach synthetic peptides at the N- or C-termini of a recombinant protein are provided. The protocols focus on protein engineering designs that involve a single ligation site. However, these protocols can be applied to

Miquel Vila-Perelló (ed.), *Expressed Protein Ligation: Methods and Protocols*, Methods in Molecular Biology, vol. 2133, https://doi.org/10.1007/978-1-0716-0434-2_5, © Springer Science+Business Media, LLC, part of Springer Nature 2020

accomplish more intricate synthetic schemes that involve the ligation of three or more fragments.

1.1 Choosing a Potential Ligation Site

The production of a protein modified with a non–genetically encodable modification (NGEM) via EPL commonly requires the reaction between a recombinant protein fragment and a peptide containing the NGEM. Given that molecular biology techniques take advantage of the cellular machinery for protein production, the recombinant fragment used in an EPL procedure can be large in size and amenable to many sequences. In contrast, the peptide fragment is typically produced by solid-phase peptide synthesis (SPPS), which limits the length of the fragment to approximately 50 residues or less [1]. Also, certain peptide sequences are prone to side reactions and therefore more challenging to synthesize than others (*see* Subheading 1.4). The generation of a synthetic peptide for EPL can therefore represent a limiting factor. Thus, it is advisable to identify a synthetic sequence that will be relatively short and amenable to high synthetic yields. Consequently, when considering EPL for the production of a semisynthetic protein, one should first perform a retrosynthetic analysis and determine which fragments of the protein product should be produced synthetically or recombinantly. As shown in Fig. 1, if the synthetic moiety is present closer to the C-terminus of the final protein product, the synthetic peptide used for EPL should be C-terminal, while the recombinant protein should be N-terminal. The opposite is true if the synthetic moiety is closer to the N-terminus of the protein product. To limit the size of the synthetic peptide needed, one should then consider the position of the ligation site between the recombinant and synthetic fragments. Because a cysteine residue is required for ligation, one must then choose the position of this residue in the protein sequence. In some cases, a cysteine may naturally be present in the protein sequence in close proximity to the NGEM. In this ideal situation, the sequence length between the ligation site and the NGEM is short, as is the peptide requiring synthesis. In this instance, ligation will also restore a wild-type cysteine residue in the protein sequence. If no cysteine is present near the NGEM, the introduction of this residue in the protein sequence may be necessary. In this scenario, one must choose which residue from the wild-type protein sequence should be mutated to a cysteine. As mentioned above, the mutation should introduce a ligation site close to the NGEM to minimize the length of the synthetic peptide produced. This mutation should also not impact the structure or function of the protein product. If the effect of this mutation is unknown prior to EPL, it is advisable to generate recombinant proteins that contain cysteine residues at positions of potential ligation sites and test for their activity and function in biochemical or biological assays. In the event that the selected cysteine

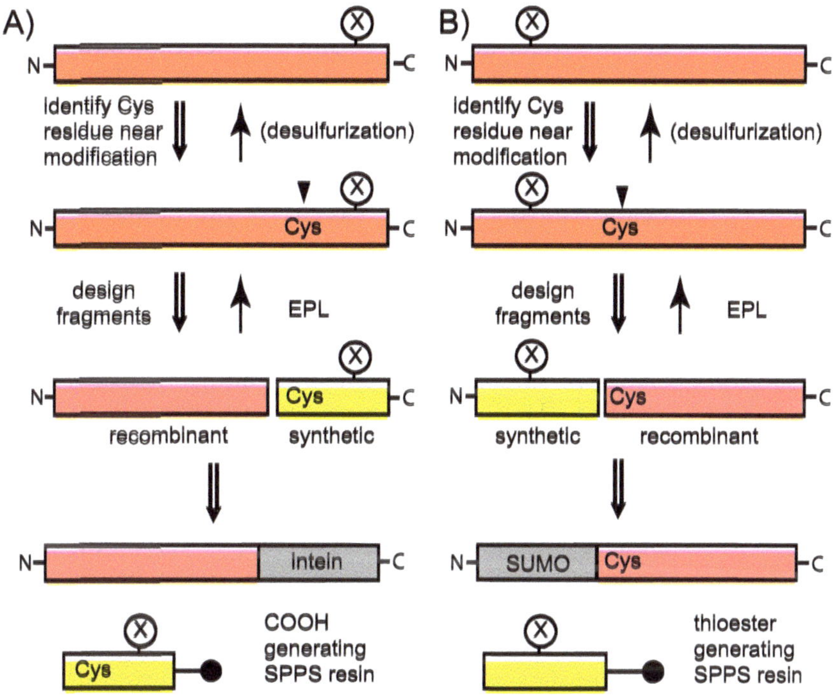

Fig. 1 Retrosynthetic analysis for the selection of a ligation site. A protein of interest, its primary structure being represented by an orange rectangle, may include an NGEM, represented by X, proximal to its N- or C-termini. (**a**) If X is close to the C-terminus of a protein (left side panel), a nearby upstream cysteine residue should be identified as a potential site of ligation. If no cysteine residue is available, an alanine residue can potentially be identified and substituted to a cysteine (later converted back into an alanine by desulfurization). The choice of a ligation site leads to two identified EPL fragments: an N-terminal recombinant fragment containing a C-terminal thioester and a synthetic peptide containing X and an N-terminal cysteine. The sequence distance between the chosen site of ligation and the modification X should be minimal, so as to reduce the length of the synthetic peptide needed. This synthetic peptide will be assembled by SPPS on a resin that generates a C-terminal carboxylic acid upon peptide cleavage. The recombinant fragment will be generated by expression of the truncated protein with an intein fusion tag at its C-terminus. (**b**) If X is near to the N-terminus of the protein, a cysteine residue will be identified downstream of X (optionally, an alanine residue is chosen, as described in **a**). The corresponding EPL fragments will consist of an N-terminal synthetic peptide and a C-terminal recombinant protein. The synthetic peptide will contain X and will be assembled by SPPS on a resin that produces a C-terminal thioester upon cleavage. The recombinant fragment will contain a fusion tag (e.g., SUMO) that can be cleaved to generate an N-terminal cysteine residue on the protein after expression and purification. As in (**a**), the sequence distance between X and the ligation site is short so as to allow the generation of a synthetic peptide that is within the length restriction of SPPS

mutations affect the activity of the protein, ligation chemistries that circumvent the involvement of a cysteine residue may be pursued (*see* Subheading 1.7).

While the size of the EPL fragments contributes to the choice of a ligation site, the reactivity between the two fragments should also be considered. The reactivity between fragments is predominantly impacted by steric hindrance caused by the side chains of the amino acids in proximity to the reactive groups (i.e., the thiolate of the cysteine residue and the C-terminal thioester). This is exemplified in several studies reporting that sterically hindering C-terminal amino acids, such as β-branched amino acids, slow the reaction kinetics [2, 3]. This is problematic because the reactive thioester-containing EPL fragment can be hydrolyzed into a nonreactive carboxylate fragment over time (the rate of this side reaction being controlled by the basicity of the buffers used for fragment production, storage, and ligation). As a result, long reaction times will often lead to a decrease in yield of the ligated product as hydrolysis out-competes ligation. To achieve optimal ligation rates, steric hindrance by amino acid side chains surrounding the thioester and the cysteine-containing constructs need to be avoided. This can be done by selecting potential ligation sites away from structured regions of the protein. For instance, choosing a ligation site in an unstructured loop or in a disordered linker region joining two protein domains may be advantageous. In contrast, a potential ligation site that is surrounded by structured elements of the protein (e.g., near alpha helices or beta-sheets) would presumably be challenging, as thioester and cysteine would be less accessible for a reaction.

Another aspect that contributes to ligation efficiency is the identity of the C-terminal amino residue of the thioester-containing fragment. Specifically, the side chain of the residue on the C-terminus of the thioester-containing fragment influences the ligation reaction, by possibly contributing steric hindrance, by affecting the electronic reactivity of the thioester carbonyl, or by promoting hydrolysis (note that temperature, concentration of reactants, catalytic thiol group used, can also affect the ligation reaction). For instance, Ghosh et al. determined that the C-terminal residues leucine, valine, isoleucine, serine, threonine, and glutamic acid yielded poor ligation efficiencies (less than 25% of the ligated product was formed) (Table 1) [4]. Similarly, Hackeng et al. reported that the ligation of two 5-mer peptides is slow and incomplete when valine, isoleucine, and proline are used at the C-terminal thioester position [2] (Table 2). Conversely, these two studies both identify histidine, tryptophan, phenylalanine, tyrosine, cysteine, methionine, alanine, and glycine as C-terminal residues favorable for rapid and efficient ligation.

1.2 Generation of Recombinant Proteins Containing a C-Terminal Thioester

Inteins are proteins that can excise themselves from a precursor protein, a multistep process referred to as splicing. Several inteins, such as Mxe GryA intein [5], Sce VMA intein [6], and Mth RIR1 intein [7], have been engineered to only undergo some of the chemistry involved in splicing. In particular, these engineered inteins act as fusion tags that can be cleaved by the addition of thiols. In turn, intein cleavage from a fused protein of interest (POI) generates a C-terminal thioester moiety that can be utilized for EPL. To utilize these thiol-inducible inteins (IN), the

Table 1
The C-terminal thioester residue impacts ligation yields of paramyosin and PB1 (native conditions)

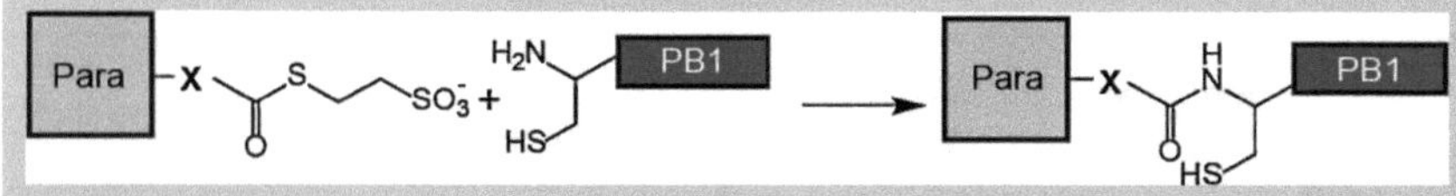

X	Yield of ligated product, %
Alanine Glutamine Histidine Phenylalanine Tryptophan Tyrosine	100
Arginine Asparagine Aspartic acid Cysteine Glycine Lysine Methionine	75
Glutamic acid Leucine Isoleucine Proline Serine Threonine Valine	<25

Paramyosin (Para) and PB1 were ligated under native conditions (folded protein) using 0.1 M Tris–HCl, 250 mM NaCl, 10 mM MESNa at 25 °C

C-terminal amino acid of POI is directly fused to the N-terminal cysteine of the IN (Fig. 2). As a result, the thiolate on the cysteine side chain can attack the carbonyl of the C-terminal residue of the POI to form a thioester between the C-terminus of the POI and the cysteine side chain of the intein, a process referred to as an N-S acyl shift. The formation of the thioester is reversible and not thermodynamically favorable. Hence, only a small population of this thioester species exists at a given time. Yet small molecule thiols, such as MESNa and MPAA, can react with the POI-IN thioester. The reaction of small molecule thiols with the POI-IN thioester results in cleavage of the intein from the POI to yield a POI thioesterified with MES or MPAA. The irreversible cleavage from the intein drives the POI-IN N-S acyl shift equilibrium toward the thioester form and complete cleavage of POI from the IN. To facilitate purification of the POI before and after cleavage of the IN fusion, a chitin binding domain (or other affinity tags) are typically added to the C-terminus of the IN fragment. After lysis of cells expressing

Table 2
The C-terminal thioester residue impacts ligation yields of LYRA and RANK peptides (denaturing conditions)

X	Time (h)	Yield of ligated product, %
Cysteine Glycine Histidine	4	100
Alanine Phenylalanine Methionine Tryptophan Tyrosine	9	100
Asparagine Aspartic acid Arginine Glutamic acid Glutamine Lysine Serine	24	100
Leucine Threonine	48	100
Valine	48	<75
Isoleucine	48	<50
Proline	48	<25

The LYRAX and CRANK peptides were ligated in 0.1 M phosphate buffer (pH 8.5) containing 6 M guanidine (denaturing conditions), 4% (v/v) benzyl mercaptan, and 4% (v/v) thiophenol at 37 °C

the POI-IN-CBD, the cell lysate can then be loaded onto a column containing chitin beads to capture POI-IN-CBD. A majority of non-CBD containing proteins from the lysate are washed away from the beads. The cleavage reaction is then initiated while POI-IN-CBD is bound to chitin beads. The POI-thioester product is then eluted, with the IN-CBD remaining bound to the column.

Several factors impact the successful purification of the POI thioesters. One factor is the efficiency of the transthioesterification reaction. As previously discussed in the context of the ligation reaction, structural or electronic effects can impact the efficiency with which a POI is cleaved from an IN. Specifically, the C-terminal amino acid of a POI can have a significant impact on cleavage efficiency. As shown in Table 3, cysteine, glycine, aspartic acid, serine, and proline are not favorable for cleavage with the Mxe

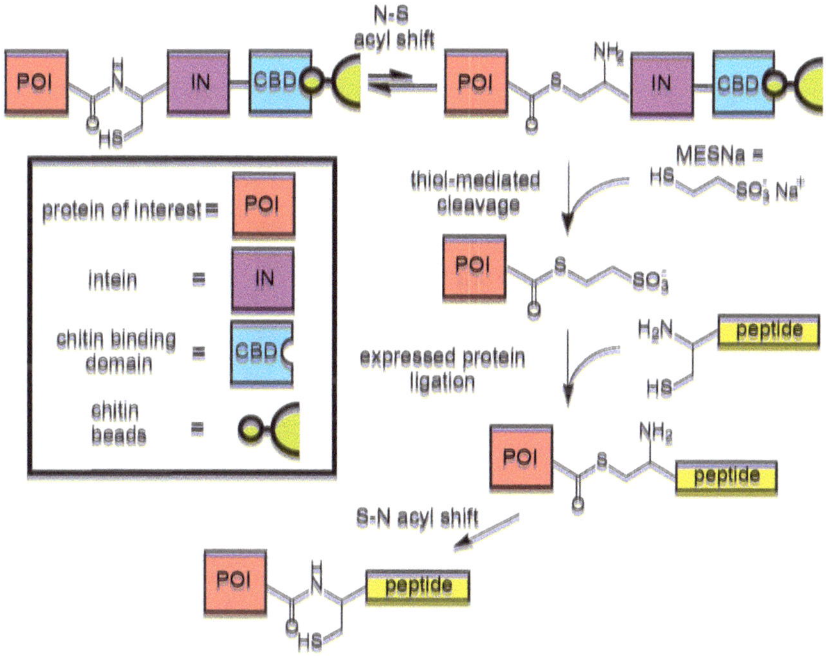

Fig. 2 Scheme for generating semisynthetic proteins by EPL. The protein of interest (POI) is fused to an engineered thiol-inducible intein (IN) and a chitin binding domain purification tag (CBD). The IN contains an N-terminal cysteine residue linked to the POI. The cysteine residue is in equilibrium between a native peptide bond and a thioester bond, the peptide bond being thermodynamically favored. Once bound to chitin beads, thiols are added to cleave the protein from the IN-CBD. Cleavage is mediated by a reaction of the added thiol, such as a 2-mercaptoethane sulfonate sodium salt (MESNa), with the protein population that is in the thioester form. To drive this reaction to completion, the thiol is added in excess to the POI. The new thioester formed as a result of this reaction is eluted from the chitin beads. A cysteine-containing peptide is then used for ligation to the POI-thioester. The N-terminal cysteine residue reacts with the POI-thioester through a transesterification reaction to form a POI-thioester–peptide linkage. The proximity of the cysteine amino group favors an S-N acyl shift. This is accompanied to the irreversible formation of a native peptide bond

GyrA intein [4]. Some of these residues may interfere with thioester formation because of their reactive side chains. For instance, aspartic acid can react with the carbonyl of a thioester to form a cleaved POI containing a C-terminal anhydride [8]. Interestingly, the C-terminal POI residues favorable for intein cleavage vary depending on the IN used (Table 4) [9]. If cleavage of the POI from the IN is expected to be poor, one may consider mutating the C-terminal residue of the POI to a residue that is favorable for cleavage. As mentioned previously, if this strategy is implemented, biochemical assays must be performed to determine if the mutation impacts protein function. Alternatively, the amount of protein produced can be increased by scaling up the process.

Another consideration for producing the thioesterified POI is the stability of the thioester group. Specifically, the cleaved POI C-terminal thioester can undergo hydrolytic degradation (the thioester being converted into a carboxylic acid). To limit the

Table 3
Mxe cleavage efficiency from paramyosin

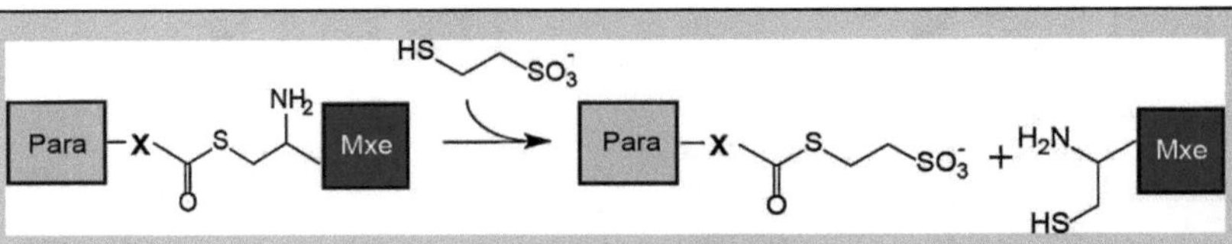

X	Yield of cleaved Para (4 °C), %	yield of cleaved Para (23 °C), %
Arginine Glutamine Histidine Leucine Lysine Methionine Phenylalanine Tyrosine	75–100	75–100
Tryptophan	50–75	75–100
Asparagine Glutamic acid Isoleucine Threonine Valine	25–50	50–75
Alanine	<25	50–75
Aspartic acid Cysteine Glycine Proline Serine	<25	<25

Cleavage was induced with 40 mM MESNa, and the reaction was allowed to proceed for 40 h. The percent yield of cleaved paramyosin (Para) was determined by mobility shift on SDS-PAGE

Table 4
Summary of favorable and unfavorable residues for thiol-induced intein cleavage

Intein	Optimal C-terminal residues for cleavage	Unfavorable residues for cleavage
Mxe GyrA	A, F, H, K, M, N, Q, T, Y; LEM	D, G, P, S
Sce VMA1	A, F, G, K, M, N, Q, W, Y; LEG	C, D, N, R, P
Mth RIR1	G, A, LEG	D, E, P

Optimal residues were determined by New England Biolabs. For more information *see* https://www.neb.com

extent of hydrolysis, the cleavage reaction is typically conducted either at room temperature (for several hours) or at 4 °C (typically overnight). The pH of the cleavage buffer is also controlled so as to, on the one hand, limit the amount of hydroxyl ion present (pH < 9), and, on the other hand, promote the deprotonation of thiols to form more nucleophilic and reactive thiolates (pH > 7). Additionally, it is worth noting that the stability of the produced POI needs to be considered, as proteins can become unstable in solution overtime (e.g., aggregation, unfolding). To avoid this issue, it is therefore often preferable to keep reaction times as short as possible.

1.3 Generation of Recombinant Proteins Containing an N-Terminal Cysteine Residue

Proteins typically contain an N-terminal methionine, a conserved residue involved in the initiation of protein synthesis. This is problematic if the ligation strategy requires a protein fragment with an N-terminal cysteine (Fig. 1). The production of proteins with an N-terminal cysteine residue is in principle possible by expressing proteins in cells and allowing intracellular peptidases to remove the N-terminal methionine (if the protein is expressed with a Met-Cys sequence at its N-terminus). However, N-terminal cysteines can react with metabolites such as pyruvate inside cells, thereby leading to fragments unreactive for EPL [10]. To circumvent this problem, one can express a POI with a fusion tag at its N-terminus. Following protein purification, this fusion tag can be cleaved selectively to create a POI with a N-terminal cysteine. One example is to fuse the N-terminus of the POI to the C-terminus of the Ssp DnaB intein [11]. The intein activity can be induced by a shift in pH from 8.5 to a lower pH (optimal range of 6.0–7.5) [11]. Notably this technique may lead to some loss of protein yield, since cleavage of the POI from the intein can occur during protein expression inside live cells. Another example of an approach based on a labile fusion tag involves fusing the protein SUMO to a Cys-POI construct (Fig. 3) [12, 13]. In this strategy, the N-terminal cysteine residue of a protein is fused to C-terminal end of His6-SUMO. After his-tag purification of the His6-SUMO-POI, the fusion protein is reacted with the SUMO protease, which recognizes a glycine–glycine motif at the C-terminus of the SUMO tag. As a result, the SUMO tag is cleaved and the released Cys-POI can be purified by loading the mixture onto a nickel column. His-tagged SUMO and SUMO protease, also labeled with His6, bind to the immobilized nickel, and the cleaved POI is then eluted from the column. The efficiency of the SUMO protease reaction depends on the accessibility of the glycine–glycine motif recognized by SUMO proteases. If structural regions of the POI hinder the glycine–glycine motif, the efficiency of the SUMO protease reaction may be reduced.

1.4 Generation of Synthetic Peptides for EPL by Solid-Phase Peptide Synthesis

Peptides are synthesized by solid-phase peptide synthesis (SPPS), a technique that relies on the stepwise conjugation of amino acids on a polymeric resin (Fig. 4) [14–16]. In essence, the C-terminus of an amino acid (aa1) is attached to the solid support. The N-terminal amine of aa1 is protected. When this protecting group is removed, the α-amine is available to react with another amino acid (aa2)

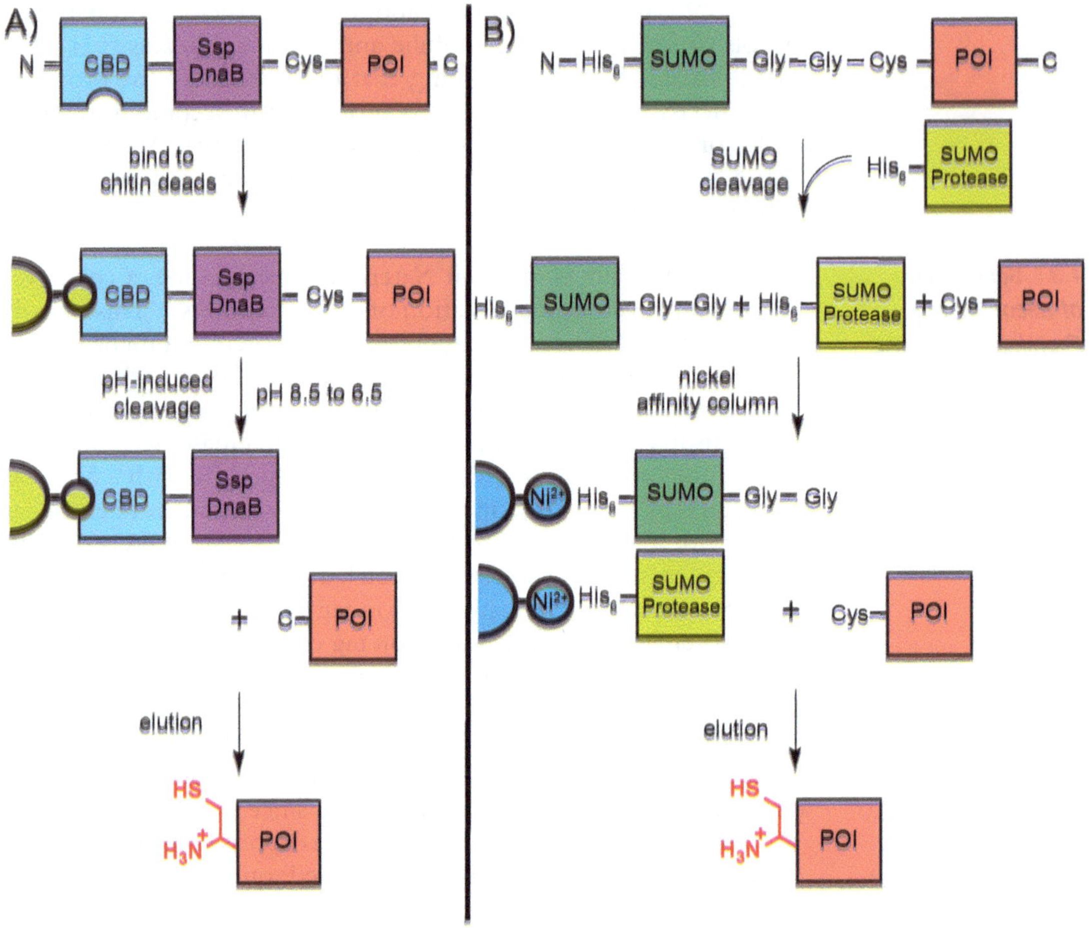

Fig. 3 Strategies for generating N-terminal cysteine-containing recombinant proteins. (**a**) The C-terminus of the Ssp DnaB intein is fused to the POI. The POI contains an N-terminal cysteine. A CBD is added to the N-terminus of the fusion protein for purification. The protein is expressed and the cell lysate is flowed through a column packed with chitin beads. The POI binds to chitin beads via the CBD while other lysate proteins are washed from the beads. A change in pH induces the cleavage of the POI from the CBD-Ssp DnaB fusion tag. The POI is eluted from beads and purified further if necessary. (**b**) SUMO is fused to the N-terminus of POI, a Gly-Gly-Cys sequence linking the two proteins. A His_6 purification tag is added to the N-terminus of SUMO. After expression, the protein is purified using affinity chromatography on nickel beads. The protein is subjected to SUMO protease, also labeled with a His_6 tag. The SUMO protease cleaves SUMO at the Gly-Gly site, leaving a POI with an N-terminal cysteine. SUMO and SUMO protease are captured using nickel beads and the desired POI is eluted from the beads

present in solution. Aa2 is also protected at its N-terminal to prevent reaction with itself and to promote the selective coupling of aa2 with aa1. Upon completion, the resin is filtered and unreacted aa2, which is typically present in excess so as to drive coupling yield to more than 98%, is simply washed away. This process is repeated until a peptide sequence of interest is assembled. The peptide is then cleaved from its solid support upon completion.

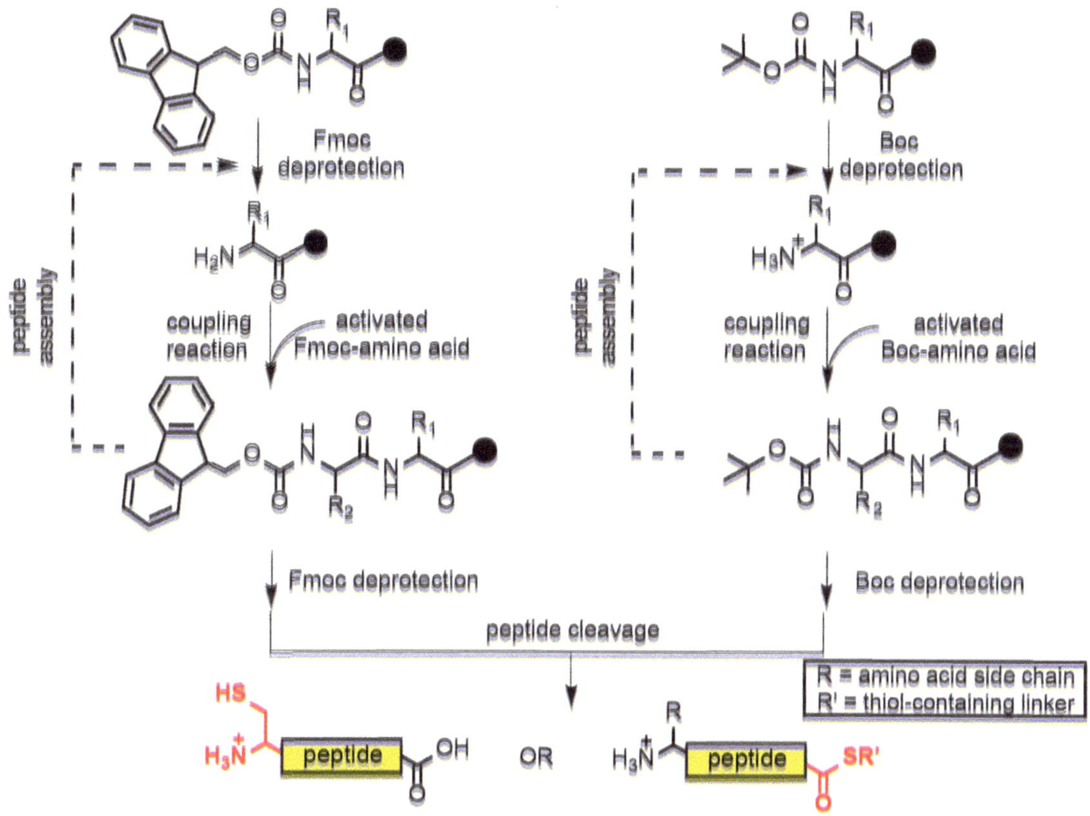

Fig. 4 Strategy for Fmoc-based and Boc-based SPPS. An amino acid is attached to a solid support (black sphere). The N-terminus of this first residue (R1 side chain) is protected by either a fluorenylmethyloxycarbonyl (Fmoc) or a tert-butyloxycarbonyl (Boc) protecting group. Fmoc is base-labile and typically removed using piperidine. Boc is acid-labile and typically removed with trifluoroacetic acid (TFA). The coupling reaction of a second protected residue (R2 side chain) to the N-terminal amine of the resin-bound peptide is initiated by addition of a mixture containing the base *N,N*-Diisopropylethylamine (DIEA) and a coupling agent (e.g., HBTU, HCTU). DIEA is used to deprotonate the amino group of the resin bound residue. The coupling agent increases the reactivity of the C-terminal carboxylate of the second amino acid to favor formation of an amide bond between the two residues. This process is repeated until a desired peptide sequence is assembled. The peptide is then cleaved from the resin, deprotection of side chains being typically concomitant to this step. For the purpose of EPL, the peptide generated can contain either an N-terminal cysteine, or a C-terminal thioester, as highlighted in red

The two common protecting groups for the α-amine are base-labile Fmoc and acid-labile Boc. The side chains of reactive amino acids also include orthogonal protection groups (opgs) that prevent side reactions. In the case of Fmoc chemistry, side chains opgs are typically resistant to the base used to cleave Fmoc (i.e., piperidine) and are instead cleaved by acids. In the case of Boc chemistry, both Boc and opgs are acid-labile. However, while Boc can be removed rapidly with 50% trifluoroacetic acid, the opgs used require harsher conditions (i.e., HF). In both Boc and Fmoc strategies, cleavage of the peptide from the solid support is often concomitant with the deprotection of side chain opgs.

SPPS is a robust strategy capable of producing a wide variety of peptides. However, some limitations exist. In particular, the efficiency of coupling reactions can decrease dramatically as the peptide increases in length. For instance, peptide chains in close proximity of the solid support can aggregate once they become long enough to reach one another. Alternatively, the peptide may fold on itself above a given length threshold, thereby making its N-terminus not accessible for a reaction. These effects, which will depend on the sequence of the target peptide and the density of peptide attached to polymer beads, can be mitigated to some extent. However, a general expectation is that their negative contribution is such that SPPS cannot produce peptides of 50 residues or more. A second limitation of SPPS relates to side reactions. These side reactions may include incomplete coupling, premature deprotection, racemization, or amino acid cyclization [17]. Each side reaction has the potential of generating a wide variety of truncated, deleted, or modified side products, making the identification and purification of the desired product challenging. When performing either Fmoc or Boc synthesis, one should therefore aim to prevent as many side products as possible. These issues have been extensively characterized, and readers are directed to excellent reviews and sources for further guidance [17–19].

The choice between Boc and Fmoc strategies will depend on a number of factors, including the sequence of the desired peptide and the nature of synthetic modifications incorporated in the peptide. For instance, in Boc synthesis, the acid hydrogen fluoride (HF) is commonly used in the final cleavage mixture to cleave the peptide from the solid support and remove side chain protecting groups. In terms of introducing NGEMs, chemical groups that react with acids (e.g., phosphorylated or glycosylated side chains) are hence incompatible with Boc-based synthesis. Another difficulty in implementing Boc-based synthesis involves the handling of HF. First, HF is a poisonous gas at room temperature. It therefore requires careful handling. To contain HF, which is used as a refrigerated liquid for Boc SPPS, specialized HF-resistant hardware (e.g., fluorocarbon-based glassware) is required [20]. Therefore, if Boc-based synthesis is the desired synthesis approach, one must consider the safety issues, cost, and setup time associated with this method. In comparison, Fmoc-based synthesis is less costly and less technically challenging. Yet Boc-based synthesis typically produces peptide thioesters in higher yields than Fmoc-based synthesis as this latter approach requires complex peptide–resin attachment strategies. The synthesis of C-terminal peptide thioesters via Boc and Fmoc chemistry are discussed below in Subheading 1.5.

For synthesizing peptides containing N-terminal cysteines, the Fmoc-based method is preferred as it is simpler to perform (i.e., more straightforward resin cleavage chemistry). For Fmoc-based synthesis, the side chain of the N-terminal cysteine can be protected with a trityl group, which is readily removed with TFA during final peptide cleavage. Ethanedithiol is utilized in the TFA cleavage mixture to prevent alkylation of the deprotected thiol group. In the case of a Boc strategy, a non-acid labile protecting group such as t-butylthio can be used. Alternatively, the Meb or Mob groups, which are removed by the final HF treatment, are also useful. If one follows the retrosynthetic analysis of Fig. 1, the peptide containing an N-terminal cysteine will include the C-terminal amino acid of the protein of interest. Thus, a Wang resin should be utilized, as it will produce a C-terminal carboxylic acid [21, 22]. If a C-terminal thioester peptide is being synthesized, then specialized resins or specialized loading of standard resins, are required. The generation of C-terminal thioester-containing peptides are described below.

1.5 Generation of C-Terminal Thioester-Containing Peptides

In order to generate peptides containing a C-terminal thioester, either Boc-based or Fmoc-based SPPS can be employed. One example of a Boc-based approach to generate a C-terminal thioester peptide involves the utilization of a trityl-associated mercaptopropionic acid–leucine resin (TAMPAL), as shown in Fig. 5. To generate TAMPAL, Boc-protected leucine is first coupled to the resin. Next, the Boc-protected group on leucine is deprotected, and S-trityl-mercaptopropionic acid is coupled to leucine. The trityl group is then deprotected and another Boc-protected amino is added. As a result, the peptide chain is anchored through a thioester linkage [23]. The thioester linkage is acid-stable, making it compatible with the repetitive exposure of TFA used to deprotect Boc. It is also stable under anhydrous HF treatment used to cleave the peptide from the resin. As a result, a peptide cleaved from the resin with HF will contain a TAMPAL thioester at its C-terminus.

In order to avoid using HF, a two-step method protocol may be followed. First, the side chains of the TAMPAL-bound peptide can be deprotected with an acid mixture consisting of trifluoromethanesulfonic acid, TFA, thioanisole, and p-cresol. Second, mercaptoethanesulfonic acid can be used to release the deprotected peptide from the TAMPAL resin by transthioesterification [24]. Recently, a different approach has been developed for the HF-free Boc synthesis of peptide thioesters [25]. In particular, Offer and coworkers have used a Merrifield resin coupled to mercaptophenylacetic acid (MPAA) as a solid support. As in the TAMPAL strategy, a peptide can then be assembled on the resin, a thioester serving as a linker. A mixture of TFA and TMSBr is then sufficient to release the peptide from the resin with a MPAA thioester at its C-terminus. Despite minor loss of peptide during TFA/-Boc cleavage (Merrifield resin being relatively labile to mild acidic conditions), Offer and coworkers were able to synthesize a phosphorylated peptide thioester using this approach.

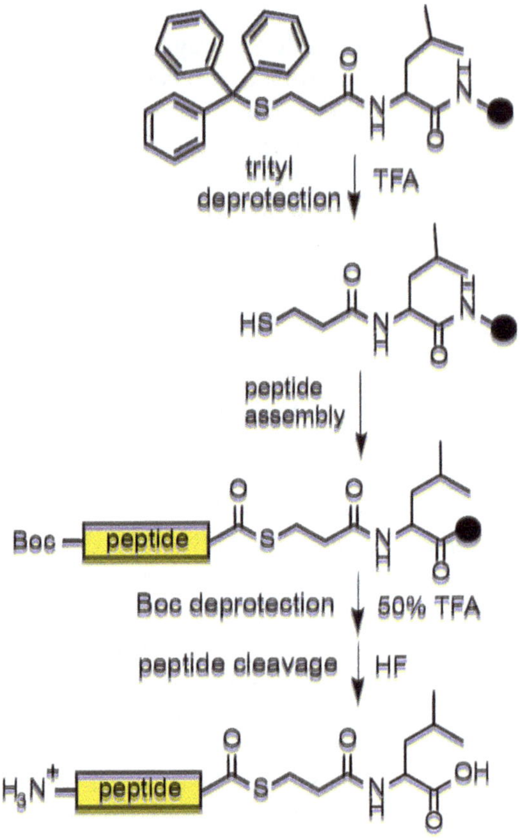

Fig. 5 Synthetic scheme for utilizing TAMPAL to generate peptide thioesters. The trityl-associated mercaptopropionic acid–leucine (TAMPAL) resin is first prepared by coupling a Boc-protected Leucine to a solid support. S-trityl mercaptopropionic acid is then coupled to N-terminus of the resin-bound leucine. Once the trityl group is removed, a Boc protected amino acid can be coupled to the free sulfhydryl to form a thioester linkage. The peptide is then elongated using Boc-based SPPS protocols. The peptide is cleaved from the resin using HF

Several Fmoc-based SPPS strategies have been developed to generate C-terminal thioesters [26]. Because thioesters are base-labile, and because Fmoc-SPPS requires repetitive base deprotection of Fmoc, the strategies used cannot utilize thioester linkers, as shown for Boc-based approaches. An example of an Fmoc-based strategy is to utilize a Dawson diaminobenzoic acid (Dbz) linker on a rink amide resin. This solid support takes advantage of the conversion of an aminoanilide group into an *N*-acylurea after peptide assembly is completed. Once the cleaved from the resin, the *N*-acylurea can be utilized as it is reactive with small molecule thiols such as MESNa, which can be added in solution to form the C-terminal thioester (Fig. 6) [27, 28]. Other Fmoc-based strategies involve using sulfonamide linkers [29], backbone amide linkers [30], Lewis

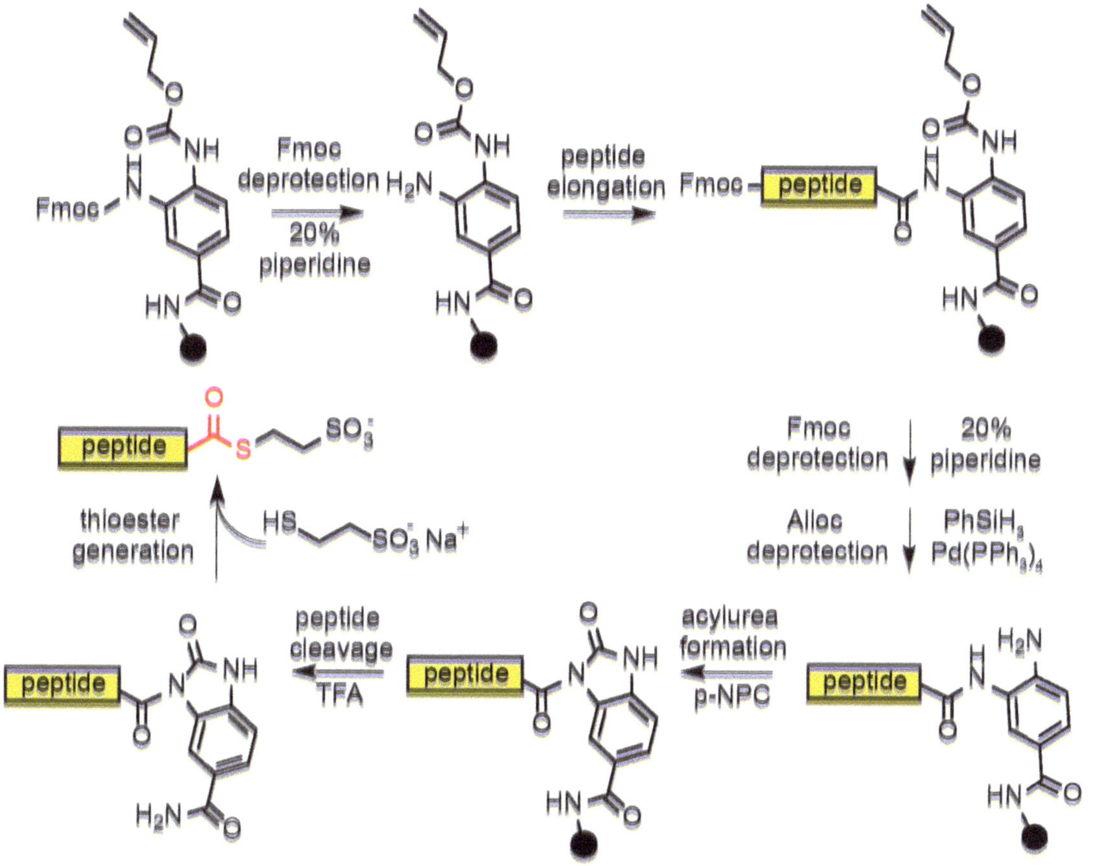

Fig. 6 Synthetic scheme for utilizing Dawson DBZ AM resin to generate peptide thioesters. The Dawson AM resin contains a 3,4 diaminobenzoic acid (DBZ) group. In this example, one of the amines is protected with allyloxy carbonyl (Alloc) and the other one is protected with Fmoc. Using the Dawson DBZ AM resin, SPPS using the Fmoc-based strategy is carried out as in Fig. 4. Once the final amino acid has been added and its Fmoc group has been removed, the Alloc group is deprotected by the addition of phenylsilane ($PhSiH_3$) and tetrakis(triphenylphosphine)palladium(0) ($Pd(PPh_3)_4$), which are mixed with the resin for 30 min at room temperature. The free amine on the DBZ linker is treated with p-nitrophenylchloroformate (p-NPC), followed by treatment with DIEA to yield an acylurea. Once the formation of the acylurea is confirmed, the peptide is cleaved. In solution, the acylurea moiety can be converted to a thioester by the addition of thiols

acid-activated cleavage [31], mercaptocarboxylethylester linker [32], aryl hydrazine linker [33], S-protected oxazolidinone linkers [34]. These methods have been described in detail in several reviews [35, 36].

1.6 Detection and Purification of the Ligated Product

In order to monitor the ligation reaction, it is necessary to identify a technique that can easily distinguish ligated product from unreacted fragments. In a simple scenario, the size of the peptide fragment is large enough to contribute significantly to the molecular weight of the ligated protein product. In such case, the ligation reaction can be monitored by SDS-PAGE, the ligated product appearing as a band above that of the unligated fragment. Ligation efficiency can be estimated by the ratio between the intensity of the Coomassie-stained bands representing the ligated and nonligated product. Monitoring the progression of the reaction over time is recommended as the ligation reaction may not go to completion. In particular, the competing hydrolysis of the thioester-fragment generates an unreactive C-terminal carboxylate. A ligation may therefore be stopped when the unligated protein fragment disappears during SDS-PAGE analysis (100% ligation efficiency) or when the amount of unligated fragment does not change substantially over time (i.e., likely presence of hydrolyzed and unreactive protein side product).

If a short peptide is utilized for EPL, the molecular weight difference between nonligated starting material and ligated product will be small and SDS-PAGE may not be adequate to distinguish these species. Increasing the percentage of the cross-linker, methylene bisacrylamide, present in the gel may first be tested to optimize the resolution of SDS-PAGE. However, if this fails, different strategies will be needed. For instance, if a fluorophore is as part of the design of the synthetic peptide used for ligation, in-gel fluorescence scanning can be used to conveniently detect the ligation product. More specically, small aliquots of the reaction mixture over time can be collected, analyzed using SDS-PAGE and scanned for fluorescence emission. An increase in fluorescence signal of a protein band should be expected as the reaction proceeds. Conversely, the reaction has likely reached completion when the fluorescence signal ceases to increase. Alternatively, one may choose to incorporate an epitope tag in a synthetic peptide. This tag may then be utilized for detection of the ligated product by western blotting. Alternatively, the ligated product could be separated from the nonligated products by reverse-phase HPLC (RP-HPLC). The RP-HPLC fractions can be analyzed by mass spectrometry to identify the HPLC peaks that correspond to ligated product and unligated fragments. Peak intensities, typically recorded from the absorbance of peptide bonds at 214 nm, can then be used to estimate ligation yields.

1.7 Conversion of a Cysteine Residue Introduced at a Ligation Site into an Alanine Residue

As mentioned in Subheading 1.1, one may replace an alanine with a cysteine in a protein sequence so as to introduce a ligation site. The cysteine residue can then be converted back into an alanine once the ligation is completed. This cysteine to alanine conversion involves a desulfurization reaction, which can be performed by using several techniques including treatment with Raney nickel or TCEP radical desulfurization [37–39]. There may be situations, however, where desulfurization may not be compatible with the POI. For instance, desulfurization is problematic if a protein contains other cysteines as all cysteine residues may be converted to alanine residues. When cysteine residues are necessary for the integrity of the protein, and a mutant cysteine residue cannot be inserted into the protein at the ideal ligation sites, one could consider using an enzyme catalyzed expressed protein ligation technique using subtiligase for the ligation of cysteine-free peptides to thioester-containing proteins [40, 41]. Another option is to synthesize cysteine surrogates to be incorporated into peptides by SPPS. These cysteine surrogates can be converted into native amino acids after performing desulfurization [42].

1.8 Choosing Native or Denaturing Conditions for EPL

Expressed protein ligation can be carried out in either nondenaturing or denaturing conditions. When possible, using folded recombinant protein fragments for EPL is usually advisable, as it circumvents the need to unfold and refold intermediates and products. However, there are instances where using denaturing conditions are necessary. This is the case when proteins are expressed in inclusion bodies in *E. coli*, when a folded protein displays a strong propensity to aggregate, or when the peptide ligated to the recombinant protein fragment is involved in the proper folding of the full-length product (in which case the recombinant EPL fragment may not be expressed in a native fold). In such cases, adding chaotropes, such as guanidinium chloride or urea, may increase the solubility of the recombinant fragment protein and subsequently increase the yield of the ligation product formed. Additionally, a recombinant fragment may display poor reactivity if the cysteine or thioester used for ligation is obstructed by the conformation of the folded protein (as previously described in Subheading 1.1). Denaturing conditions should in this case to reduce steric hindrance and facilitate the ligation reaction. Once ligation is achieved, refolding can be performed using reported techniques [43], such as dialysis [44] and dilution in the presence of chemical additives [45].

1.9 Summary of Protocols Included

Below, we provide general protocols regarding the generation of fragments useful for EPL. We outline the steps for routine Fmoc and Boc-based SPPS used to generate peptides with N-terminal cysteines or C-terminal thioesters. We also describe molecular biology protocols that yield recombinant proteins with EPL-compatible functional groups. Additionally, we provide general strategies for cloning a gene of interest (GOI) into vectors containing either pTWIN, a plasmid useful for producing fusion proteins containing Ssp DnaB, and pTXB1, a plasmid engineered for the expression of proteins fused to the intein Mxe-GyrA. The logic behind optimizing protein expression is also described. Finally, we provide protocols for performing expressed protein ligation under both native and denaturing conditions.

2 Materials

2.1 Manual Fmoc SPPS of N-Terminal Containing Cysteine Peptides on a Wang Resin

1. SPPS Vessel.
2. Quartz cuvette.
3. Calcium chloride drying tube.
4. Wang resin (0.7–1.7 mmol/g): may be purchased preloaded with an amino acid or loaded by the user.
5. Diisopropylcarbodiimide (DIPC).
6. Dimethylformamide (DMF).
7. 2% 1,8-diazabicyclo[5.4.0]undec-7-ene (DBU) in DMF (v/v).
8. 4-Dimethylaminopyridine (DMAP) powder dissolved in DMF.
9. Dichloromethane (DCM).
10. Fmoc deprotection solution: 20% Piperidine in DMF (v/v).
11. Fmoc-X-OH (X = any amino acid).
12. (6-Chloro-1-H-benzotriazole-1-yl)-1,1,3,3-tetramethylaminium hexafluorophosphate (HCTU).
13. Diisopropylethylamine (DIEA) dissolved in DMF.
14. Kaiser/ninhydrin Test reagents [46].
15. TFA-based cleavage solution: trifluoroacetic acid (TFA)/triisopropylsilane (TIS)/water (95%/2.5%/2.5% v/v). If cysteine is present, the solution should comprise 2.5% of the scavenger ethanedithiol (EDT) and the TFA should be reduced to 92.5%.
16. Acetonitrile with 0.1% TFA for use in RP-HPLC. The percent volume of acetonitrile may depend on the column used for HPLC.
17. Diethyl ether (DEE).

2.2 Manual Synthesis of C-Terminal Thioester Peptides Using Boc-SPPS

1. 4-Methylbenzhydrylamine hydrochloride (MBHA) resin.
2. DCM.
3. Boc-protected amino acid.
4. S-trityl mercaptopropionic acid.
5. *N*,*N*,*N'*,*N'*-tetramethyl-O-(1H-benzotriazol-1-yl)uranium hexafluorophosphate (HBTU).
6. DIEA.
7. Boc-deprotection solution: 50% TFA in DCM.
8. Neutralization solution: 10% triethylamine in DCM.
9. Trityl-deprotection solution: 2.5% triisopropylsilane, 2.5% water in TFA.
10. TIS.

11. (2-(1H-benzotriazol-1-yl)-1,1,3,3-tetramethyluronium hexafluorophosphate (HBTU).
12. Kaiser/ninhydrin test reagents.
13. Isopropyl alcohol (IPA).
14. Hydrogen fluoride (HF).
15. HF resistant glassware (Peptide Institute, Peptides International, Pepta Nova).

2.3 Using Dawson Dbz to Generate C-Terminal Thioester

1. Dawson Dbz AM resin.
2. Optional: Allyl chloroformate (350 mM) and DIEA (1 equivalents) in DCM.
3. Optional: Alloc deprotection solution: phenylsilane ($PhSiH_3$) (20 equivalents), and Tetrakis(triphenylphosphine)palladium (0) ($Pd(Ph_3)_4$) (0.35 equivalents) in DCM.
4. P-nitrophenylchloroformate (p-NPC) in DCM.
5. Thioester generation buffer (50 mM MESNa, 200 mM phosphate buffer, pH 7.5).
6. Same reagents as Subheading 2.1 for coupling of amino acids, deprotection of N-terminal amines, cleavage from resin, HPLC/MS analysis, etc.

2.4 Cloning of Gene of Interest into pTXB1

1. pTXB1 stock vector: Vector containing an ampicillin resistance gene, IPTG inducible lac operon, Mxe GyrA Intein-Chitin binding domain (CBD) fusion gene.
2. PCR primers to amplify the gene of interest (GOI) with restriction enzyme sites NdeI and SapI respectively (*see* **Note 1**).
3. Restriction enzymes NdeI and SapI.
4. 1× NEB 1.1 buffer: 10 mM bis-tris propane–HCl, 10 mM $MgCl_2$, 100 μg/mL BSA, pH 7.0 at 25 °C. If another restriction enzyme is used in place of Nde1, a different NEB buffer may be needed.
5. Spin columns.
6. Agarose.
7. T4 DNA Ligase.
8. 1× T4 DNA ligase buffer: 50 mM Tris–HCl, 10 mM $MgCl_2$, 10 mM dithiothreitol (DTT) 1 mM ATP, pH 7.5 at 25 °C.
9. DH5α competent cells.
10. Sterilized Luria broth (LB) and Luria agar (LA), autoclaved before use.
11. 1000× stock (100 mg/mL) of ampicillin or carbenicillin (*see* **Note 2**).

2.5 Cloning of Gene of Interest into pTWIN

1. pTWIN1 stock vector: Vector containing an ampicillin resistance gene, IPTG inducible lac operon, Mxe GyrA Intein, Ssp DnaB intein, and Chitin binding domain (CBD) genes.
2. PCR primers containing gene of interest (GOI) with Sap1 and Pst1 (*see* **Note 3**).
3. 1× Cutsmart Buffer: 50 mM potassium acetate, 20 mM tris–acetate, 10 mM magnesium acetate, 100 μg/mL BSA, pH 7.9.
4. Same materials of in Subheading 2.4.

2.6 Protein Expression

1. BL21-DE3 cells: certain BL21 cells have been engineered for specific purposes and may be more optimal depending on the POI needed to be produced.
2. Isopropyl β-D-1-thiogalactopyranoside (IPTG).
3. Lysis buffer: 20 mM Na-HEPES, 500 mM NaCl (can be modulated to increase protein solubility), 1 mM EDTA, 1× Halt Protease Inhibitor, pH 8.5.
4. SDS-PAGE gels (12% acrylamide recommended).
5. 5× SDS-PAGE loading dye: 0.05% bromophenol blue, 10% sodium dodecyl sulfate, 30% glycerol, 250 mM tris pH 6.8. The loading dye may be supplemented with reducing agent to prevent disulfide bonds (especially nonnative), but in the case of the Mxe-GyrA intein, thiol-based reducing agents (dithiothreitol) may cause partial cleavage between POI and intein.

2.7 Thiol-Mediated Mxe-GyrA Intein Cleavage to Generate a C-Terminal Thioester-Containing Protein

1. Chitin beads stored in 20% ethanol.
2. Wash buffer: 20 mM Na-HEPES, 500 mM (variable) NaCl, pH 8.5.
3. Mxe-GyrA cleavage buffer: 10 mM MESNa, 20 mM Na-HEPES, 500 mM NaCl (variable), pH 8.5.
4. Recommended storage buffer: 5 mM bis-tris, pH 6, 0.25 M NaCl.

2.8 Cleavage of Ssp DnaB Intein to Generate N-Terminal Cysteine on a Recombinant Protein

1. Cell lysis buffer (same composition as above, but ensure that the pH is at 8.5 or higher).
2. Ssp DnaB cleavage buffer: 20 mM Na-HEPES (recommended for pH 6.8–8.2), 500 mM (variable) NaCl, pH 6–7 (buffering system can be changed to accommodate pH range 6.0–6.8).

2.9 Ligation of C-Terminal Thioester and N-Terminal Cysteine-Containing Fragments

1. 10× Ligation buffer for native conditions: 0.1 M MESNa, 1 M Na-HEPES, 5 M NaCl (may be modulated to aid in solubilization of the protein), pH variable (*see* **Note 4**).
2. Denaturing ligation buffer: 0.1 M MESNa, 1 M Na-HEPES, 5 M NaCl, 6 M guanidine hydrochloride. Purge this buffer with ultrapure nitrogen for 20 min or use commercially available vacuum degassers.

3. C-terminal thioester-containing fragment.
4. N-terminal cysteine-containing fragment.
5. Dialysis buffer 1: 100 mM Na-HEPES, 500 mM NaCl, 4 M guanidine hydrochloride, pH 8.5.
6. Dialysis buffer 2: 100 mM Na-HEPES, 500 mM NaCl, 2 mM guanidine hydrochloride, pH 8.5.
7. Storage buffer: 100 mM Na-HEPES, 500 mM NaCl, pH 8.5. The components of this buffer can be changed to fit the needs of the experiment (e.g., pH 7.4 for assays at physiological pH).

3 Methods

3.1 Manual Fmoc SPPS of N-Terminal Containing Cysteine Peptides on a Wang Resin

The Wang resin is functionalized with *p*-alkoxy-benzyl alcohol. The first reaction of the synthetic scheme to utilize the Wang resin is to load the first amino acid onto the resin. This involves linking the carboxylic acid of the peptide to the primary alcohol on the Wang resin to form an ester bond. A general method used to carry out this esterification consists in converting the amino acid into a reactive anhydride with DIPC and catalyzing the reaction with a base called DMAP [47]. While this method is compatible with most amino acids, it is not ideal for amino acids that are prone to base-catalyzed racemization. In these specific cases, other esterification methods may be pursued [21]. One example is the use of 2,4,6-masitylene-sulfonyl-3-nitro-1,2,4-triazelide (MSNT) in the presence of 1-methylimidazole (MeIm), a method previously used to successfully load C-terminal glycosylated serine and cysteine amino acids onto a Wang resin with minimal racemization [48]. Alternatively, Wang resins preloaded with amino acids can be utilized. The loading capacity (mmol of peptide growth site per g of resin) of preloaded resin will be indicated by manufacturers. The loading capacity of Wang resin loaded by users should be determined after the esterification reaction, as it relates to the yield with which the first amino acid was successfully attached onto the resin. This can be performed by cleaving the Fmoc group of the attached amino acid with DBU and then, analyzing the concentration of released dibenzofulvene (the product of Fmoc cleavage) using UV spectrophotometry. The measured loading capacity of the resin can then be used to calculate the amount of the reactants needed for the coupling reaction. The amount of reagents weighed will be dependent on the desired equivalency of the coupling reagents. SPPS is carried out by iterative cycles of deprotection of the N-terminal amine of a resin-bound peptide and coupling of the next amino acid to the deprotected amine (as previously described in more detail in Subheading 1.4). Note that synthesis can be paused at the end of a coupling step, as all functional groups are then protected and unreactive. Conversely, the coupling step should be carried out immediately after Fmoc deprotection to avoid side reactions involving the deprotected and reactive N-terminal amine of the growing peptide.

3.1.1 Loading the Wang Resin

1. Weight out an amount of Wang resin (often available at 0.6–1.7 mmol/g loading capacity) suitable to obtain a peptide amount for EPL and future experiments (*see* **Note 5**).
2. Place the resin in a SPPS vessel, cover the resin with DMF, and swell the resin in this solvent for at least 1 h. The amount of volume needed to just cover the resin will be referred to in this section as 1 column volume (CV). Use this time to prepare reagents needed for the next step.
3. In a scintillation vial, dissolve the Fmoc-amino acid (10 equivalents) in DCM (the solvent should be dry and degassed with N_2). Then add DIPC (5 equivalents).
4. Transfer the solution to a calcium chloride drying tube, then stir the mixture for 20 min at 0 °C. The calcium chloride drying tube should prevent moisture from accumulating, which could cause side products to form.
5. Remove DCM using a rotary evaporator. Dissolve in DCM, using the minimal volume possible and add to the SPPS vessel.
6. Dissolve DMAP (1 equivalent) in DMF (~1–2 CV) and add to the SPPS vessel. Let the reaction proceed for 1 h. Mix by bubbling nitrogen into the SPPS vessel.
7. Remove a small amount of the resin (usually 20 mg is sufficient) and place in a 10 mL flask. Add 2 mL of 2% DBU in DMF and mix for 30 min. In a separate flask, also add 2 mL of 2% DBU in DMF without resin. Dilute both solutions to 10 mL in acetonitrile.
8. Record the absorbance at 304 nm to measure free dibenzofulvene, which is proportional to amount of Fmoc groups cleaved from the resin.
9. Use the following equation to determine the concentration of amino acid loaded onto the resin. Mmol/g = $(Abs_{sample} - Abs_{ref}) \times (16.4\ mg^{-1}$ of resin) (*see* **Note 6**).

3.1.2 Coupling of Amino Acids to Loaded Wang Resin

1. Add the Fmoc deprotection solution (piperidine in DMF, 20% v/v) to the vessel. Bubble nitrogen gas to mix the resin with Fmoc deprotection solution for 5 min. Wash resin with one CV of DMF and remove the solvent using a vacuum line. Repeat to a total of three washes, washing with one column volume each time. Remove the solvent, add one CV of the Fmoc deprotection solution to the SPPS vessel, and mix for an additional 15 min. Wash three times with 1 CV of DMF.

2. Add the coupling solution (mixture of Fmoc-aa-OH/HCTU/DIEA in DMF) to the SPPS vessel, which should be added in 4.0/3.9/10 equivalents (relative to the loading of the resin) respectively. Typically Fmoc-aa-OH and the HCTU are dissolved together, as they are commonly available as powders. Once these powders are dissolved, then DIEA (commonly available in liquid form) is added to the solution. Mix the solution in the SPPS vessel by bubbling nitrogen into the vessel for at least 4 h.
3. Use the Kaiser/ninhydrin test to detect the presence of free amines, an indication of an incomplete coupling reaction (*see* **Note 7**) [46].
4. In a synthesis where side reactions are possible and expected, one may decide to perform test cleavages followed by high pressure liquid chromatography (HPLC)/mass spectrometry (MS) analysis to follow the progress of the synthesis and for possible troubleshooting. If so, collect small aliquots of resin (10–20 mg). Then, cleave the peptide by adding the TFA-based cleavage solution to the small resin aliquot and shake for 1 h. The TFA-based cleavage solution comprises TFA/TIS/Water (95%/2.5%/2.5% v/v). If a trityl-protected cysteine residue is present in the peptide sequence, the solution should include 2.5% of the scavenger ethanedithiol (EDT), and the percentage of TFA should be decreased to 92.5%. This TFA-based cleavage solution will be the same for cleavage of the final product. Precipitate the peptide in diethyl ether. Spin down the pellet in a centrifuge and decant the solvent. If peptide did not precipitate, the ether solution can be removed by using a rotary evaporator or evaporation of the solution by N_2 drying. Then, the peptide can be extracted in water and then lyophilized. Suspend the product in an appropriate solvent, based on the hydrophobicity or hydrophilicity of the molecule (e.g., water and 0.1% TFA for peptides that have an overall positive charge) (*see* **Note 8**).
5. Determine the relative purity of the desired product by HPLC (*see* **Note 9**) and verify that the target peptide is within the sample by mass spectrometry. It is then advisable to optimize the HPLC gradient and determine if the product can be successfully separated.
6. When the peptide is successfully synthesized, perform a full cleavage using the TFA-based cleavage solution for 3 h.
7. Precipitate with ether as before and centrifuge to remove the solvent from the pellet. If the peptide does not precipitate, the ether solution can be removed as previously described (**step 4**).

8. Purify on a preparative HPLC column. From the fractions collected, run small amounts on an analytical HPLC to estimate purity.
9. Lyophilize pure samples and store final peptide at −80 °C (*see* **Note 10**).

3.2 Manual Synthesis of C-Terminal Thioester Peptides Using Boc-SPPS

Several methods have been used to prepare C-terminal thioesters. When designing the peptide, ensure that none of the protecting groups require base deprotection, as this will cleave the thioester. An extensive list of amino acids protecting groups have been described in great detail [49]. Herein, we describe the protocol used with the TAMPAL resin, a method established by Hackeng et al. [2, 50]. Ideally, one should use a small scale amount of MBHA resin, confirm the preparation of the resin is successful in high yields, and then use a larger scale. In this protocol, we will use a 2 mmol resin scale as an example:

1. Swell 2 mmol of MBHA resin in DMF for 1 h.
2. Meanwhile, prepare a coupling solution containing 4 mmol of Boc-protected leucine, 3.6 mmol of HBTU, and 6 mmol of DIEA. Dissolve in DMF as previously described in Subheading 3.1.2 (**step 2**). Remove DMF from the reaction vessel and add the coupling solution to the SPPS vessel.
3. Perform a Kaiser/ninhydrin test to determine if the reaction has gone to completion.
4. Deprotect the Boc-protected amine by mixing the resin with the Boc-deprotection solution (50%TFA in DCM) for 5 min. After washing three CVs with DCM, repeat the addition of the Boc-deprotection solution and allow the reaction to proceed for 20 min.
5. Wash the peptide twice with one column volume of DCM (2 column volumes total). Repeat the washes but instead use IPA. This should remove a majority of the deprotection solution (*see* **Note 11**).
6. To neutralize any residual TFA, add the neutralization solution (10%TEA in DCM) for 10 min. Wash the resin with DCM. Repeat this step once.
7. Prepare a coupling solution consisting of 3 mmol of S-trityl mercaptopropionic acid, 2.7 mmol of HBTU, and 6 mmol of DIEA in DCM for 15 min.
8. Use the Kaiser/ninhydrin test to determine if the reaction has gone to completion. Store a small aliquot of trityl-deprotected resin for use downstream HPLC analysis.

9. Deprotect the trityl group by adding 1 CV of the trityl deprotection solution (2.5% triisopropylsilane and 2.5% water in TFA). Allow the reaction to proceed for 1 min. Wash 3× with DCM, 1 column CV for each wash. Repeat this step once.
10. Continue coupling amino acids to the peptide chain by repeating **steps 2–6** of this section until the peptide is complete.
11. Once the peptide is ready for final cleavage, transfer the resin to a HF-compatible container. Performing HF-cleavage requires specialized equipment and specific handling to avoid exposure to HF (*see* **Note 12**).

3.3 Using Dawson Dbz Resin to Generate a C-Terminal Thioester-Containing Peptide

One example of utilizing Fmoc-based SPPS to generate a C-terminal thioester on synthetic peptides involves the use of a resin functionalized with 3,4 diaminobenzoic acid (Dbz) (the 3-amine is protected with Fmoc) [28]. After deprotection of the Fmoc-protected 3-amine, the first amino acid is coupled to the resin using an equivalent, so as to derivatize only one of two Dbz amines present. From this point forward, routine Fmoc-based SPPS is carried out. Once the desired peptide sequence has been assembled, the nonprotected Dbz amine is utilized to form an acylurea called N-acyl-benzimidazolinone (Nbz). In the presence of thiols, the Nbz undergoes thiolysis, yielding a thioester. When working with this resin, it is imperative to perform the coupling reaction in conditions where only one of the amines is coupled with an amino acid. Otherwise, the other Dbz amine cannot be utilized to form Nzb. Without Nzb, formation of the thioester is not possible. Before performing the first coupling reaction, one must consider that each amino acid differs in reactivity with the Dbz amine. Below, we describe conditions that have been optimized for coupling the first amino acid onto the resin.

It is important to note that the utilizing the Dawson resin is only one of many options for Fmoc-based synthesis of peptides containing C-terminal thioesters. Essentially, other Fmoc-based methods to generating thioesters may be more suitable for synthesizing the peptide of interest. For instance, this method is not ideal for glycine rich sequences, as they lack bulky side chains that prevent the Dbz free amine (either 3 or 4) from being acylated. Peptides of considerable length are also not ideal as the free Dbz amine is repetitively exposed to activated amino acids. Thus in these cases, one may elect to pursue a different method. Alternatively, one of the two Dbz amines can be protected with an Alloc or Me group prior to starting amino acid coupling reactions. These synthesis schemes circumvent the problem of amine over acylation [27]. Moreover, while the Alloc-Dbz strategy requires an additional deprotection step (Pd(PPh3)4 and phenylsilane) prior to urea formation, the MeDbz strategy does not [51].

Table 5
Recommended coupling conditions for the coupling reaction of the first amino acid to the Dawson DBZ resin

Amino acids	Fmoc-amino acid (mmole)	Activator (mmole)	DIEA (mmol)	Recommended activator	Perform double coupling
Arginine Isoleucine Threonine Valine Proline	0.6	0.6	0.9	HATU	X
Glycine	0.6	0.6	N/A	HOBT	
Other amino acids	0.6	0.6	0.9	HCTU	

3.3.1 Loading of the First Amino Acid onto the Dawson Dbz Resin

1. Weigh out the desired amount of Dawson Dbz AM resin. Transfer the resin to an SPPS vessel, cover the resin with DCM and swell for 1 h. Wash the resin 3× with DMF.
2. It is recommended, albeit optional, to protect the Dbz 4-amine with Alloc, while the 3-amine is Fmoc protected. Prepare a mixture containing allyl chloroformate (350 mM) and DIEA (1 equivalents) in DCM and place on a nutator for 24 h at room temperature. Use the Kaiser/ninhydrin test to determine if the Dbz amine was successfully protected.
3. Remove the Fmoc group from the 3-Dbz-amine with the Fmoc-deprotection solution for 5 min. Wash the resin 3× with DMF. Add Fmoc-deprotection solution for 15 min and wash with DMF.
4. The coupling of the first amino acid has been previously optimized. Use Table 5 to perform the first amino acid coupling. After adding the components of the reactions, mix for 1 h. For Ile, Val, Thr, Pro, and Arg, add reagents twice (1 h × 2) before performing a Kaiser/ninhydrin test to check for reaction completion. For glycine couplings use Fmoc-Gly-OPfp.
5. Once complete perform a small scale test cleavage using a small amount of resin. Cleave with the TFA-based cleavage solution for 30 min and analyze the product using HPLC.
6. Collect fractions and analyze the molecular weight by MS before continuing synthesis.

3.3.2 Alloc Deprotection

1. If Alloc was utilized to protect a Dbz 4-amine perform the following, otherwise skip to Subheading 3.3.3:
2. Swell the resin in DCM for 1 h.
3. Quickly flush DCM with argon to remove any dissolved oxygen.

4. Add 1 CV of $PhSiH_3$ (20 eq.) and $Pd(PPh_3)_4$ (0.35 eq.)
5. Mix for 30 min at room temperature.
6. Repeat the deprotection cycle until deprotection is detected as complete (e.g., HPLC analysis of small scale cleavage).

3.3.3 Generation of NBZ (Fig. 6, ***Step 5****)*

1. Swell the resin for 1 h in DCM.
2. Treat the resin with 1 column volume of p-NPC (5 equivalents) in DCM for 45 min at room temperature.
3. Wash the resin with three CVs of DCM.
4. Swell the resin for 1 h in DMF.
5. Remove DMF from the SPPS vessel and add DIEA (0.5 M in DMF).

3.3.4 Thioester Generation

1. Remove the Fmoc group, as previously described (piperidine cleavage).
2. Cleave the peptide from the resin as previously described (TFA cleavage).
3. Lyophilize the peptide and resuspend in buffer containing 200 mM phosphate buffer (pH 7.5) and 50 mM MESNa pH 7.5 for 1 h. If solubility is an issue, chaotropic agents such as 6 M guanidinium hydrochloride can be added to the resuspension buffer. Alternatively 250 mM of MPAA can be used in place of MESNa, as MPAA may speed up the reaction [52].
4. Monitor the completion of the reaction by RP-HPLC and mass spectrometry.

3.4 Cloning of Gene of Interest into pTXB1 Vector

1. Amplify the gene of interest using primers that allow for the addition of the restriction enzyme sites NdeI and SapI to the 5′-start of the target sequence and at the 3′-end of the target sequence respectively (*see* **Note 13**).
2. Purify the DNA fragment by spin columns. Alternatively, DNA can be purified by phenol–chloroform extraction (*see* **Note 14**).
3. After PCR, check for the presence of the amplicon by loading at least 10 ng of product into an agarose gel (*see* **Note 15**).
4. Place approximately 1 μg of the stock vector pTXB1 (previously prepared by a mini-prep procedure) and 1 μg of purified PCR product in separate tubes. Prepare the restriction digestion reaction mixture to be added to each tube: 1× NEB1.1 buffer, 1 unit of NdeI and SapI (1 unit is the amount of enzyme needed to digest 1 μg of DNA in 1 h), and bring to a total volume of 50 μL with distilled water that has been filtered through a 0.22 μm membrane. Digest the vector for 1 h at

37 °C. Analyze the DNA by loading 5 μL of the digestion reaction into an agarose gel. Load the linearized and noncut stock vector as controls. If the digest is successful, purify the fragment by agarose gel extraction (*see* **Note 16**).

5. Perform a ligation reaction between the digested amplicon and the digested pTXB1 vector. Prepare the ligation reaction mixture: 1× T4 DNA ligase buffer, T4 DNA ligase, and the desired molar ratio of the vector and the insert DNA in 20 μL of nuclease-free water. As a starting condition, a 1:3 vector-to-insert ratio is commonly utilized as an initial attempt at ligation, but this reaction can be optimized further by testing other ratios ranging from 1:1 to 1:7. Let the reaction proceed overnight at 16 °C.
6. Add the 1–5 μL of ligation mixture into DH5α cells.
7. Plate cells on agar plates containing the containing 100 μg/mL of carbenicillin (diluted 1:1000 from 1× stock).
8. Pick single colonies and inoculate them in separate culture tubes containing 5 mL of LB and carbenicillin (100 μg/mL) to an OD_{600} of 1.0.
9. With cultures that have successfully grown to $OD_{600} = 1.0$, perform a mini-prep procedure, which can be conveniently performed with commercially available kits. Once the DNA has been purified, use sequencing to ensure that the ligation was successful (mutations are rare but may have occurred during DNA replication).

3.5 Cloning of Gene of Interest into pTWIN

1. Follow the same cloning strategy as with pTXB1 with the following exceptions:
2. Use Sap1 and Pst1 cloning site for PCR amplification.
3. For restriction enzyme digest, use the cutsmart buffers, which are commonly supplied with NEB enzymes.

3.6 Protein Expression

Conditions that are optimal for the expression of proteins in *E.* coli are protein-dependent, as there are several factors that impact the production of a soluble protein (e.g., folding maturation time, protein toxicity to *E. coli*). For each protein, it is ideal to optimize the expression conditions to achieve a maximum amount of soluble protein possible, as a portion of the POI may be lost in each of the downstream protein purification steps. Before embarking on expression on the liter scale and subsequent purification, it is advisable to perform a screen for ideal expression conditions on the milliliter scale, as samples will be easier to handle and reagents will be saved. Thus, we recommend the following procedure as an example for screening for conditions ideal for protein expression in *E. coli* for a given construct:

1. Transform 1 pg to 10 ng of vector containing the GOI in a pTXB1 or pTWIN plasmid into BL1-DE3 cells. BL1-DE3 cells are standard *E. coli* competent cells that have been optimized for protein production (*see* **Note 17**).
2. Screen conditions optimal for protein expression using milliliter scale volumes. The conditions we recommend for the first round of screening are the following: 15 °C for 24 h, 25 °C for 24 h, 30 °C for 5–6 h, 37 °C for 3–4 h (times indicated corresponding to expression time post-IPTG induction). Screening these different conditions using 1 mL cultures of *E. coli* is convenient, as they can be readily cultured in commonly available 15 mL disposable culture tubes and can be readily transferred to Eppendorf tubes for lysis. In order to screen for the conditions listed above, we recommend taking the following steps:
   (a) First, transform the plasmid DNA containing the GOI.
   (b) Pick a single colony from an agar plate.
   (c) Inoculate the single colony in 5 mL LB media containing carbenicillin at a final concentration of 100 μg/mL (*see* **Note 18**).
   (d) Grow the culture to an optical density at 600 nm (OD_{600}) of 0.6. This density is commonly used as *E. coli* are still growing exponentially and have not exhausted nutrients in the media (*see* **Note 19**).
   (e) Induce expression by adding IPTG to achieve a final concentration of 50 μM. Since IPTG initiates the expression of the lac operon on the plasmid containing the GOI, the amount of protein expression can be modulated by the concentration of added IPTG. Further optimization of protein expression by screening various IPTG concentrations can also be pursued.
   (f) Transfer the culture into four separate 1 mL aliquots.
   (g) Take each 1 mL aliquot and culture them in different incubators corresponding to the conditions specified above.
   (h) Prepare cells for analysis on SDS-PAGE (*see* **Note 20**).

Once conditions are optimized, protein expression should be performed in several liters worth of bacterial culture. As described before, the ideal OD_{600} for IPTG induction is 0.6. In order to reach this optical density ideal for IPTG induction ($OD_{600} = 0.6$) in a time-efficient manner on the liter scale, it is ideal to first seed the large volume of culture media with a starter culture of cells in log-phase growth:

1. To prepare the starter culture, take an aliquot from the previously prepared glycerol stock, which should predominantly contain the plasmid from the original single-picked colony.
2. The starter culture should be grown to an OD_{600} of approximately 1.0.
3. In a 4 L flask, add 1 L of autoclaved LB supplemented with carbenicillin to achieve a final concentration of 100 μg/mL. Add 30 mL of the starter culture to the 1 L flask. Several 1 L cultures can be utilized to scale up the protein production.
4. Grow the bacteria to an $OD_{600} = 0.6$ and induce expression by adding IPTG to achieve a final concentration of 50 μM.
5. Transfer the 1 L culture to a tube suited for centrifugation at 13,000 × *g*. Spin the cultures at 13,000 × *g* at 4 °C. Decant the LB.
6. Lyse cells (*see* **Note 21**) and centrifuge (13,000 × *g* at 4 °C for 20 min) to separate soluble proteins from insoluble material, including inclusion bodies.
7. A recombinant POI may be present in the supernatant or in the pellet (as an inclusion body), which is determined by SDS-PAGE analysis of these fractions. If the protein is present in the supernatant, the purification process can continue to the steps described below. If the protein is present in the pellet fraction, the protein may be recovered by following a denaturation/refolding protocol, as described [43].

3.7 Thiol-Mediated Mxe-GyrA Intein Cleavage to Generate a C-Terminal Thioester-Containing Protein

Once a POI-IN-CBD has been expressed in *E. coli* and recovered from a cell lysate, it will be incubated with chitin beads to capture the protein. As it is often difficult to determine the amount of overexpressed POI-IN-CBD present in a lysate (a complex mixture of proteins), the amount of chitin beads needed to completely isolate the POI from the cell lysate is unknown and must be determined empirically. A common starting point consists of using approximately 20 mL of chitin beads per 1 L culture (the loading capacity of chitin bead is 2 mg protein/mL of beads). The amount of beads is then increased if excess protein is found in flow-through fractions. Record the amount of chitin beads needed for optimal loading of the POI and use this information for subsequent preparations or for scaling up the protein production.

1. The chitin beads are typically stored in 20% ethanol. After transferring the beads to a column (column 1). The volume of beads transferred into the column is considered as the column volume, and will be referred herein as CV.
2. Remove the ethanol by washing the beads with 3 column volumes (CV) of wash buffer.

3. Load the cell lysate onto the column at a flow rate of no more than 1 mL/min to allow for the protein to bind to the chitin beads. Alternatively, incubate the lysate with a slurry of chitin beads with gentle shaking overnight and then flow the mixture in a column (*see* **Note 22**).
4. Collect a sample of the flow through and load onto SDS-PAGE. Check for a decrease in the band at the molecular weight of POI-IN-CBD. If there is still a detectable amount of protein in the flow through continue to **step 4**. If not, proceed to **step 5**.
5. In a separate column (column 2), pack 20 mL of chitin beads and run the fractions collected from column 1 through column 2. Collect the flow through and determine if the band corresponding to the POI-IN-CBD is absent or decreases in intensity. Repeat until POI-IN-CBD is fully captured.
6. Wash the column with 5–10 CV of washing buffer. Add 1 CV at a time and collect the corresponding flow-through fraction (1CV fraction, 5–10 fractions total).
7. Analyze aliquots of the wash fractions and of the washed beads by SDS-PAGE. Prepare the chitin beads for SDS-PAGE by mixing with 5× SDS-Loading dye. Then boil the sample to ensure that the POI has dissociated from chitin beads. Confirm successful binding of POI-IN-CBD by determining if a band at the expected molecular weight is present in the bead sample. An additional band corresponding to IN-CBD, indicative of in situ cleavage, may be present. Confirm successful washing of the column by determining that the successive wash fraction becomes devoid of protein content (*see* **Note 23**).
8. Take a sample (100 μL) of POI-IN-CBD bound chitin beads. This sample will be used to compare with a sample of chitin beads pulled after the cleavage reaction has proceeded for a given time.
9. Equilibrate the column by flowing through 3 CV of cleavage buffer (collect fractions).
10. Stop the flow and place the column in a 4 °C cold room overnight. Alternatively, one can choose to place the column at room temperature to accelerate the reaction.
11. Collect the eluant and wash the column with 3 column volumes of the cleavage buffer.
12. Take 100 μL of beads for loading in SDS-PAGE, as before.
13. Analyze the bead samples and the elution fractions by SDS-PAGE. The elution fractions should display a protein band corresponding to the expected molecular weight for POI-thioester. The beads sample obtained post-cleavage should show bands corresponding to uncleaved POI-IN-

CBD and IN-CBD (*see* **Note 23**). The ratio of these two bands can be used to determine the extent of cleavage achieved. If cleavage is incomplete, fresh cleavage buffer can be added to the chitin column and cleavage time can be extended (*see* **Note 24**). It is also possible that the cleaved protein has become insoluble (*see* **Note 25**).

14. As thioesters are prone to hydrolysis over time, the eluted POI-thioester should ideally be used immediately for ligation with a peptide. However, if storage is necessary, use a low pH storage buffer to prevent hydrolysis of the thioester, flash freeze in liquid nitrogen, and store in a −80 °C freezer.
15. If the expressed protein will be ligated under denaturing conditions (*see* Subheading 1.8 for rationale), the sample is dialyzed against deionized water, and subsequently lyophilized [53]. For additional purification (if necessary, based on SDS-PAGE analysis), the lyophilized protein is resuspended in HPLC buffers and purified by HPLC. Pure protein fractions are collected, combined, and lyophilized. For ligation, the purified lyophilized fragment should be resuspended in denaturing ligation buffer, as described in Subheading 3.9.2.

3.8 Cleavage of Ssp DnaB Intein to Generate N-Terminal Cysteine on a Recombinant Protein

This method utilizes an intein (Ssp DnaB) as a N-terminal fusion tag that can self-cleave to liberate a POI with an N-terminal cysteine. To facilitate protein purification, a chitin binding domain is placed N-terminal of the Ssp DnaB. Similarly to the Mxe-GyrA system, the CBD-IN-POI is incubated with chitin beads. Instead of inducing cleavage of the intein by thiols, the Ssp DnaB intein cleavage is induced by a pH change from 8.5 to 6.0–7.0. Once cleaved from the intein, the Cys-POI is released from the chitin beads and ready for EPL. This method can be rendered ineffective if several precautions are not taken. First, since the cleavage reaction is initiated at pH 7.0 or lower, premature cleavage may take place in cells during protein expression. It is therefore advisable to limit expression time to minimize this problem (we recommend monitoring expression and Ssp DnaB cleavage by SDS-PAGE while optimizing both expression time and temperature on a small expression scale). Additionally, the buffer used to lyse cells should be at a pH of at least 8.5 to avoid intein cleavage during protein purification. Finally, it is important to assess the solubility of the protein and the activity of the protein after the pH shift (for intein cleavage and buffer exchange to EPL reaction conditions). If the protein is susceptible to any of these pitfalls, other techniques may be pursued. This includes the SUMO fusion approach described in Subheading 1.3 [12].

1. Express the Ssp DnaB intein and POI fusion and prepare cell lysates according to protocols described in Subheading 3.6, the only difference being that the lysis buffer used will be at pH 8.5

to minimize self-cleavage of the intein. Pack a column with chitin beads and load the supernatant of the cell lysate (bind and wash as in Subheading 3.7).

2. Before initiating cleavage, take a sample of chitin beads, load onto SDS-PAGE, and confirm that CBD-Ssp DnaB-POI has bound to the chitin beads.
3. Induce cleavage of Ssp DnaB intein by first equilibrating with 3 CV of Ssp DnaB cleavage buffer (recommended: 20 mM Na-HEPES, 500 mM (variable) NaCl, pH 7).
3. Allow for the cleavage reaction to proceed overnight at room temperature.
4. Collect the eluant from the column. Wash with 3 CV and take a sample of the chitin beads. If the protein is not soluble, prepare the protein as previously described (Subheading 3.7 **step 11**).
5. Analyze the eluant by SDS-PAGE to determine purity. Also, analyze a sample of the chitin beads to determine if the protein was cleaved from the resin.
6. Perform a buffer exchange with the ligation buffer to prepare the protein sample for the ligation step (native conditions) or to store the protein. Alternatively, the protein can be prepared for ligation under denaturing conditions (*see* Subheading 3.7 **step 15**).

3.9 Ligation of C-Terminal Thioester and N-Terminal Cysteine-Containing Fragments

When preparing fragments for EPL one should consider that the reactive groups (e.g., N-terminal cysteine and C-terminal thioester) can be rendered nonreactive, which ultimately will lead to a decrease in the overall yield of ligation. For instance, cysteines are prone to oxidation, which may result in disulfide bond formation (e.g peptide/protein dimerization) during storage. To mitigate this issue, oxygen should be purged from samples with nitrogen or argon prior to storage at −80 °C. In addition, samples can also be reduced post-storage by using immobilized TCEP (iTCEP), a reducing agent bound to beads that can simply be filtered out prior to beginning a ligation reaction. When using such reducing agents, avoid allowing the reduction reaction to proceed for extended periods of time and/or at temperatures of 37 °C or above, as the cysteine can get desulfurized or cleaved [54, 55].

As previously mentioned, the thioester moiety can be hydrolyzed over time, rendering it nonreactive toward the cysteine-containing fragment. Hence, the ligation reaction will result in low yields if a majority of the thioester moiety is hydrolyzed prior to the reaction. To maximize ligation yields, one should take preventative strategies to retain the thioester moiety. One strategy is to use the protein-thioester for ligation rapidly after bacterial expression and purification (i.e., avoiding storage). If this is not possible or convenient, storing the protein in a low pH buffer (6.5 or below) may be preferable, as thioesters are more susceptible to hydrolysis at higher pH. Additionally, excess thiol agents can be supplemented to the storage buffer so as to compete against the hydrolysis reaction of the thioester. MESNa (10 mM) is commonly used for this purpose.

When performing the EPL reaction, one should also consider that the stoichiometry of the fragments and the concentration of the overall reactants can be modulated to increase yield. In terms of stoichiometry, the EPL reaction is a bimolecular reaction and applying one of the fragments in excess of the other will drive the equilibrium toward formation of the ligated product. Because peptides can be obtained in relatively large quantities, and because small peptides can be easily separated from a larger protein (e.g., by size exclusion chromatography), the synthetic peptide fragment is usually added in excess (depending on the solubility of the peptide) over the recombinant protein fragment.

Another strategy to increase ligation yield is to increase the overall concentration of both reactants. As discussed in Subheading 1.8, a higher concentration of reactants (usually in the millimolar concentration range) can be achieved under denaturing conditions, as compared to native conditions. Thus, a larger amount of the ligated product can be achieved (if the protein can be successfully refolded and retain activity). There are also other scenarios that may require denaturing conditions. For instance, denaturing conditions should be considered when the C-terminal amino acid of the thioester-containing fragment leads to inefficient ligation and cannot be mutated. By performing the ligation reaction under denaturing conditions, some of the steric hindrance associated with the C-terminal amino acid of the thioester containing fragment is alleviated. Specifically, the side chain of the C-terminal amino acid is less likely to interact with other surrounding residues. Also, one must consider utilizing denaturing conditions if the peptides are not soluble at native conditions. In any of these scenarios, it is advisable to first produce the full-length proteins and determine if the protein can be efficiently refolded and retain its activity. Below we provide general protocols for performing EPL at both denaturing conditions and conditions that support the native fold of the POI.

3.9.1 Expressed Protein Ligation Using Nondenaturing Conditions

1. Prepare a 10× stock of ligation buffer (nondenaturing) that is purged of gases.
2. Dissolve a lyophilized peptide stock in ligation buffer to make a concentrated stock (when added to the final reaction mixture, the concentration of peptide should be at least 0.5 mM) (*see* **Note 26**).

3. If the protein is not already in ligation buffer, either dialyze or dilute the protein in ligation buffer. Concentrate the protein stock if necessary. When added to the final reaction mixture, the recommended concentration of protein should be at least 0.01 mM.
4. Prepare the ligation reaction by adding the 10× ligation buffer (from **step 1**), the peptide (from **step 2**), the protein (from **step 3**), and water (enough volume of water to give a final concentration of 1× ligation buffer) to a tube. We recommend using at least 0.5–1 mM of peptide and at least 0.01 mM of protein as a starting point.
5. Gently apply inert gas to purge oxygen from the environment and to prevent oxidation of the cysteine-containing peptide.
6. Place the mixture of protein, peptide, and 1× ligation buffer either at room temperature or in the 4 °C cold room (*see* **Note 27**).
7. Throughout the course of the experiment, take small aliquots of the ligation reaction at different time points. For example, collect a few microliters of solution every hour for reactions performed at room temperature, or every 6 h if at 4 °C. These samples will be loaded onto an SDS-PAGE gel for analysis. The amount of sample taken depends on the concentration of EPL fragments present in the ligation reaction and on the sensitivity of the method used for detection (i.e., 1 μL may be sufficient if fluorescence is used; 10 μL may be necessary for Coomassie staining). Establish when the intensity of the band corresponding to a ligated product reaches a plateau with time. This indicates that the reaction will not proceed further (note that this does not mean that 100% product formation is achieved, as product formation yield depends on the competing thioester hydrolysis reaction).
8. As mentioned earlier, the purification of the ligated product will vary depending on the EPL fragments used, their sizes, and their chemical properties. Optimizing conditions for the ligation reaction is recommended as achieving 100% yield (i.e., disappearance of the unligated recombinant protein fragment) will likely simplify the purification scheme.

3.9.2 Expressed Protein Ligation Using Denaturing Conditions

EPL using denaturing conditions is typically performed using a high concentration of protein and peptide fragments (millimolar concentration), as compared to native conditions (micromolar concentration). Using a high concentration of reactants for EPL can potentially lead to a large quantity of ligated product. However, the overall yield of desired product formation will also depend on the efficiency of refolding the ligated protein. When designing a strategy for efficient refolding, one must decide whether or not to remove nonligated fragments, which may interfere with the refolding process. Essentially, refolding the ligated protein can either be performed on the ligation mixture (potentially containing starting materials) or on purified and denatured ligated product. When performing refolding on the ligation mixture, nonligated protein fragments may interfere with the refolding of the ligated product. In principle, while it requires additional steps, purification on the ligation mixture before refolding is therefore recommended [56]. Below we present one approach for ligation of denatured fragments and there subsequent refolding. It is important to note that many refolding methods are available (*see* Subheading 1.8) and that this step will likely require optimization:

1. Purge the denaturing ligation buffer with nitrogen or argon for 20 min. Note that unlike the ligation buffer used in Subheading 3.9.1, the denaturing ligation buffer is prepared at a 1× concentration, as guanidine hydrochloride would not be soluble at a 10× concentration.
2. Dissolve a lyophilized peptide stock in denaturing ligation buffer to make a concentrated stock (*see* **Note 28**).
3. Dissolve a lyophilized protein stock in denaturing ligation buffer (*see* **Note 28**).
4. Prepare the reaction mixture by diluting the peptide and protein fragments stocks in denaturing ligation buffer. The final concentration of the peptide and protein fragments should be 1–2 and 0.5–1 mM, respectively.
5. Follow the progress of the reaction as in Subheading 3.9.1.
6. In the event that the reaction has progressed to near completion (little to no unligated protein fragment remaining), one may choose dialysis to purify the ligated protein product from the excess peptide remaining. This method can be advantageous as purification and refolding can be performed simultaneously.
7. Transfer the ligation reaction to a cassette with a membrane MWCO that is large enough to let the peptide escape but small enough to trap the protein (*see* **Note 29**).
8. Perform at least three buffer exchanges, each time decreasing the amount of guanidine hydrochloride (e.g., from 4 to 2 M and 0 M guanidine). Dialyze three times using the storage buffer (*see* **Note 30**). The sample should be kept in each buffer for at least 12 h (*see* **Note 31**).
9. After each dialysis step, observe if a precipitate forms. If a precipitate is present, use a centrifuge (13,000 × *g*) to remove the precipitate before the next dialysis step. Determine the concentration of protein remaining in the supernatant (absorbance, Bradford, fluorescence if labeled with the fluorophore

etc.). In the event that the protocol used leads to a significant loss of protein, repeat and optimize dialysis, varying for instance buffers concentration and dialysis time. Please note that the protein precipitate formed may be resuspended in denaturing buffer and that this material can in principle be used for additional purification efforts.

10. Using a protein activity assay specific to your protein of interest, establish if the refolded protein is functional. Alternatively, use biophysical techniques (e.g., NMR, circular dichroism) to determine whether the ligated and refolded protein shares structural similarities with a wild-type protein standard.

4 Notes

1. Recommended design of these primers can be found in manuals provided by New England Biolabs. The manual can be found at the NEB website: https://www.neb.com/.
2. Carbenicillin is more stable to heat and pH in culture medium as compared to ampicillin. It is also inactivated by beta-lactamases more slowly than ampicillin. Carbenicillin selection is therefore more stringent than ampicillin selection and reduces the chance that bacteria will grow in the absence of expression plasmid [57].
3. Recommended design for primers can also be found at the in manuals posted on NEB website (link in **step 1**).
4. The ideal pH for the ligation buffer is one that favors the thiolate form of the cysteine (deprotonated state) and one that prevents hydrolysis of the thioester. Ideally, these conditions correspond to a basic pH of 8–9. This may not be applicable for many cytosolic proteins as they may be optimally soluble at physiological pH. For these reasons, a pH of 7–8 is recommended. A pH below 7 should be avoided as this will favor protonation of the cysteine residue, rendering it less nucleophilic and less reactive. In contrast, a basic pH above pH 9 will contribute to accelerating the hydrolysis rate of the thioester. Another concern is oxidation of the cysteine. Thus, when preparing this buffer, degas by purging with ultrapure nitrogen for 20 min or use commercially available vacuum degassers.
5. If it is known that the peptide synthesis has low yield, it may be necessary to compensate by using more resin.
6. The equivalencies of the coupling reagents used in the next steps should be adjusted based upon the concentration of peptide loaded onto the resin.

7. If the solution is left for more than 5 min, the Boc or 4-methyltrityl (MTT) protected epsilon amine of lysine and the Fmoc on the alpha-amine of the amino acid, may be cleaved. As a result, the solution may turn blue, yielding a false positive. The intensity of the blue color is often amino acid dependent. Additionally, N-terminal proline and secondary amines may yield a brown or yellow color. The TNBS test is an alternative method to the Kaiser/ninhydrin test [58].
8. Common HPLC solvents for peptide synthesis consist of two solutions (e.g., water/0.1% TFA and acetonitrile/water/0.1% TFA). 0.1% TFA is present to protonate acidic groups on peptides and residual silanols on the C18 beads to ensure the separation is based on hydrophobicity. Additionally, the acidic pH protects the HPLC stationary phase from damage by bases. Solubility of the peptide in these buffers should be assessed prior to HPLC analysis, as particulates can damage the HPLC system.
9. Peptide bonds absorb at 214 nm and this signal can be used to monitor peptides by HPLC. The Fmoc group has a unique absorbance spectrum, with distinctive peaks at 265 and 300 nm. In some cases, it may be advantageous to leave the Fmoc group on the peptide to further facilitate analysis (e.g., distinguish peptides that are growing chains from peptides that are truncated and capped, therefore lacking Fmoc). Conversely, if the peptide already contains a group with an identifiable absorbance spectrum (e.g., aromatic amino acids or fluorophore), the Fmoc group can be removed. This is especially the case if the solubility of the peptide hinders the use of reverse phase HPLC.
10. Peptides will be obtained as TFA salts. When using peptides for experiments, buffer the solvent system to prevent TFA-induced changes in pH. Additionally, cysteine and methionine residues are prone to oxidation. Store these peptides under oxygen-free conditions.
11. As a result of Boc-deprotection by TFA, residual trifluoroacetate salts may still be present. Thus, a neutralization procedure must be performed.
12. Anhydrous HF is reactive with glass, so the resin must be transferred to a container that is compatible with HF. Appropriate setup for the anhydrous HF cleavage apparatus have been described in great detail in [20].
13. As discussed earlier, the C-terminal amino acid of the POI impacts the cleavage efficiency. One should consider that at this stage if the C-terminal amino acid of the POI could be changed by introducing a mutation in the appropriate codon.

14. Molecular cloning protocols are described in Molecular cloning: a laboratory manual, by Sambrook et al. [59].
15. If there are multiple PCR products, either redesign the primers to improve specificity of the target sequence or purify by agarose gel extraction. Alternatively, other methods to prepare the GOI may be pursued (as described in Molecular cloning: a laboratory manual, by Sambrook et al. [59]).
16. If the digest is incomplete, it is advisable to either add more enzyme to accelerate the reaction or extend the incubation time.
17. Depending on the protein, one may consider choosing modified BL21-DE3 cells. For instance, if a protein causes toxicity to *E. coli* cells, one may choose to utilize BL21(DE3) pLysE cells that repress low-levels transcription prior to IPTG induction or pLysS that reduces basal expression post-induction, respectively [60].
18. Optimal aeration for *E. coli* is required for optimal growth. This is achieved when the volume of the culture does not exceed 10% of the container. In this case, we recommend using at least a 50 mL tube to culture 5 mL of *E. coli* to provide adequate room for mixing and aeration.
19. At this stage, a glycerol stock can be prepared to be used for large scale expression, as described in 3.07. For this, transfer an aliquot of the culture to ice to prevent further growth. Prepare a glycerol stock by combining 500 μL of culture with 500 μL of 50% (v/v) glycerol. Store at −80 °C.
20. A volume of 10 μL of cell sample is added to 5× SDS-PAGE loading dye. The samples are boiled at 100 °C for 10 min to lyse cells and denature proteins.
21. There are several methods of lysis of bacterial cells, including sonication and disruption with a French pressure cell press. If using sonication, it is important to control the heating that accompanies this process as this may lead to protein unfolding (cells are placed on ice and a temperature probe is used to insure that the sonication regimen does not cause sharp temperature increases).
22. Flow rates that exceed 1 mL/min may disrupt CBD binding.
23. The presence of unidentified protein bands in the bead and/or final wash aliquot are indicative of incomplete washing. In this event, repeat the washing of the column. Conversely, the presence of POI-IN-CBD in wash fractions is indicative of washing steps that are too stringent and disruptive to CBD binding.
24. It is possible that a population of intein present is partially unfolded and therefore unable to mediate N-to-S acyl transfer and cleavage.

25. If the cleaved protein is insoluble, keep record for future purifications. If another purification attempt is required, after repeating the cleavage of the POI from the chitin beads, it is then ideal to wash the beads with wash buffer containing denaturants [61] or organic solvents [62] to promote solubility of the cleaved fragment.
26. Note that lyophilized peptides may contain TFA salts as a result of the purification process. After preparing the peptide stock solution, add a small amount (about 2 μL) of the solution to a pH strip to ensure the pH of the peptide stock solution is not acidic. Adjust the pH if it has fallen below 7.0. Addition of a low pH peptide solution to the ligation mix will otherwise make the protein precipitate upon addition to the ligation mixture.
27. The choice of temperature used is based on the stability of the protein and on the rate of hydrolysis of the thioester present.
28. As in Subheading 3.9.1, ensure the pH of the stock solutions and reaction mixture are within the desired range of 7–8.5 (precipitation of protein is less of a concern under denaturing conditions; however, the reactivity of the cysteine residue and the rate of hydrolysis of the thioester remain important considerations).
29. This can be performed by dialysis or a protein concentrator.
30. When dialyzing, use large volumes of dilution buffer and relatively small volumes of protein sample in the dialysis tubing to maximize the dilution of the reagents that escape the dialysis cassette.
31. The dialysis method presented may need to be optimized. For instance, it may be more favorable to gradually decrease the guanidine concentration in smaller increments (e.g., 0.5 M each time, instead of decreasing 2 M). The time for each dialysis step may also need to be increased. If optimization does not yield a properly folded protein, other refolding techniques may be pursued.

References

1. Dawson PE, Kent SB (2000) Synthesis of native proteins by chemical ligation. Annu Rev Biochem 69:923–960. https://doi.org/10.1146/annurev.biochem.69.1.923
2. Hackeng TM, Griffin JH, Dawson PE (1999) Protein synthesis by native chemical ligation: expanded scope by using straightforward methodology. Proc Natl Acad Sci U S A 96 (18):10068–10073
3. Marcaurelle LA, Mizoue LS, Wilken J, Oldham L, Kent SB, Handel TM, Bertozzi CR (2001) Chemical synthesis of lymphotactin: a glycosylated chemokine with a C-terminal mucin-like domain. Chemistry 7(5):1129–1132
4. Ghosh I, Considine N, Maunus E, Sun L, Zhang A, Buswell J, Evans TC Jr, Xu MQ (2011) Site-specific protein labeling by intein-mediated protein ligation. Methods Mol Biol 705:87–107. https://doi.org/10.1007/978-1-61737-967-3_6
5. Telenti A, Southworth M, Alcaide F, Daugelat S, Jacobs WR Jr, Perler FB (1997)

The Mycobacterium xenopi GyrA protein splicing element: characterization of a minimal intein. J Bacteriol 179(20):6378–6382

6. Chong S, Mersha FB, Comb DG, Scott ME, Landry D, Vence LM, Perler FB, Benner J, Kucera RB, Hirvonen CA, Pelletier JJ, Paulus H, Xu M-Q (1997) Single-column purification of free recombinant proteins using a self-cleavable affinity tag derived from a protein splicing element. Gene 192 (2):271–281. https://doi.org/10.1016/S0378-1119(97)00105-4
7. Evans TC Jr, Benner J, Xu MQ (1999) The in vitro ligation of bacterially expressed proteins using an intein from Methanobacterium thermoautotrophicum. J Biol Chem 274 (7):3923–3926
8. Minteer CJ, Siegart NM, Colelli KM, Liu X, Linhardt RJ, Wang C, Gomez AV, Reitter JN, Mills KV (2017) Intein-promoted cyclization of aspartic acid flanking the intein leads to atypical N-terminal cleavage. Biochemistry 56 (8):1042–1050. https://doi.org/10.1021/acs.biochem.6b00894
9. Biolabs NE (2019) Impact vectors and applications. Biolabs NE, Ipswich. https://www.neb.com/tools-and-resources/selection-charts/impact-vectors-and-applications
10. Gentle IE, De Souza DP, Baca M (2004) Direct production of proteins with N-terminal cysteine for site-specific conjugation. Bioconjug Chem 15(3):658–663. https://doi.org/10.1021/bc049965o
11. Mathys S, Evans TC, Chute IC, Wu H, Chong S, Benner J, Liu XQ, Xu MQ (1999) Characterization of a self-splicing mini-intein and its conversion into autocatalytic N- and C-terminal cleavage elements: facile production of protein building blocks for protein ligation. Gene 231(1–2):1–13
12. Malakhov MP, Mattern MR, Malakhova OA, Drinker M, Weeks SD, Butt TR (2004) SUMO fusions and SUMO-specific protease for efficient expression and purification of proteins. J Struct Funct Genom 5(1):75–86. https://doi.org/10.1023/B:JSFG.0000029237.70316.52
13. Peroutka Iii RJ, Orcutt SJ, Strickler JE, Butt TR (2011) SUMO fusion technology for enhanced protein expression and purification in prokaryotes and eukaryotes. Methods Mol Biol 705:15–30. https://doi.org/10.1007/978-1-61737-967-3_2
14. Merrifield RB (1963) Solid phase peptide synthesis. I. The synthesis of a tetrapeptide. J Am Chem Soc 85(14):2149–2154. https://doi.org/10.1021/ja00897a025
15. Chan WC (2004) Fmoc solid phase peptide synthesis : a practical approach. University Press, Oxford
16. Palomo JM (2014) Solid-phase peptide synthesis: an overview focused on the preparation of biologically relevant peptides. RSC Adv 4 (62):32658–32672. https://doi.org/10.1039/C4RA02458C
17. Yang Y (2016) Side reactions in peptide synthesis. Elsevier, Amsterdam
18. Fields GB (2002) Introduction to peptide synthesis. Curr Protoc Protein Sci 18:Unit-18.11. https://doi.org/10.1002/0471140864.ps1801s26
19. Amblard M, Fehrentz J-A, Martinez J, Subra G (2005) Fundamentals of modern peptide synthesis. In: Howl J (ed) Peptide synthesis and applications. Humana, Totowa, NJ, pp 3–24. https://doi.org/10.1385/1-59259-877-3:003
20. Muttenthaler M, Albericio F, Dawson PE (2015) Methods, setup and safe handling for anhydrous hydrogen fluoride cleavage in Boc solid-phase peptide synthesis. Nat Protoc 10:1067. https://doi.org/10.1038/nprot.2015.061. https://www.nature.com/articles/nprot.2015.061#supplementary-information
21. Ganesan A (2003) Wang resin. Encyclopedia of reagents for organic synthesis Wiley, Hoboken, NJ. https://doi.org/10.1002/047084289X.rn00168
22. Wang S-S (1973) p-Alkoxybenzyl alcohol resin and p-alkoxybenzyloxycarbonylhydrazide resin for solid phase synthesis of protected peptide fragments. J Am Chem Soc 95(4):1328–1333. https://doi.org/10.1021/ja00785a602
23. Camarero JA, Cotton GJ, Adeva A, Muir TW (1998) Chemical ligation of unprotected peptides directly from a solid support. J Pept Res 51(4):303–316
24. Gates ZP, Dhayalan B, Kent SB (2016) Obviation of hydrogen fluoride in Boc chemistry solid phase peptide synthesis of peptide-(alpha)thioesters. Chem Commun (Camb) 52 (97):13979–13982. https://doi.org/10.1039/c6cc07891e
25. Raz R, Burlina F, Ismail M, Downward J, Li J, Smerdon SJ, Quibell M, White PD, Offer J (2016) HF-free boc synthesis of peptide thioesters for ligation and cyclization. Angew Chem Int Ed Engl 55(42):13174–13179. https://doi.org/10.1002/anie.201607657
26. Noisier AFM, Albericio F (2015) Advance in ligation techniques for peptide and protein synthesis. Amino Acids, Peptides and Proteins. R Soc Chem 39:1–20. https://doi.org/10.1039/9781849739962-00001

27. Mahto SK, Howard CJ, Shimko JC, Ottesen JJ (2011) A reversible protection strategy to improve Fmoc-SPPS of peptide thioesters by the N-acylurea approach. Chembiochem 12 (16):2488–2494. https://doi.org/10.1002/cbic.201100472
28. Blanco-Canosa JB, Dawson PE (2008) An efficient Fmoc-SPPS approach for the generation of thioester peptide precursors for use in native chemical ligation. Angew Chem Int Ed Engl 47(36):6851–6855. https://doi.org/10.1002/anie.200705471
29. Shin Y, Winans KA, Backes BJ, Kent SBH, Ellman JA, Bertozzi CR (1999) Fmoc-based synthesis of peptide-αthioesters: application to the total chemical synthesis of a glycoprotein by native chemical ligation. J Am Chem Soc 121 (50):11684–11689. https://doi.org/10.1021/ja992881j
30. Alsina J, Yokum TS, Albericio F, Barany G (1999) Backbone amide linker (BAL) strategy for N(alpha)()-9-fluorenylmethoxycarbonyl (Fmoc) solid-phase synthesis of unprotected peptide p-nitroanilides and thioesters(1). J Org Chem 64(24):8761–8769
31. Swinnen D, Hilvert D (2000) Facile, Fmoc-compatible solid-phase synthesis of peptide C-terminal thioesters. Org Lett 2 (16):2439–2442. https://doi.org/10.1021/ol0060836
32. Botti P, Villain M, Manganiello S, Gaertner H (2004) Native chemical ligation through in situ O to S acyl shift. Org Lett 6(26):4861–4864. https://doi.org/10.1021/ol0481028
33. Camarero JA, Hackel BJ, de Yoreo JJ, Mitchell AR (2004) Fmoc-based synthesis of peptide alpha-thioesters using an aryl hydrazine support. J Org Chem 69(12):4145–4151. https://doi.org/10.1021/jo040140h
34. Ohta Y, Itoh S, Shigenaga A, Shintaku S, Fujii N, Otaka A (2006) Cysteine-derived S-protected oxazolidinones: potential chemical devices for the preparation of peptide thioesters. Org Lett 8(3):467–470. https://doi.org/10.1021/ol052755m
35. Behrendt R, White P, Offer J (2016) Advances in Fmoc solid-phase peptide synthesis. J Pept Sci 22(1):4–27. https://doi.org/10.1002/psc.2836
36. Camarero JA, Mitchell AR (2005) Synthesis of proteins by native chemical ligation using Fmoc-based chemistry. Protein Pept Lett 12 (8):723–728
37. Yan LZ, Dawson PE (2001) Synthesis of peptides and proteins without cysteine residues by native chemical ligation combined with desulfurization. J Am Chem Soc 123(4):526–533
38. Dawson PE (2011) Native chemical ligation combined with desulfurization and deselenization: a general strategy for chemical protein synthesis. Israel J Chem 51(8–9):862–867. https://doi.org/10.1002/ijch.201100128
39. Wan Q, Danishefsky SJ (2007) Free-radical-based, specific desulfurization of cysteine: a powerful advance in the synthesis of polypeptides and glycopolypeptides. Angew Chem Int Ed Engl 46(48):9248–9252. https://doi.org/10.1002/anie.200704195
40. Henager S, Chu N, Chen Z, Bolduc D, Dempsey D, Hwang Y, Wells J, Cole P (2016) Expressed protein ligation using subtiligase, Nat Methods. 2016 13(11):925–927 https://doi.org/10.1038/nmeth.4004
41. Henager SH, Chu N, Chen Z, Bolduc D, Dempsey DR, Hwang Y, Wells J, Cole PA (2016) Enzyme-catalyzed expressed protein ligation. Nat Methods 13(11):925–927. https://doi.org/10.1038/nmeth.4004
42. Wong CT, Tung CL, Li X (2013) Synthetic cysteine surrogates used in native chemical ligation. Mol BioSyst 9(5):826–833. https://doi.org/10.1039/c2mb25437a
43. Yamaguchi H, Miyazaki M (2014) Refolding techniques for recovering biologically active recombinant proteins from inclusion bodies. Biomol Ther 4(1):235–251. https://doi.org/10.3390/biom4010235
44. Zhi W, Landry SJ, Gierasch LM, Srere PA (1992) Renaturation of citrate synthase: influence of denaturant and folding assistants. Protein Sci 1(4):522–529. https://doi.org/10.1002/pro.5560010407
45. Yamaguchi H, Miyazaki M, Briones-Nagata MP, Maeda H (2010) Refolding of difficult-to-fold proteins by a gradual decrease of denaturant using microfluidic chips. J Biochem 147 (6):895–903. https://doi.org/10.1093/jb/mvq024
46. Kaiser E, Colescott RL, Bossinger CD, Cook PI (1970) Color test for detection of free terminal amino groups in the solid-phase synthesis of peptides. Anal Biochem 34(2):595–598
47. Neises B, Steglich W (1978) Simple method for the esterification of carboxylic acids. Angew Chem Int Ed Engl 17(7):522–524. https://doi.org/10.1002/anie.197805221
48. Blankemeyer-Menge B, Nimtz M, Frank R (1990) An efficient method for anchoring fmoc-anino acids to hydroxyl-functionalised solid supports. Tetrahedron Lett 31 (12):1701–1704. https://doi.org/10.1016/S0040-4039(00)88858-9
49. Isidro-Llobet A, Álvarez M, Albericio F (2009) Amino acid-protecting groups. Chem Rev 109

(6):2455–2504. https://doi.org/10.1021/cr800323s

50. Southworth MW, Amaya K, Evans TC, Xu MQ, Perler FB (1999) Purification of proteins fused to either the amino or carboxy terminus of the Mycobacterium xenopi gyrase A intein. Biotechniques 27(1):110–120. https://doi.org/10.2144/99271st04
51. Blanco-Canosa JB, Nardone B, Albericio F, Dawson PE (2015) Chemical protein synthesis using a second-generation N-acylurea linker for the preparation of peptide-thioester precursors. J Am Chem Soc 137(22):7197–7209. https://doi.org/10.1021/jacs.5b03504
52. Johnson ECB, Kent SBH (2006) Insights into the mechanism and catalysis of the native chemical ligation reaction. J Am Chem Soc 128(20):6640–6646. https://doi.org/10.1021/ja058344i
53. Kamei A, Hauser PS, Beckstead JA, Weers PMM, Ryan RO (2012) Expressed protein ligation-mediated template protein extension. Protein Expr Purif 83(2):113–116. https://doi.org/10.1016/j.pep.2012.03.014
54. Wang Z, Rejtar T, Zhou ZS, Karger BL (2010) Desulfurization of cysteine-containing peptides resulting from sample preparation for protein characterization by mass spectrometry. Rapid Commun Mass Spectrom 24(3):267–275. https://doi.org/10.1002/rcm.4383
55. Liu P, O'Mara BW, Warrack BM, Wu W, Huang Y, Zhang Y, Zhao R, Lin M, Ackerman MS, Hocknell PK, Chen G, Tao L, Rieble S, Wang J, Wang-Iverson DB, Tymiak AA, Grace MJ, Russell RJ (2010) A tris (2-carboxyethyl) phosphine (TCEP) related cleavage on cysteine-containing proteins. J Am Soc Mass Spectrom 21(5):837–844. https://doi.org/10.1016/j.jasms.2010.01.016
56. Spriestersbach A, Kubicek J, Schafer F, Block H, Maertens B (2015) Purification of His-tagged proteins. Methods Enzymol 559:1–15. https://doi.org/10.1016/bs.mie.2014.11.003
57. Traub WH, Leonhard B (1995) Heat stability of the antimicrobial activity of sixty-two antibacterial agents. J Antimicrob Chemother 35(1):149–154
58. Sarin VK, Kent SB, Tam JP, Merrifield RB (1981) Quantitative monitoring of solid-phase peptide synthesis by the ninhydrin reaction. Anal Biochem 117(1):147–157
59. Green MR, Sambrook J (2012) Molecular cloning: a laboratory manual, 4th edn. Cold Spring Harbor Laboratory Press, Cold Spring Harbor, NY
60. Studier FW (1991) Use of bacteriophage T7 lysozyme to improve an inducible T7 expression system. J Mol Biol 219(1):37–44
61. Virdee S, Macmillan D, Waksman G (2010) Semisynthetic Src SH2 domains demonstrate altered phosphopeptide specificity induced by incorporation of unnatural lysine derivatives. Chem Biol 17(3):274–284. https://doi.org/10.1016/j.chembiol.2010.01.015
62. Zuberek J, Wyslouch-Cieszynska A, Niedzwiecka A, Dadlez M, Stepinski J, Augustyniak W, Gingras A-C, Zhang Z, Burley SK, Sonenberg N, Stolarski R, Darzynkiewicz E (2003) Phosphorylation of eIF4E attenuates its interaction with mRNA 5′ cap analogs by electrostatic repulsion: intein-mediated protein ligation strategy to obtain phosphorylated protein. RNA 9(1):52–61. https://doi.org/10.1261/rna.2133403

Chapter 6

Peptide Hydrazides as Thioester Equivalents for the Chemical Synthesis of Proteins

Yu Wang and Yi-Ming Li

Abstract

The chemical synthesis of proteins allows for the precise control of structural information at the atomic level, overcoming the limits of protein expression. Peptide hydrazides are widely used as thioester equivalents in the total chemical synthesis and semisynthesis of proteins as they can be easily prepared using standard solid phase peptide synthesis (SPPS) and recombinant peptide techniques. Via treatment with $NaNO_2$ and subsequent thiolysis, peptide hydrazides can be rapidly converted to peptide thioesters, which then selectively react with recombinant protein containing an N-terminal cysteine (Cys) to form a native peptide bond, thereby linking the two peptide segments without isolating any intermediates.

Key words Chemical synthesis of proteins, Total chemical synthesis, Semisynthesis, Peptide hydrazides, Peptide thioesters

1 Introduction

The chemical synthesis of proteins enables the acquisition of proteins that are difficult to obtain through biological expression, such as post-translationally modified proteins or toxic proteins [1]. However, SPPS is limited to peptides of 50 amino acids or fewer, and the synthesis of larger peptides or proteins requires the use of peptide segment condensation strategies [2]. One popular and well-developed strategy is native chemical ligation (NCL), wherein a C-terminal peptide thioester and N-terminal Cys are selectively reacted [3, 4]. Both total synthetic and recombinantly peptides can be used in NCL [3, 5, 6], and its tremendous utility has been repeatedly demonstrated through the total and semisynthesis of a variety of proteins [7–9]. The combination of synthetic peptides with recombinant proteins greatly expands the reach of protein chemical synthesis, allowing for the generation of complex semisynthetic proteins [10, 11]. However, NCL is predicated on the efficient preparation of peptide thioester intermediates. Although Boc-SPPS can yield peptide thioesters, it requires the use of the

Miquel Vila-Perelló (ed.), *Expressed Protein Ligation: Methods and Protocols*, Methods in Molecular Biology, vol. 2133,
https://doi.org/10.1007/978-1-0716-0434-2_6, © Springer Science+Business Media, LLC, part of Springer Nature 2020

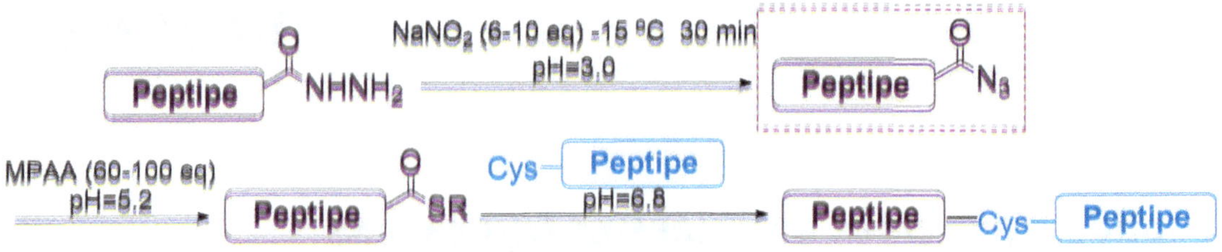

Fig. 1 Native chemical ligation of peptide hydrazides

highly corrosive hydrogen fluoride (HF) for deprotection, which has limited its widespread application [12]. The most commonly used method is the Fmoc-SPPS method which still has difficulty in obtaining peptide thioesters [13]. In order to overcome this problem, extensive studies have been conducted on how to synthesize peptide thioesters or their equivalents by Fmoc-SPPS [14–22].

Here, we describe the strategy of peptide hydrazide-based NCL [23–26]. The method involves a chemoselective reaction between a C-terminal peptide hydrazide and an N-terminal Cys to produce a native peptide bond (Fig. 1). Notably, peptide hydrazides can be readily prepared not only in Boc-SPPS or Fmoc-SPPS [27, 28], but also through recombinant expression technology [29–34]. Therefore, peptide hydrazides can be used in both total synthesis and semisynthesis of proteins, and therefore have been extensively used in the preparation of many different types of proteins [35–41]. Here we provide protocols for the total synthesis of centruroides suffusus suffusus II (CssII) [23] (*see* Subheading 3.1) and semisynthesis of acid stress chaperone HdeA (HdeA) [32] (*see* Subheading 3.2) by hydrazide-based NCL.

2 Materials

Prepare all solutions with ultrapure water and analytical grade reagents. Reagents or solvents are stored at room temperature, unless otherwise stated. All organic solvents should be processed in a chemical fume hood.

2.1 Total Synthesis of CssII by Hydrazide-Based NCL

2.1.1 Preparation of 2-Cl-(Trt)-$NHNH_2$ Resin

1. 2-Cl-(Trt)-Cl resin: 0.34 mmol/g. Store at 4 °C.
2. Dichloromethane (DCM).
3. *N*,*N*-dimethylformamide (DMF).
4. Peptide synthesis vessel.
5. Constant temperature shaker.
6. 5% (v/v) NH_2NH_2: 5% (v/v) NH_2NH_2 in DMF (*see* **Note 1**).
7. 5% (v/v) MeOH: 5% (v/v) MeOH in DMF.

2.1.2 Synthesis of Peptide Fragments of CssII

1. Rink Amide-AM Resin: 0.32 mmol/g. Store at 4 °C.
2. Wang Resin: 0.8 mmol/g. Store at 4 °C.
3. Acetonitrile (ACN): HPLC grade.
4. Standard Fmoc-protected amino acids: Fmoc-Ala-OH, Fmoc-Arg(Pbf)-OH, Fmoc-Asn(Trt)-OH, Fmoc-Asp(OtBu)-OH, Fmoc-Cys(Trt)-OH, Fmoc-Gln(Trt)-OH, Fmoc-Glu(OtBu)-OH, Fmoc-Gly-OH, Fmoc-His(Trt)-OH, Fmoc-Ile-OH, Fmoc-Leu-OH, Fmoc-Lys(Boc)-OH, Fmoc-Met-OH, Fmoc-Phe-OH, Fmoc-Pro-OH, Fmoc-Ser(tBu)-OH, Fmoc-Thr(tBu)-OH, Fmoc-Trp(Boc)-OH, Fmoc-Tyr(tBu)-OH, Fmoc-Val-OH.
5. 2-(6-Chloro-1H-benzotriazole-1-yl)-1,1,3,3-tetramethylaminiumhexafluorophosphate (HCTU).
6. 1-[Bis(dimethylamino)methylene]-1H-1,2,3-triazolo-[4,5-b]pyridiniumhexafluorophosphate 3-oxide (HATU).
7. *N*,*N*-Diisopropylethylamine (DIEA).
8. Trifluoroacetic acid (TFA).
9. 20% (v/v) piperidine: 20% (v/v) piperidine in DMF (*see* **Note 2**).
10. TFA cleavage mixture: 20 mL of TFA/H_2O/phenol/TIPS (v/v/wt/v) = 88:5:5:2. Mix 17.6 mL of TFA, 1 mL of H_2O, 0.4 mL of TIPS and 1 g of phenol to prepare 20 mL of TFA cleavage mixture (*see* **Note 3**).
11. 50% (v/v/v) ACN: ACN/H_2O/TFA (v/v/v) = 50:50:0.1.
12. Analytical and semipreparative HPLC system: *see* Table 1.

Table 1
Analytical and semipreparative HPLC conditions

	Analytical	Semipreparative
Column	C4 reverse-phase column (4.6 × 250 mm, pore size: 300 Å, particle size: 5 μm)	C18 reverse-phase column (10 × 250 mm, pore size: 300 Å, particle size: 10 μm)
Solvents	A: deionized H_2O containing 0.1% (v/v) TFA; B: ACN containing 0.08% (v/v) TFA	A: deionized H_2O containing 0.1% (v/v) TFA; B: ACN containing 0.08% (v/v) TFA
Flow rate	1.0 ml/min	3.0 ml/min
Gradient	Alterable, depending on the peptides or reactions	Alterable, depending on the peptides or reactions
Wavelength	210 and 254 nm	210 and 254 nm

2.1.3 Synthesis of CssII(1–24)-$NHNH_2$ by the Native Chemical Ligation of Peptide Hydrazides

1. 100 mL Dewar flask.
2. pH meter.
3. 2 M NaOH solution: 2 M NaOH in water (*see* **Note 4**).
4. 2 M HCl solution: 2 M HCl in water. Add 3.33 mL of 12 M HCl to 16.67 mL H_2O in a 50 mL tube and mix (*see* **Note 5**).
5. 0.2 M NaH_2PO_4 solution: 0.2 M NaH_2PO_4, 6 M Gn·HCl in water, pH 3.0–3.1. 3.12 g $NaH_2PO_4 \cdot 2H_2O$ and 57.4 g Gn·HCl are placed in a 100 mL volumetric flask, dissolve in water and adjust to final volume. Adjust the pH to 3.0–3.1 and filter the solution through a 0.22 μm microporous membrane filter.
6. 0.2 M $NaNO_2$ solution: 0.2 M $NaNO_2$ in 0.2 M NaH_2PO_4 solution (*see* **Note 6**).
7. 0.2 M 4-mercaptophenylacetic acid (MPAA) solution: 0.2 M MPAA in 0.2 M NaH_2PO_4 solution. Adjust the pH to 5.2 and filter the solution through a 0.22 μm microporous membrane filter.
8. 0.6 M Tris(2-carboxyethyl)phosphine hydrochloride (TCEP·HCl) solution: 0.6 M TCEP in 0.2 M NaH_2PO_4 solution. Adjust the pH to 7.0 and filter the solution through a 0.22 μm microporous membrane filter (*see* **Note 7**).
9. Analytical and semipreparative HPLC system: *See* Table 1.

2.1.4 Synthesis of CssII(1–40)-$NHNH_2$ by the Native Chemical Ligation of Peptide Hydrazides

Same as the materials in Subheading 2.1.3.

2.1.5 Synthesis of CssII(1–66)-NH_2 by the Native Chemical Ligation of Peptide Hydrazides

Same as the materials in Subheading 2.1.3.

2.2 Semisynthesis of HdeA by Hydrazide-Based Expressed Protein Ligation

2.2.1 Synthesis of α-Hydroxy Acid BKOH

1. Amberlyst®15 wet ion-exchange column (H+ form).
2. Lysine (compound 1).
3. Di-tert-butyl dicarbonate (Boc_2O).
4. 1,4-dioxane.
5. CH_2Cl_2.
6. $MgSO_4$.
7. 10% (v/v) H_2SO_4 solution: 10% (v/v) H_2SO_4 in water. Add 9.8 mL of 98% H_2SO_4 to 89.2 mL H_2O in to a 250 mL graduated cylinder and mix (*see* **Note 8**).

8. 3.86 M $NaNO_2$ solution: 3.86 M $NaNO_2$ in water.
9. 6 M urea solution: 6 M urea in water.
10. 1 M HCl solution: 1 M HCl in water. Add 1 mL of 12 M HCl and 11 mL H_2O to a 25 mL graduated cylinder and mix.
11. 8% (v/v) ammonia solution: 8% (v/v) ammonia in water.
12. 0.2%(w/v) Ninhydrin solution: 0.2% (w/v) Ninhydrin in C_2H_5OH.
13. 1 M NaOH solution: 1 M NaOH in water.
14. 50% methanol (v/v) solution: 50% methanol (v/v) in diethyl ether.

2.2.2 Genetic Incorporation of α-Hydroxy Acid BKOH

1. Constant temperature incubator.
2. Falcon tube.
3. Ni-NTA resin.
4. Centrifugal filter: 3000 MW cutoff.
5. Plasmid PylRS-tRNA$^{Pyl}_{CUA}$ pair: Plasmid PylRS-tRNA$^{Pyl}_{CUA}$ pair from the laboratory of Peng R. Chen, Peking University.
6. Plasmid HdeA-V58TAG-myc-His_6: The plasmids expressing HdeA-myc-His_6-V58BKOH proteins His-tagged at the C terminus are generated using HindIII-XhoI restriction sites in pBAD. In order to introduce the unnatural amino acid BKOH at position 58, the V58 codon is mutated to TAG.
7. *E. coli* DH10B cells.
8. 50 mg/mL ampicillin solution: 50 mg/mL ampicillin in water. Store at −20 °C. Ampicillin solution and LB medium are used in a ratio of 1:500.
9. 50 mg/mL chloramphenicol solution: 50 mg/mL chloramphenicol in water. Store at −20 °C. Chloramphenicol solution and LB medium are used in a ratio of 1:1000.
10. LB medium: 1%(w/v) NaCl, 0.5% (w/v) yeast extract, 1% (w/v) tryptone in water. 10 g NaCl, 5 g yeast extract and 10 g tryptone are placed in a 1 L Erlenmeyer flask, dissolve in water and adjust to final volume. The solution is autoclaved at 121 °C for 20 min.
11. LB agar plate: 1%(w/v) NaCl, 0.5% (w/v) yeast extract, 1% (w/v) tryptone, 0.5% (w/v) agar in water. 2 g NaCl, 1 g yeast extract, 2 g tryptone, and 1 g agar are placed in a 200 mL Erlenmeyer flask, dissolve in water and adjust to final volume. The solution is autoclaved at 121 °C for 20 min. When culture medium of agar is cooled, add 0.2 mL of 50 mg/mL ampicillin solution and 0.1 mL of 50 mg/mL chloramphenicol solution to it. Add about 30 mL of medium to petri dish, after cooling, Store at 4 °C.

12. 1 M BKOH solution: 1 M BKOH in C_2H_5OH. Store at −20 °C. BKOH solution and LB medium are used in a ratio of 1:1000.
13. 0.8 M IPTG solution: 0.8 M IPTG in water. Store at −20 °C. IPTG solution and LB medium are used in a ratio of 1:1600.
14. 20% (v/v) arabinose solution: 20% (v/v) arabinose in water. Stored at −20 °C. Arabinose solution and LB medium are used in a ratio of 1:100.
15. Lysis buffer: 20 mM Tris–HCl, 300 mM NaCl in water, pH 7.4. 0.24 g Tris base and 1.75 g NaCl are placed in a 100 mL volumetric flask, dissolve in water and adjust to final volume. Adjust the pH to 7.4 and filter the solution through a 0.22 μm microporous membrane filter.
16. Washing buffer: 20 mM Tris–HCl, 300 mM NaCl, 35 mM imidazole in water, pH 7.4. 0.24 g Tris base, 0.24 g imidazole, and 1.75 g NaCl are placed in a 100 mL volumetric flask, dissolve in water and adjust to final volume. Adjust the pH to7.4 and filter the solution through a 0.22 μm microporous membrane filter.
17. Elution buffer: 20 mM Tris–HCl, 300 mM NaCl, 250 mM imidazole in water, pH 7.4. 0.24 g Tris base, 1.7 g imidazole, and 1.75 g NaCl are placed in a 100 mL volumetric flask, dissolve in water and adjust to final volume. Adjust the pH to7.4 and filter the solution through a 0.22 μm microporous membrane filter.

2.2.3 Hydrazinolysis of Protein Oxoester

1. Hydrazine buffer: 6 M Gn·HCl, 0.2 M Na_2HPO_4, 4% (v/v) $NH_2NH_2 \cdot H_2O$ in water, pH 9.0. 0.716 g $Na_2HPO_4 \cdot 12H_2O$, 5.74 g Gn·HCl and 0.4 mL $NH_2NH_2 \cdot H_2O$ are placed in a 10 mL volumetric flask. Dissolve in water and adjust to final volume. Adjust the pH to 9.0 and filter the solution through a 0.22 μm microporous membrane filter.
2. 5%(w/v) TCEP solution: 6 M Gn·HCl, 0.2 M Na_2HPO_4, 5% TCEP in water, pH 7.0. 0.716 g $Na_2HPO_4 \cdot 12H_2O$, 5.74 g Gn·HCl, and 0.5 g TCEP are placed in a 10 mL volumetric flask, dissolve in water and adjust to final volume. Adjust the pH to 7.0 and filter the solution through a 0.22 μm microporous membrane filter.
3. Analytical and semipreparative HPLC system: *see* Table 1.

2.2.4 Synthesis of HdeA(V58C-89)-OH

Same as the materials in Subheading 2.1.2.

2.2.5 Synthesis of HdeA by Native Chemical Ligation of Peptide Hydrazides

Same as the materials in Subheading 2.1.3.

3 Methods

3.1 Total Synthesis of CssII by Hydrazide-Based NCL

The sequential synthesis of N-terminus to C-terminus in NCL is poorly developed, mainly due to the difficulties associated with the masking of thioesters. The hydrazide peptide does not require masking, since the addition of MPAA is prior to the addition of the second peptide not only converts the first peptide azide to the peptide thioester but also removes the unreacted $NaNO_2$, which allows the hydrazide of the second peptide to be retained for the next ligation.

CssII has 66 amino acid residues including eight cysteine residues [42]. We identified Cys residues at positions 12, 25, and 41 as suitable ligation sites. In order to highlight the advantages of the peptide hydrazide N-to-C ligation strategy, we decided to use four fragments to synthesize the protein CssII [23] (Fig. 2). In this

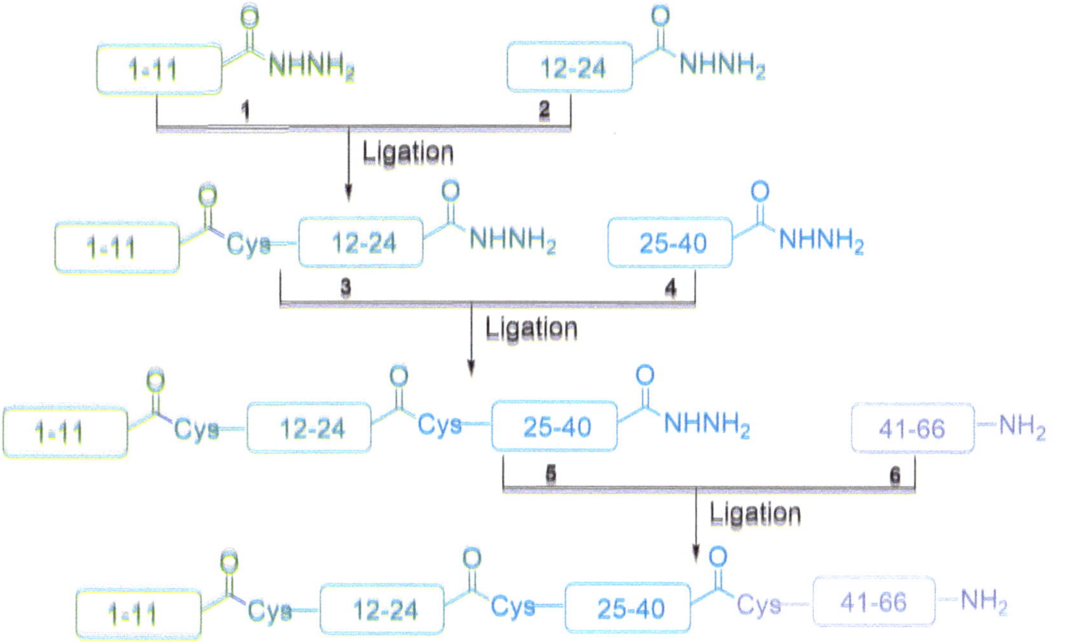

7(CssII)

H-Lys-Glu-Gly-Tyr-Leu-Val-Ser-Lys-Ser-Thr-Gly-Cys-Lys-Tyr-Glu-Cys-Leu-Lys-Leu-Gly-Asp-Asn-Asp-Tyr-Cys-Leu-Arg-Glu-Cys-Lys-Gln-Gln-Tyr-Gly-Lys-Ser-Ser-Gly-Gly-Tyr-Cys-Tyr-Ala-Phe-Ala-Cys-Trp-Cys-Thr-His-Leu-Tyr-Glu-Gln-Ala-Val-Val-Trp-Pro-Leu-Pro-Asn-Lys-Thr-Cys-Asn-NH_2

Fig. 2 The N-to-C sequential ligation strategy was tested for the synthesis of CssII

section we provide protocols for the synthesis of the four following peptides: CssII(1–11)-$NHNH_2$, CssII(12–24)-$NHNH_2$, CssII (25–40)-$NHNH_2$ and CssII(41–66)-NH_2.

3.1.1 Preparation of 2-Cl-(Trt)-$NHNH_2$ Resin

1. Weigh 0.735 g (0.25 mmol) of 2-Cl-(Trt)-Cl resin (0.34 mmol/g) into a peptide synthesis vessel.
2. Add 5 mL of DMF to the peptide synthesis vessel, gently shake it for 10 s, and then drain it. Repeat two more times.
3. Add 5 mL of DCM to the peptide synthesis vessel, gently shake it for 10 s, and then drain it. Repeat two more times.
4. Swell the resin in 6 mL of 50% (v/v) DMF/DCM for 30 min, then drain it.
5. Add 4 mL of 5% (v/v) NH_2NH_2 solution to the resin for hydrazination (Fig. 3). Shake for 30 min using a constant temperature shaker at 30 °C, then drain it.
6. Repeat **steps 2** and **3**.
7. Add 5 mL 5% (v/v) MeOH solution to the peptide synthesis vessel. Shake for 10 min using a constant temperature shaker at 30 °C, then drain it (*see* **Note 9**).
8. Repeat **steps 2** and **3**. Directly use the resin for the next coupling step (*see* **Note 10**).

3.1.2 Synthesis of Peptide Fragments of CssII

Here we provide protocols for preparing peptide hydrazides by Fmoc-SPPS. After coupling, deprotection, circulation and cleavage operations, a peptide hydrazide is obtained (Fig. 4). This protocol is used for synthesis of CssII(1–11)-$NHNH_2$, CssII(12–24)-$NHNH_2$,CssII(25–40)-$NHNH_2$ and CssII(41–66)-NH_2.

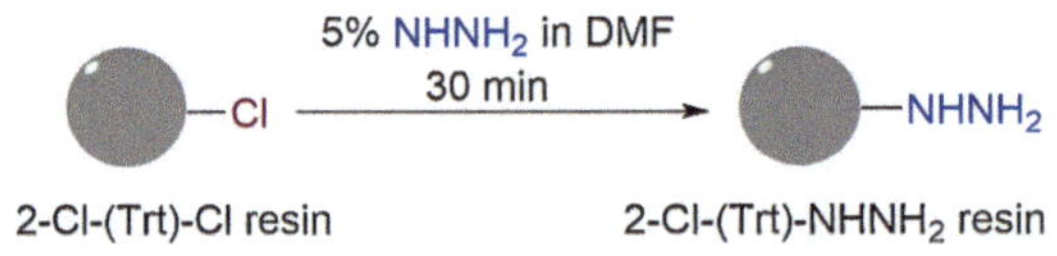

Fig. 3 Preparation of 2-Cl-(Trt)-$NHNH_2$ resin

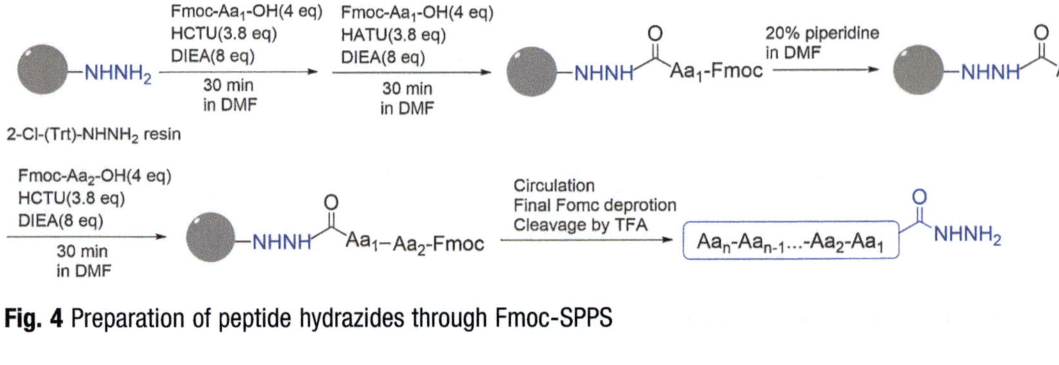

Fig. 4 Preparation of peptide hydrazides through Fmoc-SPPS

Synthesis of CssII(1–11)-$NHNH_2$, CssII(12–24)-$NHNH_2$ and CssII(25–40)-$NHNH_2$ are performed in 2-Cl-(Trt)-$NHNH_2$ resin. Synthesis of CssII(41–66)-NH_2 is performed on a Rink Amide-AM Resin (0.781 mg, 0.25 mmol).

1. Dissolve 1 mmol of the desired Fmoc-protected amino acid and 0.393 g HCTU (0.95 mmol) in 5 mL of DMF(*see* **Note 11**).
2. Add 0.331 mL DIEA (2 mmol) to the above solution. Mix the solution using a vortex to activate the amino acid for 0.5–1 min (*see* **Note 12**).
3. Add the activated amino acid to the resin. Shake for 30 min using a constant temperature shaker at 30 °C, then drain it.
4. If the amino acid derivative to be coupled is sterically hindered (e.g., Fmoc-Cys(Trt)-OH, Fmoc-His(Trt)-OH, Fmoc-Ile-OH, Fmoc-Asn(Trt)-OH, Fmoc-Gln(Trt)-OH, Fmoc-Arg (Pbf)-OH, Fmoc-Thr(tBu)-OH, Fmoc-Val-OH) or coupled after a Pro residue, the amino acid needs to be coupled again using HATU (0.361 g, 0.95 mmol), instead of HCTU, using the procedure described above (**steps 1–3**).
5. Add 5 mL of DMF to the peptide synthesis vessel, gently shake it for 10 s, and then drain it. Repeat two more times.
6. Add 5 mL of DCM to the peptide synthesis vessel, gently shake it for 10 s, and then drain it. Repeat two more times (*see* **Note 13**).
7. Add 4 mL of 20% (v/v) piperidine to the resin. Shake for 5 min using a constant temperature shaker at 30 °C, then drain it.
8. Add 4 mL of 20% (v/v) piperidine to the resin. Shake for 15 min using a constant temperature shaker at 30 °C, then drain it.
9. Repeat **steps 5** and **6**.
10. After completing the Fmoc deprotection above, go back to step1 to perform the coupling of the next amino acid. Repeat **steps 1** through **8** (circulation) until all amino acid residues have been coupled.
11. Once chain elongation is completed use DCM for the last washing. Then vacuum-dry the resin for 5 min (*see* **Note 14**).
12. Add 15 mL of TFA cleavage mixture (2–3 mL per 100 mg of resin) to the dried resin.
13. Shake for 2–3 h using a constant temperature shaker at 30 °C, then the cleavage mixture is filtered into a 50 mL centrifuge tube, and the reaction vessel and resin are washed with 1 mL of TFA and collected into the same centrifuge tube.
14. Remove most of TFA by bubbling nitrogen through the solution (*see* **Note 15**).

15. Cool the ether to 0 °C and add to the above solution to precipitate crude peptides (*see* **Note 16**), centrifuge at high speed for 1 min, and the solvent is discarded. Repeat the above steps and wash the crude peptide again with ether.
16. Weigh the crude peptide (typically around 50 mg) and dissolve in 4 mL of 50% ACN. The solution is filtered through a 0.22 μm microporous membrane filter (*see* **Note 17**).
17. Dilute 1 μL of the above solution with 19 μL of 50% ACN and analyze it by analytical HPLC (For CssII(1–11)-$NHNH_2$ and CssII(12–24)-$NHNH_2$ use Gradient A: a linear gradient of 2–40% B over 28 min, then a linear gradient of 40–100% B over 20 min. For CssII(25–40)-$NHNH_2$ and CssII(41–66)-NH_2 use Gradient B: a linear gradient of 10–14% B over 10 min, then a linear gradient of 14–35% B over 38 min, then a linear gradient of 35–100% B over 10 min). Fractions corresponding to the main peak are collected, and the target peptide mass is confirmed by ESI-MS (Fig. 5).

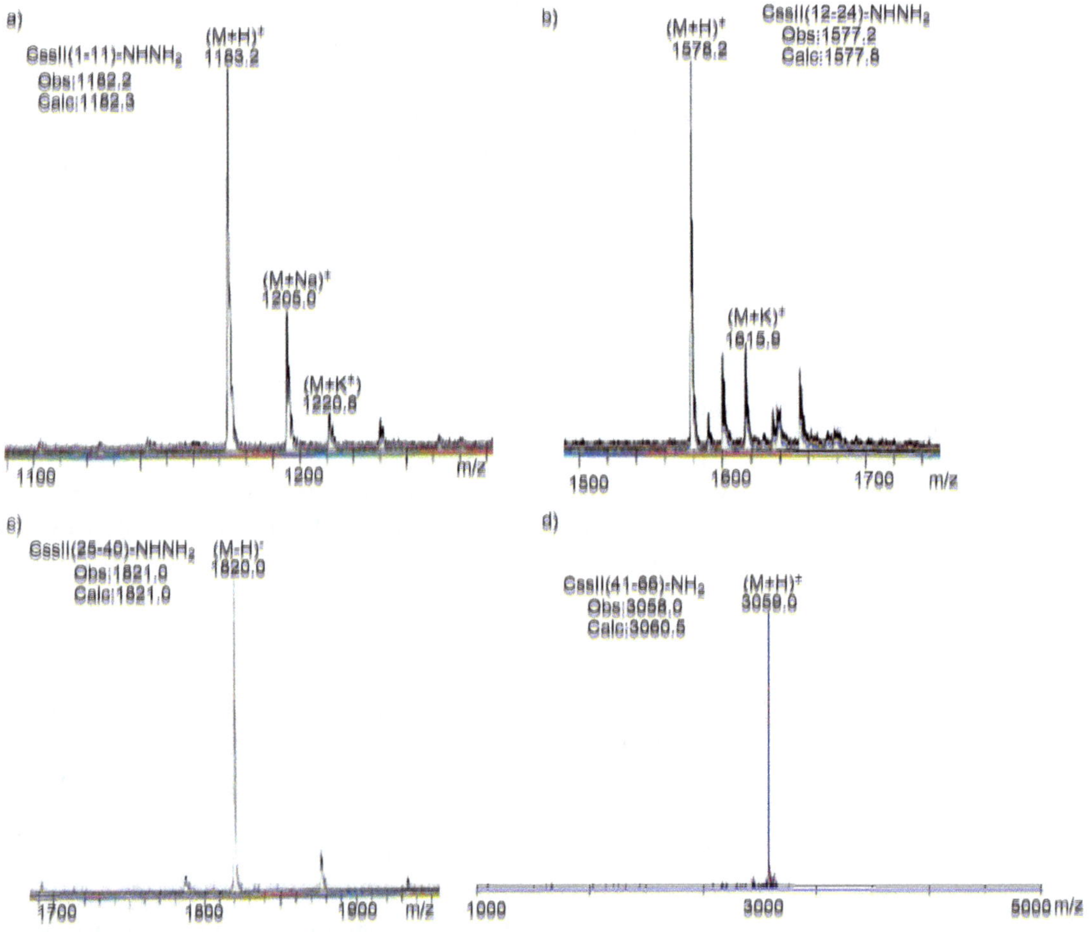

Fig. 5 (**a**) ESI-MS of purified CssII(1–11)-$NHNH_2$, (**b**) CssII(12–24)-$NHNH_2$, (**c**) CssII(25–40)-$NHNH_2$, (**d**) CssII (41–66)-NH_2

18. Purify the crude peptide by semipreparative HPLC (using the gradients above), collect pure fractions, and lyophilize.

3.1.3 Synthesis of CssII(1–24)-NHNH$_2$ by the Native Chemical Ligation of Peptide Hydrazides

1. Prepare a −15 °C ice/salt bath, mix 12.5 g of sodium chloride and 50 g of crushed ice in a 100 mL Dewar flask.
2. Dissolve 5.9 mg of CssII(1–11)-NHNH$_2$ (5 μmol) in 1.64 mL of 0.2 M NaH_2PO_4 solution, and place the above reaction tube in the above ice salt bath. Precool for 5 min.
3. To oxidize the peptide hydrazide to the corresponding acyl azide, add 151 μL of 0.2 M $NaNO_2$ to the above solution and stir the solution at −15 °C for 20 min (*see* **Note 18**).
4. To convert the peptide acyl azide to the peptide thioester, the above reaction tube is taken out from the ice salt bath, add 1.64 mL of 0.2 M MPAA to the above solution, adjust the pH to 5.2 with 2 M NaOH solution, and stir 30 min at room temperature (*see* **Note 19**).
5. To achieve ligation of the two peptide segments, add 7.9 mg of CssII(12–24)-NHNH$_2$ (5 μmol) to the above solution, adjust the pH to 6.8 with 2 M NaOH solution, stir at room temperature for 2 h (*see* **Note 20**).
6. Add 1 mL of 0.6 M TCEP solution to the above reaction solution (*see* **Note 21**), and the reaction is monitored by analytical HPLC (Fig. 6) (Gradient C: a linear gradient of 5–10% B

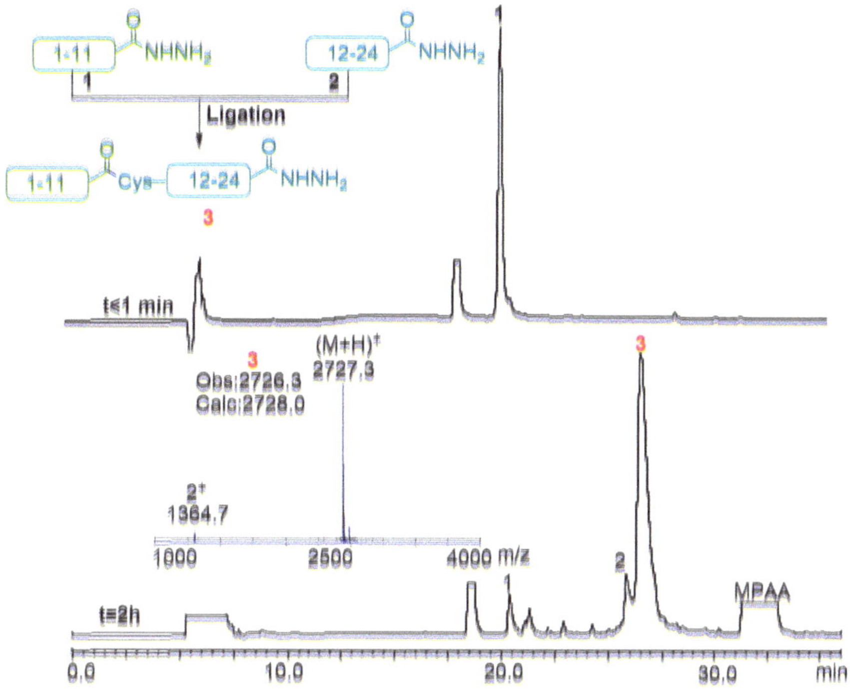

Fig. 6 Hydrazide-based native chemical ligation between 1 and 2

over 2 min, then a linear gradient of 10–70% B over 30 min). Fractions corresponding to the main peak are collected, and the target peptide mass is confirmed by ESI-MS (Fig. 6).

7. Purify the product by semipreparative HPLC (Gradient C), collect and lyophilize. A measure of 9.0 mg of purified CssII (1–24)-$NHNH_2$ will be obtained.

3.1.4 Synthesis of CssII(1–40)-$NHNH_2$ by the Native Chemical Ligation of Peptide Hydrazides

1. Prepare a −15 °C ice–salt bath, mix 12.5 g of sodium chloride and 50 g of crushed ice in a 100 mL Dewar flask.
2. Dissolve 13.7 mg of CssII(1–24)-$NHNH_2$ (5 μmol) in 1.64 mL of 0.2 M NaH_2PO_4 solution, and place the above reaction tube in the above ice salt bath. Precool for 5 min.
3. To oxidize the peptide hydrazide to the corresponding acyl azide, add 164 μL of 0.2 M $NaNO_2$ to the above solution and stir the solution at −15 °C for 20 min.
4. To convert the peptide acyl azide to the peptide thioester, the above reaction tube is taken out from the ice salt bath, add 1.64 mL of 0.2 M MPAA to the above solution, adjust the pH to 5.2 with 2 M NaOH solution, and stir 30 min at room temperature.
5. To achieve ligation of the two peptide segments, add 27.4 mg of CssII(24–40)-$NHNH_2$ (15 μmol) to the above solution, adjust the pH to 6.8 with 2 M NaOH solution, stir at room temperature for 12 h.
6. Add 1 mL of 0.6 M TCEP solution to the above reaction solution, and the reaction is monitored by analytical HPLC (Fig. 7) (Gradient D: a linear gradient of 10–14% B over 2 min, then a linear gradient of 14–70% B over 30 min). Fractions corresponding to the main peak are collected, and the target peptide mass is confirmed by ESI-MS (Fig. 7).
7. Purify the product by semipreparative HPLC (Gradient D), collect and lyophilize. A measure of 11 mg of purified CssII (1–40)-$NHNH_2$ will be obtained.

3.1.5 Synthesis of CssII(1–66)-NH_2 by the Native Chemical Ligation of Peptide Hydrazides

1. Prepare a −15 °C ice–salt bath, mix 12.5 g of sodium chloride and 50 g of crushed ice in a 100 mL Dewar flask.
2. Dissolve 4.6 mg of CssII(1–40)-$NHNH_2$ (1.0 μmol) in 0.69 mL of 0.2 M NaH_2PO_4 solution, and place the above reaction tube in the above ice salt bath. Precool for 5 min.
3. To oxidize the peptide hydrazide to the corresponding acyl azide, add 34.5 μL of 0.2 M $NaNO_2$ to the above solution and stir the solution at −15 °C for 20 min.
4. To convert the peptide acyl azide to the peptide thioester, the above reaction tube is taken out from the ice salt bath, add 0.69 mL of 0.2 M MPAA to the above solution, adjust the pH

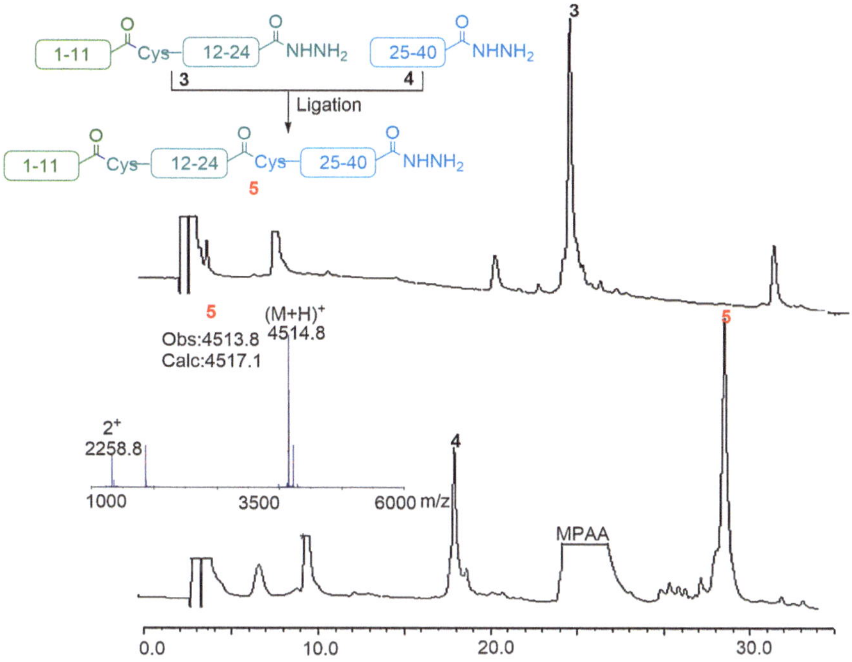

Fig. 7 Hydrazide-based native chemical ligation between 3 and 4

to 5.2 with 2 M NaOH solution, and stir 30 min at room temperature.

5. To achieve ligation of the two peptide segments, add 9.17 mg of CssII(41–66)-NH_2 (3 μmol) to the above solution, adjust the pH to 6.8 with 2 M NaOH solution, stir at room temperature for 12 h.
6. Add 0.2 mL of 0.6 M TCEP solution to the above reaction solution, and the reaction is monitored by analytical HPLC (Fig. 8) (Gradient D). Fractions corresponding to the main peak are collected, and the target peptide mass is confirmed by ESI-MS (Fig. 8).
7. Purify the product by semipreparative HPLC (Gradient D), collect, and lyophilize. A measure of 3.79 mg of purified CssII(1–66)-NH_2 will be obtained.

3.2 Semisynthesis of HdeA by Hydrazide-Based Expressed Protein Ligation

Peptide hydrazides can be prepared through recombinant expression technology and used for protein-based expressed protein ligation (EPL), allowing for the generation of complex semisynthetic proteins. Here we provide protocols for the recombinant protein α-hydrazides produced via genetic incorporation of α-hydroxy acids.

3.2.1 Synthesis of α-Hydroxy Acid BKOH

1. Dissolve 2 g of compound 1 (Fig. 9) (13.6 mmol) in 25 mL of 10% H_2SO_4, stir at 40 °C for 5 min.

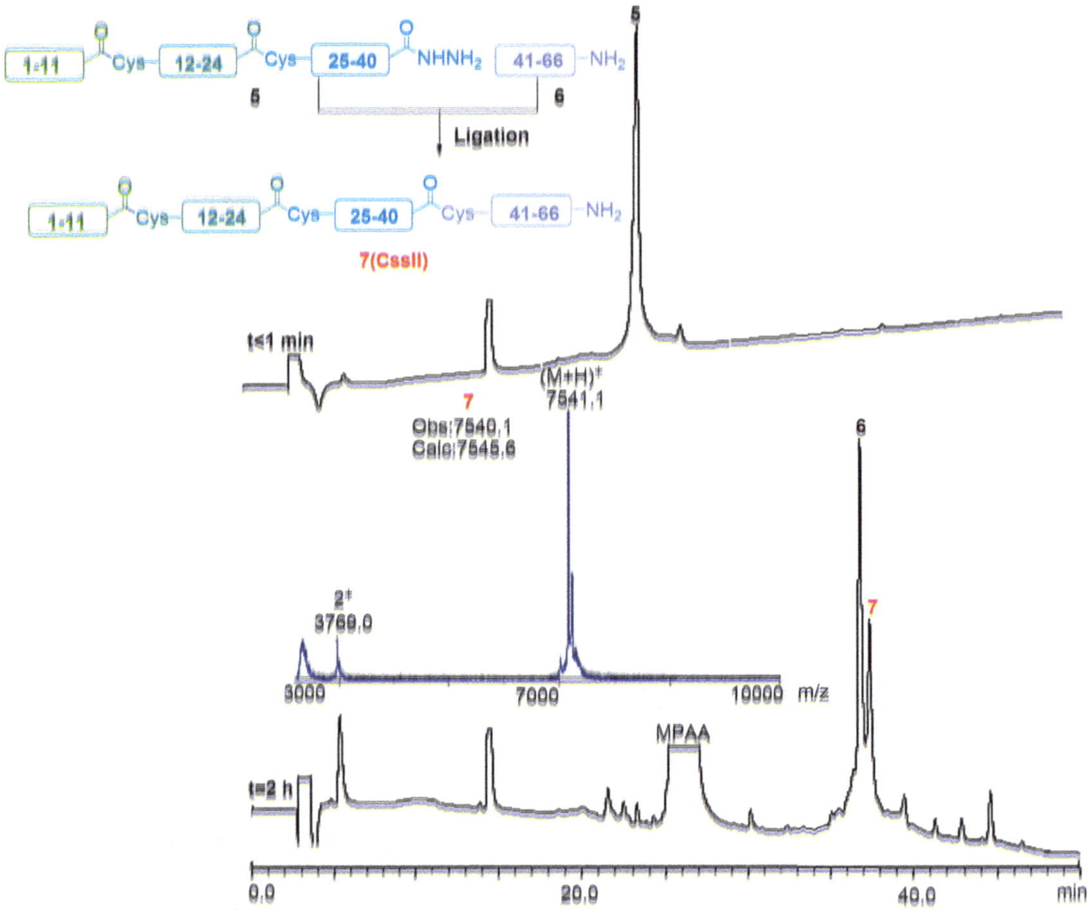

Fig. 8 Hydrazide-based native chemical ligation between 5 and 6

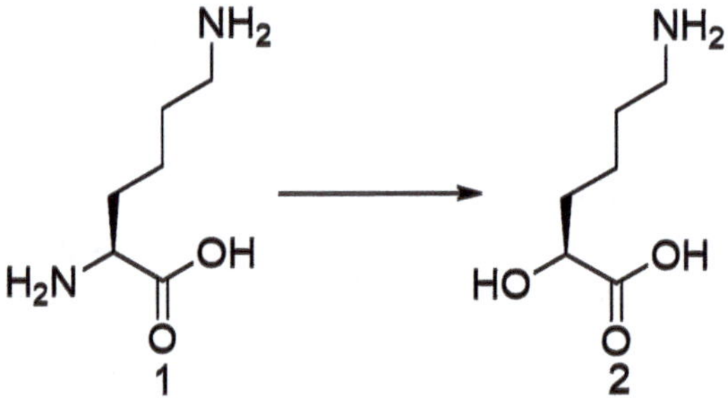

Fig. 9 Synthesis of compound 2

2. Slowly add 12 mL of 3.86 M $NaNO_2$ solution to the above solution, heat the solution to 50 °C and stir for 3 h.
3. Add 7.72 mL of 6 M urea solution to the above solution to decompose the remaining $NaNO_2$.

4. Add 10 mL of 1 M HCl to Amberlyst®15 wet ion-exchange column (H+ form), then the reaction mixture is transferred to it.
5. Add 5 bed volumes of deionized water to wash out impurities.
6. Add 3 bed volumes of 8% ammonia solution to elute the column, collect 1 bed volume fractions, perform ninhydrin test on the fractions and pool together ninhydrin-positive fractions. Rotaevaporate the pooled fractions until a yellowish semisolid is obtained.
7. Place the yellowish semisolid in a 25 mL round bottom flask, add a small amount of 50% methanol (v/v) solution, heat the mixture to boiling, and keep adding 0.5 mL of 50% methanol (v/v) solution until solid is completely dissolved. Allow the solution to stand at room temperature to crystalize a yellow needle-like solid, compound 2 (Fig. 9). Perform ^{1}H NMR (300 MHz, D_2O): δ (ppm) 4.24 (dd, 4H), 2.94 (m, 2H), 1.67 (m, 4H), 1.43 (m, 2H). ^{13}C NMR (75 MHz, D_2O): δ (ppm) 162.3, 71.9, 39.2, 33.6, 27.4, 21.7. ESI-MS (positive) 148.9 (observed, M + H) 147.1 (calculated, M).
8. Dissolve 0.6 g of compound 2 (4.08 mmol) in a mixture of 10 mL of 1 M NaOH and 10 mL of 1,4-dioxane.
9. Add 1.12 mL of Boc_2O (4.90 mmol, 1.2 equivalent) to the above solution and stir at room temperature for 3 h.
10. Add 20 mL diethyl ether to extract the reaction mixture, collect the aqueous fraction and adjust the aqueous layer pH to 2.0 with 1 M HCl (*see* **Note 22**).
11. Extract the aqueous fraction with CH_2Cl_2, collect organic layer and dried it over $MgSO_4$, evaporate to dryness. Typically 0.52 g of a colorless liquid compound 3 (Fig. 10) (BKOH) are obtained (51% yield). Perform ^{1}H NMR (300 MHz, D_2O): δ (ppm) 4.23 (dd, 1H), 3.03 (m, 2H), 1.80–1.72 (m, 1H), 1.72–1.64 (m, 1H), 1.50–1.42 (m, 4H), 1.38 (s, 9H). ^{13}C NMR

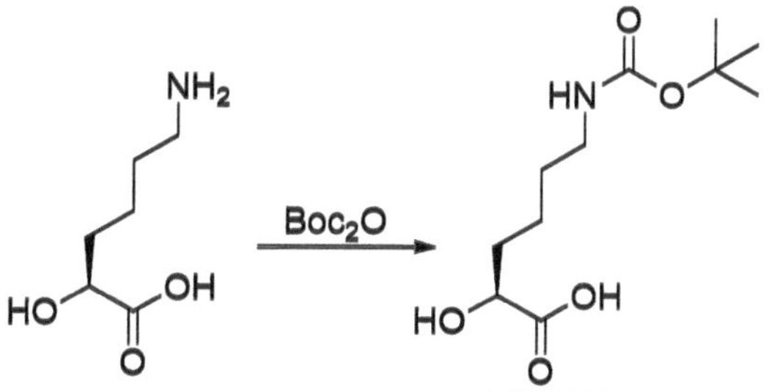

2 3 **BKOH**

Fig. 10 Synthesis of compound 3

(75 MHz, D_2O): δ (ppm) 178.2, 158.3, 80.7, 70.1, 39.8, 33.0, 28.6, 27.7, 21.5. ESI-MS (negative) 246.1 (observed, M-H) 493.6 (observed, 2 M-H) 247.3 (calculated, M).

3.2.2 Genetic Incorporation of α-Hydroxy Acid BKOH

1. Dissolve 0.4 μg of plasmid PylRS-tRNA$^{Pyl}_{CUA}$ pair in 40 μL H_2O. Dissolve 0.4 μg of plasmid HdeA-V58TAG-myc-His_6 in 40 μL H_2O.
2. Take 100 μL of *E. coli* DH10B competent cells (stored at −80 °C) and thaw on ice for 30 min, then add 2 μL plasmid PylRS-tRNA$^{Pyl}_{CUA}$ pair solution and 2 μL plasmid HdeA-V58TAG-myc-His_6 solution, mix gently and hold the ice for 30 min.
3. Heat shock the cells for 1 min in a 42 °C water bath, then quickly put the cells into the ice, let stand for 2 min.
4. Add 400 μL of LB medium to the cells, culture at 37 °C (200 rpm) for 40 min.
5. Apply the LB medium to an LB agar plate containing 0.1 mg/mL and chloramphenicol 0.05 mg/mL, and culture overnight at 37 °C.
6. Take a single colony from LB agar plate and inoculate 5 mL LB medium containing 0.1 mg/mL and chloramphenicol 0.05 mg/mL in Falcon tube, then culture at 37 °C (200 rpm) for 12 h.
7. Dilute culture from **step 6** 1:100 in LB medium containing 0.1 mg/mL ampicillin and 0.05 mg/mL chloramphenicol, and grow cells to $OD_{600} = 0.5$ at 37 °C (200 rpm).
8. Add BKOH to a final concentration of 1 mM and continue to shake for 30 min to allow the cells to completely absorb compound 3.
9. Protein expression is induced by the addition of 0.2% (v/v) arabinose and 0.5 mM IPTG. Grow cells, at 30 °C (200 rpm) for 12 h.
10. The cells are harvested by centrifugation at 10,000 × *g* for 10 min, and then resuspend in lysis buffer.
11. Bacterial lysate after sonication is centrifuged at 16,000 × *g* for 30 min. Transfer the supernatant to the Ni-NTA column (Histrap 5 mL, add 15 mL of lysis buffer to equilibrate the column).
12. The column is washed with 30 mL of washing buffer (*see* **Note 23**) and the target protein is eluted with 15 mL of elution buffer (*see* **Note 24**).
13. Concentrate the eluted protein using a centrifugal filter (3000 MW cutoff). Purify the concentrate by RP-HPLC (Fig. 11) (Gradient E: a linear gradient of 10–14% B over

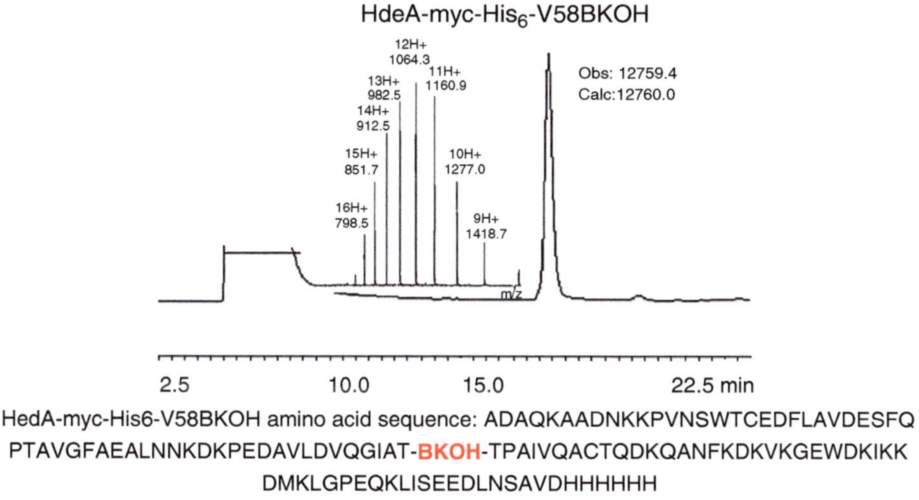

Fig. 11 Determination of HdeA-myc-His6-V58BKOH

2.5 min, then a linear gradient of 14% to 70% B over 20 min) and lyophilize for use.

3.2.3 Hydrazinolysis of Protein Oxoester

1. Dissolve 12.7 mg (1 mM) of HdeA-myc-His$_6$-V58BKOH in 1 mL of hydrazine buffer (Fig. 12a), stir at room temperature for 2 h.
2. Mix the reaction buffer with an equal volume of pH-neutral 5% TCEP solution and analyze by analytic HPLC (Fig. 12b) (Gradient D). Fractions corresponding to the main peak are collected, and the molecular weight of target peptide is confirmed by ESI-MS (Fig. 12c).
3. Purify the product by semipreparative HPLC (Gradient D), collect and lyophilize. 1 L cells can get 1–2 mg of purified peptide hydrazide HdeA(1–57)-$NHNH_2$.

3.2.4 Synthesis of HdeA(V58C-89)-OH

1. Perform the steps in Subheading 3.1.2 using 0.25 mmol (315 mg) of Wang Resin (0.8 mmol/g) to obtain HdeA (V58C-89)-OH. The peptide is analyzed by analytical HPLC (gradient D) and confirmed by ESI-MS (Fig. 13).

3.2.5 Synthesis of HdeA by Native Chemical Ligation of Peptide Hydrazides

1. Prepare a −15 °C ice/salt bath, mix 12.5 g of sodium chloride and 50 g of crushed ice in a 100 mL Dewar flask.
2. Dissolve 6.11 mg HdeA(1–57)-$NHNH_2$ (1.0 μmol, obtain in **step 3**, Subheading 3.2.3) in 0.69 mL of 0.2 M NaH_2PO_4 solution, and place the above reaction tube in the above ice salt bath. Precool for 5 min.
3. To oxidize the peptide hydrazide to the corresponding acyl azide, add 34.5 μL of 0.2 M $NaNO_2$ to the above solution and stir the solution at −15 °C for 20 min.

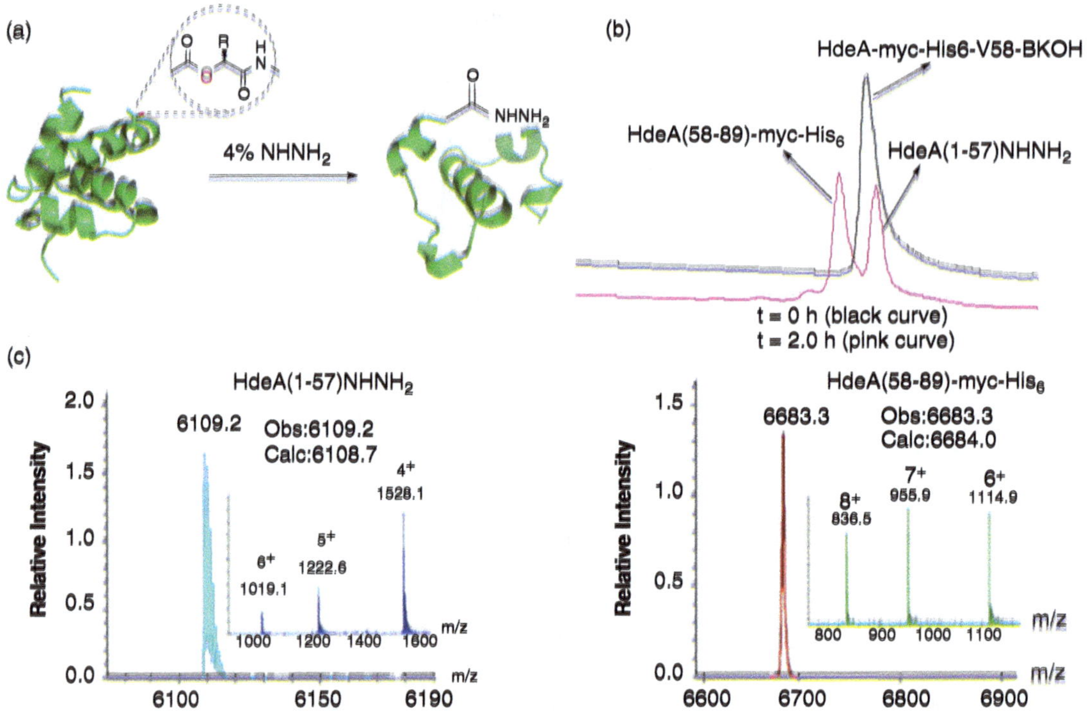

Fig. 12 (**a**) Hydrazinolysis of HdeA-myc-His6-V58BKOH. (**b**) Analytical HPLC traces of hydrazinolysis of HdeA-myc-His6-V58BKOH. (**c**) ESI-MS spectra of the two segments HdeA-myc-His6-V58-BKOH after hydrazinolysis

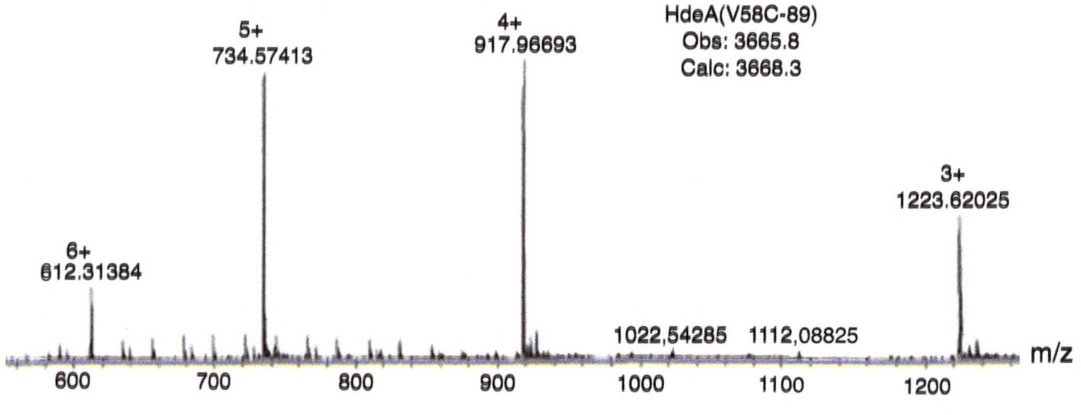

Fig. 13 ESI-MS of purified HdeA(V58C-89)

4. To convert the peptide acyl azide to the peptide thioester, the above reaction tube is taken out from the ice salt bath, add 0.69 mL of 0.2 M MPAA to the above solution, adjust the pH to 5.2 with 2 M NaOH solution, and stir for 30 min at room temperature.
5. To achieve ligation of the two peptide segments, add 4.43 mg HdeA(V58C-89)-OH (1.2 μmol) to the above solution, adjust

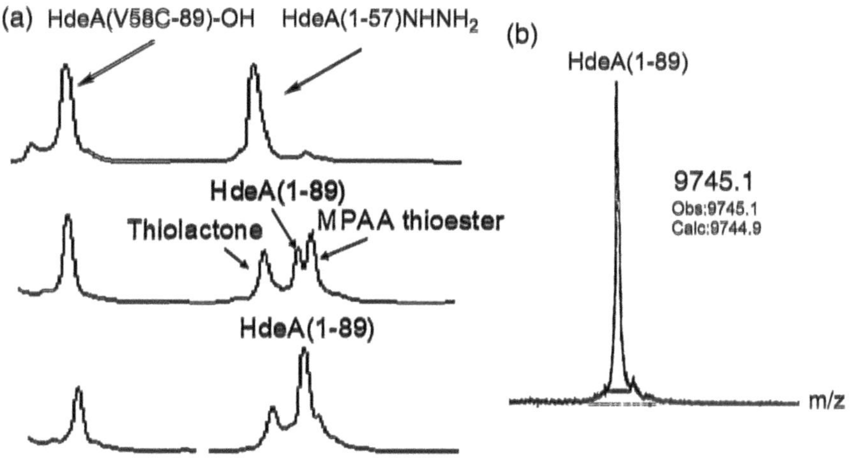

Fig. 14 (**a**) Analytical HPLC trace for reaction mixture before and after ligation. (**b**) ESI-MS analysis of ligation product HdeA

the pH to 6.8 with 2 M NaOH solution, stir at room temperature for 4 h.

6. Add 0.2 mL of 0.6 M TCEP solution to the above reaction solution, and the reaction is monitored by analytical HPLC (Fig. 14a) (Gradient D). Fractions corresponding to the main peak are collected, and the target peptide mass is confirmed by ESI-MS (Fig. 14b).
7. Purify the product by semipreparative HPLC (Gradient D), collect and lyophilize. A measure of 2.71 mg of purified HdeA will be obtained.

4 Notes

1. $NH_2NH_2 \cdot H_2O$ is highly toxic. Wear a lab coat, gloves, and lab goggles when preparing this solution.
2. Piperidine is highly flammable and toxic. When preparing the solution, wear a lab coat, gloves, and lab goggles, and work in a chemical fume hood.
3. TFA is strongly corrosive and toxic. Phenol is toxic and corrosive. When preparing solution of these, wear a lab coat, gloves, and lab goggles, and work in a chemical fume hood. Freshly prepare the solution before use.
4. NaOH solids and their solutions are highly corrosive. Wear a lab coat, gloves when preparing the solution. The solution should be prepared freshly before use because NaOH will react with CO_2 after prolonged exposure to air.

5. 12 M HCl solutions are highly corrosive and volatile. When preparing the solution, wear a lab coat, gloves, and lab goggles, and work in a chemical fume hood.
6. The solution should be prepared freshly before use because $NaNO_2$ may decompose after prolonged storage.
7. The solution should be prepared freshly before use because TCEP will oxidize after prolonged exposure to air.
8. Slowly pour 98% H_2SO_4 into the water and keep stirring. Ninety-eight percentage of sulfuric acid solutions are highly corrosive. Wear a lab coat, gloves, and lab goggles when preparing the solution.
9. This step is important because it ensures that unreacted sites on the resin are capped.
10. The 2-Cl-(Trt)-$NHNH_2$ resin should be prepared freshly before use because 2-Cl-(Trt)-$NHNH_2$ resin will be unstable for long periods of storage.
11. Stick to the amounts of amino acid and HCTU indicated, it is important to ensure that the concentration of HCTU is slightly lower than the concentration of amino acid building blocks.
12. The activation of amino acids is exothermic and results in the solution turning yellow, but these phenomena are not always observed. Activation time should not exceed 5 min.
13. As long as the Fmoc protecting group is not removed, the synthesis can be interrupted after the end of the amino acid condensation cycle and the resin can be stored in DMF overnight.
14. Thoroughly clean and dry the resin before the cleavage step to avoid side reactions.
15. The volume of the concentrated cleavage mixture should be less than 5 mL.
16. The volume of diethyl ether should be at least ten times that of the concentrated cleavage mixture.
17. Filter the crude peptide solution with a hydrophobic membrane filter to remove insoluble impurities, otherwise the impurities will block HPLC lines.
18. Make sure the temperature drops to about -15 °C during oxidation. In order to avoid side reactions, the oxidation reaction should not exceed 30 min.
19. The addition of MPAA not only converts the azide group to a thioester group, but also removes unreacted HNO_2.
20. The pH value is important for this series of reactions, ensure pH adjustments are accurate.

21. TCEP is used to reduce disulfide by-products and is a necessary step before HPLC purification.
22. Adjust pH to 2.0 to change the product from a carboxylate to a carboxylic acid. Carboxylic acid is more soluble in CH_2Cl_2 for easy extraction.
23. The wash buffer contains a low concentration of imidazole for washing other proteins. If too much washing buffer is added, some of the target protein may also be washed out.
24. The elution buffer contains a high concentration of imidazole, which can be competitively combined with Ni to elute the target protein from Ni-NTA column.

References

1. Kent SBH (2009) Total chemical synthesis of proteins. Chem Soc Rev 38:338–351
2. Kochendoerfer GG, Kent SBH (1999) Chemical protein synthesis. Curr Opin Chem Biol 3:665–671
3. Dawson PE, Muir TW, Clark-Lewis I, Kent SBH (1994) Synthesis of proteins by native chemical ligation. Science 266:776–779
4. Johnson ECB, Kent SBH (2006) Insights into the mechanism and catalysis of the native chemical ligation reaction. J Am Chem Soc 128:6640–6646
5. Muir TW, Sondhi D, Cole PA (1998) Expressed protein ligation: a general method for protein engineering. Proc Natl Acad Sci 95 (12):6705–6710
6. Evans TC Jr, Benner J, Xu MQ (1998) Semisynthesis of cytotoxic proteins using a modified protein splicing element. Protein Sci 7 (11):2256–2264
7. McGinty RK, Kim J, Chatterjee C, Roeder RG, Muir TW (2008) Chemically ubiquitylated histone H2B stimulates hDot1L-mediated intranucleosomal methylation. Nature 453 (7196):812
8. Wang P, Dong SW, Shieh JH, Peguero E, Hendrickson R, Moore MA, Danishefsky SJ (2013) Erythropoietin derived by chemical synthesis. Science 342:1357–1360
9. Scheuermann JC, de Ayala Alonso AG, Oktaba K, Ly-Hartig N, McGinty RK, Fraterman S, Müller J et al (2010) Histone H2A deubiquitinase activity of the Polycomb repressive complex PR-DUB. Nature 465:243–247
10. Chu N, Salguero AL, Liu AZ, Chen Z, Dempsey DR, Ficarro SB, Cole PA et al (2018) Akt kinase activation mechanisms revealed using protein semisynthesis. Cell 174(4):897–907
11. Haj-Yahya M, Lashuel HA (2018) Protein semisynthesis provides access to tau disease-associated post-translational modifications (PTMs) and paves the way to deciphering the tau PTM code in health and diseased states. J Am Chem Soc 140:6611–6621
12. Schnolzer M, Alewood P, Jones A, Alewood D, Kent SBH (2007) In situ neutralization in Boc-chemistry solid phase peptide synthesis. Int J Pept Res Ther 13:31–44
13. Mende F, Seitz O (2011) 9-Fluorenylmethoxycarbonyl-based solid-phase synthesis of peptide α-thioesters. Angew Chem Int Ed 50:1232–1240
14. Ingenito R, Bianchi E, Fattori D, Pessi A (1999) Solid phase synthesis of peptide C-terminal thioesters by Fmoc/t-Bu chemistry. J Am Chem Soc 121:11369–11374
15. Kawakami T, Aimoto S (2007) Peptide ligation using a building block having a cysteinyl prolyl ester (CPE) autoactivating unit at the carboxy terminus. Chem Lett 36:76–77
16. Blanco-Canosa JB, Dawson PE (2008) An efficient Fmoc-SPPS approach for the generation of thioester peptide precursors for use in native chemical ligation. Angew Chem Int Ed 47:6851–6855
17. Tofteng AP, Sorensen KK, Conde-Frieboes KW, Hoeg-Jensen T, Jensen KJ (2009) Fmoc solid-phase synthesis of C-terminal peptide thioesters by formation of a backbone pyroglutamyl imide moiety. Angew Chem Int Ed 48:7411–7414
18. Kang J, Richardson JP, Macmillan D (2009) 3-Mercaptopropionic acid-mediated synthesis of peptide and protein thioesters. Chem Commun 2009:407–409
19. Tsuda S, Shigenaga A, Bando K, Otaka A (2009) N-to-S acyl-transfer-mediated synthesis

of peptide thioesters using anilide derivatives. Org Lett 11:823–826
20. Zheng JS, Cui HK, Fang GM, Xi WX, Liu L (2010) Chemical protein synthesis by kinetically controlled ligation of peptide O-esters. Chembiochem 11:511–515
21. Ollivier N, Dheur J, Mhidia R, Blanpain A, Melnyk O (2010) Bis(2-sulfanylethyl)amino native peptide ligation. Org Lett 12:5238–5241
22. Zheng JS, Chang HN, Wang FL, Liu L (2011) Fmoc synthesis of peptide thioesters without post-chain-assembly manipulation. J Am Chem Soc 133:11080–11083
23. Fang GM, Li YM, Shen F, Huang YC, Li JB, Lin Y, Cui HK, Liu L (2011) Protein chemical synthesis by ligation of peptide hydrazides. Angew Chem Int Ed 50:7645–7649
24. Fang GM, Wang JX, Liu L (2012) Convergent chemical synthesis of proteins by ligation of peptide hydrazides. Angew Chem Int Ed 51:10347–10350
25. Zheng JS, TangS QYK, Wang ZP, Liu L (2013) Chemical synthesis of proteins using peptide hydrazides as thioester surrogates. Nat Protoc 8(12):2483
26. Zheng JS, Tang S, Huang YC, Liu L (2013) Development of new thioester equivalents for protein chemical synthesis. Acc Chem Res 46:2475–2484
27. Wang SS (1997) p-Alkoxybenzyl alcohol resin and p-alkoxybenzyloxycarbonylhydrazide resin for solid phase synthesis of protected peptide fragments. J Am Chem Soc 95:1328–1333
28. Chang JK, Shimizu M, Wang SS (1976) Fully automated solid phase synthesis of protected peptide hydrazides on recycling hydroxymethyl resin. J Org Chem 41:3255–3258
29. Kalia J, Raines RT (2006) Reactivity of intein thioesters: appending a functional group to a protein. Chembiochem 7:1375–1383
30. Yi L, Sun H, Wu YW, Triola G, Waldmann H, Goody RS (2010) A highly efficient strategy for modification of proteins at the C terminus. Angew Chem Int Ed 49:9417–9421
31. Thom J, Anderson D, McGregor J, Cotton G (2011) Recombinant protein hydrazides: application to site-specific protein PEGylation. Bioconjug Chem 22:1017–1020
32. Li YM, Yang MY, Huang YC, Li YT, Chen PR, Liu L (2012) Ligation of expressed protein α-hydrazides via genetic incorporation of an α-hydroxy acid. ACS Chem Biol 7:1015–1022
33. Li YM, Li YT, Pan M, Kong XQ, Huang YC, Hong ZY, Liu L (2014) Irreversible site-specific hydrazinolysis of proteins by use of sortase. Angew Chem Int Ed 53:2198–2202
34. Adams AL, Cowper B, Morgan RE, Macmillan D et al (2013) Cysteine promoted C-terminal hydrazinolysis of native peptides and proteins. Angew Chem Int Ed 52:13062–13066
35. Tang S, Si YY, Wang ZP, Mei KR, Chen X, Cheng JY, Zheng JS, Liu L (2015) An efficient one-pot four-segment condensation method for protein chemical synthesis. Angew Chem Int Ed 54:5713–5717
36. Wang Z, Xu W, Liu L, Zhu TF (2016) A synthetic molecular system capable of mirror-image genetic replication and transcription. Nat Chem 8:698–704
37. Pan M, Gao S, Zheng Y, Tan X, Lan H, Tan X, Sun D, Lu L, Wang T, Zheng Q, Huang Y, Wang J, Liu L (2016) Quasi-racemic X-ray structures of K27-linked ubiquitin chains prepared by total chemical synthesis. J Am Chem Soc 138:7429–7435
38. Tang S, Liang LJ, Si YY, Gao S, Wang JX, Liang J, Mei Z, Zheng JS, Liu L (2017) Practical chemical synthesis of atypical ubiquitin chains by using an isopeptide-linked Ub isomer. Angew Chem Int Ed 56:13333–13337
39. Li YT, Yi C, Chen CC, Lan H, Pan M, Zhang SJ, Huang YC, Guan CJ, Li YM, Yu L, Liu L (2017) A semisynthetic Atg3 reveals that acetylation promotes Atg3 membrane binding and Atg8 lipidation. Nat Commun 8:14846
40. Pan M, Zheng QY, Ding S, Zhang LJ, Qu Q, Wang T, Chen CL, Mei ZQ, Liu L et al (2019) Chemical protein synthesis enabled mechanistic studies on the molecular recognition of K27-linked ubiquitin chains. Angew Chem 131(9):2653–2657
41. Chu GC, Pan M, Li JB, Liu SL, Zuo C, Tong ZB, Tian CL, Li YM, Liu L et al (2019) Cysteine-aminoethylation-assisted chemical ubiquitination of recombinant histones. J Am Chem Soc 141:3654–3663
42. Martin MF, Perez LGG, Ayeb ME, Kopeyan C, Bechis G, Jover E, Rochat H (1987) Purification and chemical and biological characterizations of seven toxins from the Mexican scorpion, Centruroides suffusus suffusus. J Biol Chem 262:4452–4459

Chapter 7

Native Chemical Ligation via N-Acylurea Thioester Surrogates Obtained by Fmoc Solid-Phase Peptide Synthesis

Judith Palà-Pujadas and Juan B. Blanco-Canosa

Abstract

Native chemical ligation (NCL) enables the direct chemical synthesis and semisynthesis of proteins of different sizes and compositions, streamlining the access to proteins containing posttranslational modifications (PTMs). NCL assembles peptide fragments through the chemoselective reaction of a C-terminal α-thioester peptide, prepared either by chemical synthesis or via intein-splicing technology, and a recombinant or synthetic peptide containing an N-terminal Cys. Whereas the generation of C-terminal α-thioester proteins can be achieved via the recombinant fusion of the sequence of interest to an intein domain, chemical methods can also be used for synthetically accessible proteins. The use of Fmoc solid-phase peptide synthesis (Fmoc-SPPS) to obtain α-thioester peptides requires the development of novel strategies to overcome the lability of the thioester bond toward piperidine Fmoc-removal conditions. These new synthetic methods enable the easy introduction of PTMs in the thioester fragment. In this chapter, we describe an approach for the synthesis and use of C-terminal α-N-acylbenzimidazolinone (Nbz) and α-N-acyl-N′-methylbenzimidazolinone (MeNbz) peptides in NCL. Following stepwise peptide elongation, acylation with *p*-nitrophenylchloroformate and cyclization affords the Nbz/MeNbz peptides. The optimization of the coupling conditions allows the chemoselective incorporation of the C-terminal amino acid (aa) on the 3,4-diaminobenzoyl (Dbz) and prevents undesired diacylations of the resulting *o*-aminoanilide. Following synthesis, these Nbz/MeNbz peptides undergo NCL straightforwardly at neutral pH catalyzed by the presence of arylthiols. Herein, we apply the Nbz technology solid phase synthesis, NCL-mediated cyclization and folding of the heterodimeric RTD-1 defensin, an antimicrobial peptide isolated from the rhesus macaque leukocytes.

Key words Native chemical ligation (NCL), Peptide-thioesters, N-acylurea peptides, Peptide *o*-aminoanilides

1 Introduction

1.1 C-Terminal Peptide Thioesters in NCL: Fmoc Approaches

Native chemical ligation is a chemoselective, epimerization-free amide bond formation reaction between two unprotected peptides, one of them functionalized with a thioester group at the C-terminal side and the other featuring an N-terminal Cys (Fig. 1a) [1–6]. Both fragments react in aqueous solution under mild conditions

Miquel Vila-Perelló (ed.), *Expressed Protein Ligation: Methods and Protocols*, Methods in Molecular Biology, vol. 2133, https://doi.org/10.1007/978-1-0716-0434-2_7,

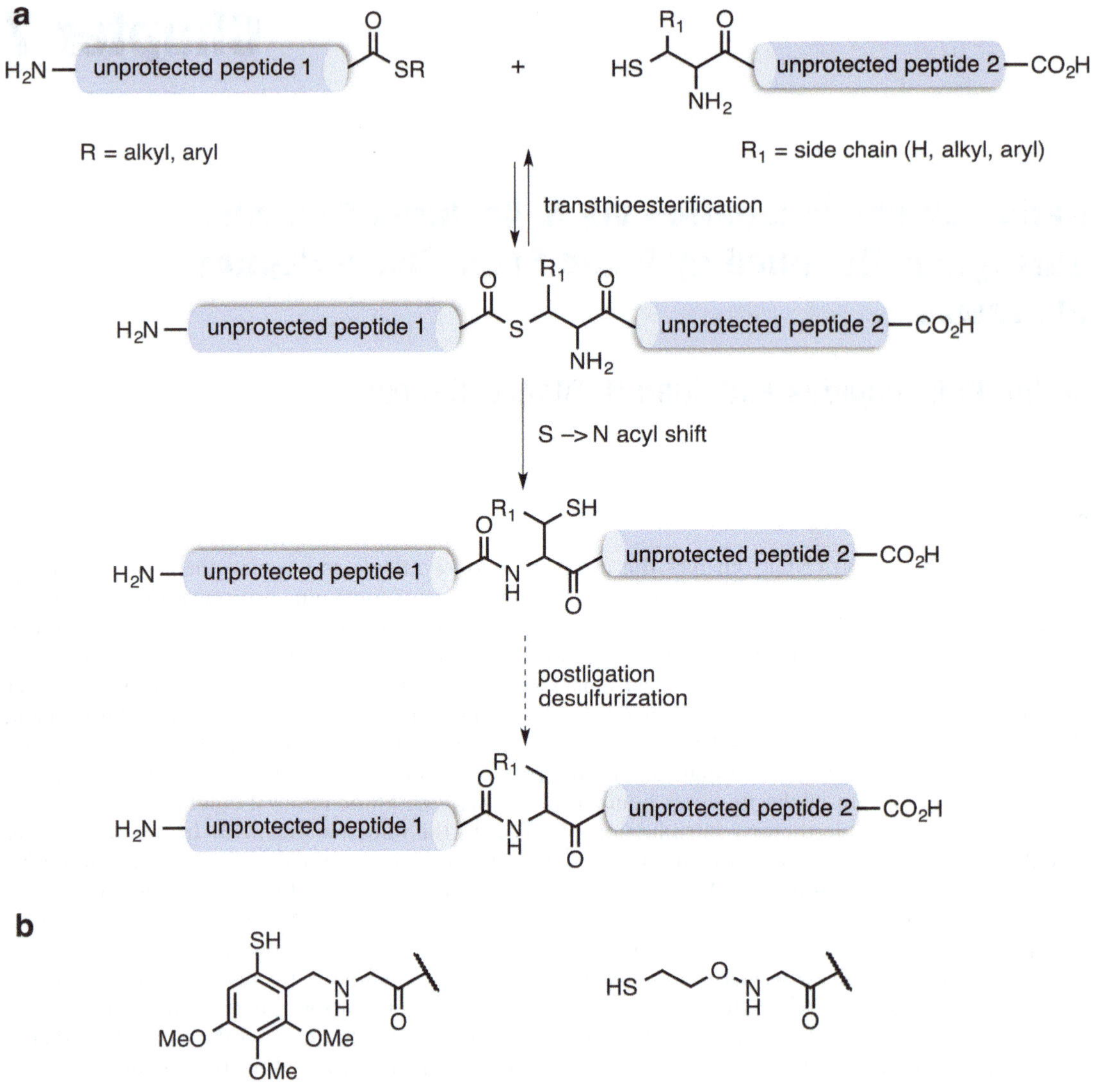

Fig. 1 (**a**) Native chemical ligation (NCL) between an N-terminal Cys (or a Cys surrogate residue) and a C-terminal α-thioester peptide. (**b**) Common Gly *N*-auxiliaries used in NCL

(pH 6.5–7.5, 25–37 °C) and the use of chaotropic agents such as guanidine hydrochloride, urea or sodium dodecyl sulfate, improves the solubility of the fragments and hence the overall efficiency. Despite the reaction is constrained to the presence of Cys at the ligation site, continuous work on the field has allowed to expand it to other residues. Therefore, it is currently possible to carry out NCL at other amino acids furnished with a thiol group at ß or γ-position [7, 8], as well as at Gly through the use of *N*-auxiliaries (Fig. 1b) [9, 10]. Following NCL, the natural amino acid can be restored by desulfurization [7, 11–13]. More importantly, the efficient preparation of peptide α-thioesters by Fmoc-SPPS, which is compatible with PTMs such as glycosylations and phosphorylations, has presented a major synthetic challenge due to the lability of the thioester bond toward aminolysis by piperidine. A variety of different approaches have been devised including: (1) the Kenner's sulfonamide safety-catch linker [14, 15], (2) activation of protected peptides in solution [16], (3) the backbone amide linker (BAL) [17], (4) solid phase side-chain anchoring bearing orthogonal α-carboxy protection [18–20], (5) O-to-S and N-to-S intramolecular acyl transfer reactions [21–23], (6) activation of C-terminal peptide hydrazides for acyl-azide to S transfer [24, 25], (7) oxidation of *o*-aminoanilides to generate active benzotriazoles [26–28]. Common to the majority of these methods is the utilization of inactivated precursors that, upon activation, can be transformed into thioesters in the presence of thiol additives. Both, the coupling of the C-terminal amino acid to the linker and the thioesterification are critical steps, since they usually involve the conversion of the precursor into a high energy-rich intermediate that is susceptible to epimerization [29]. Thus, there have been many efforts to find strategies that involve mildly activated intermediates that subsequently undergo transthioesterification to generate the desired thioesters. The first generation Nbz and later the second generation MeNbz approaches were introduced with the aim of facilitating the synthesis of α-thioester peptides via simple and robust methods overcoming limitations of previous approaches.

1.2 N-Acylurea Peptides as Thioester Surrogates

The Nbz [30–32] and the MeNbz [33] method provide a new strategy for the synthesis of C-terminal α-thioester peptides. Using the most simple and powerful reaction in peptide synthesis, the acylation of an amine, followed by a 5-member ring intramolecular cyclization, the Nbz/MeNbz peptides furnish a straightforward and simple route for the preparation of these key intermediates in NCL.

Nbz and MeNbz peptides are readily synthesized from the commercially available 3-(Fmoc-amino)-4-aminobenzoic acid (Fmoc-Dbz) and 3-(Fmoc-amino)-4-(methylamino)benzoic acid (Fmoc-MeDbz) or from Fmoc-Dbz and Fmoc-MeDbz resins (Fig. 2). Starting from a resin derivatized with a Rink, PAL (peptide amide linker) or a similar TFA cleavable linker, first a Gly is introduced as a spacer before the coupling of the Fmoc-Dbz or Fmoc-MeDbz (resin **1**). Following Fmoc-deprotection, the C-terminal amino acid is incorporated using standard peptide coupling conditions (HBTU, HATU or DIC/HOOBt). Achieving selective acylation of the C-terminal amino acid (Table 1) avoids undesired diacylations of the Dbz derivative. On the other hand, the Me group of the MeDbz linker endows a natural protection to the unprotected *o*-(methylamino)anilide offering a wider range of applications which includes microwave assisted SPPS and difficult Gly-rich sequences [33]. The anilide bond is stable under standard

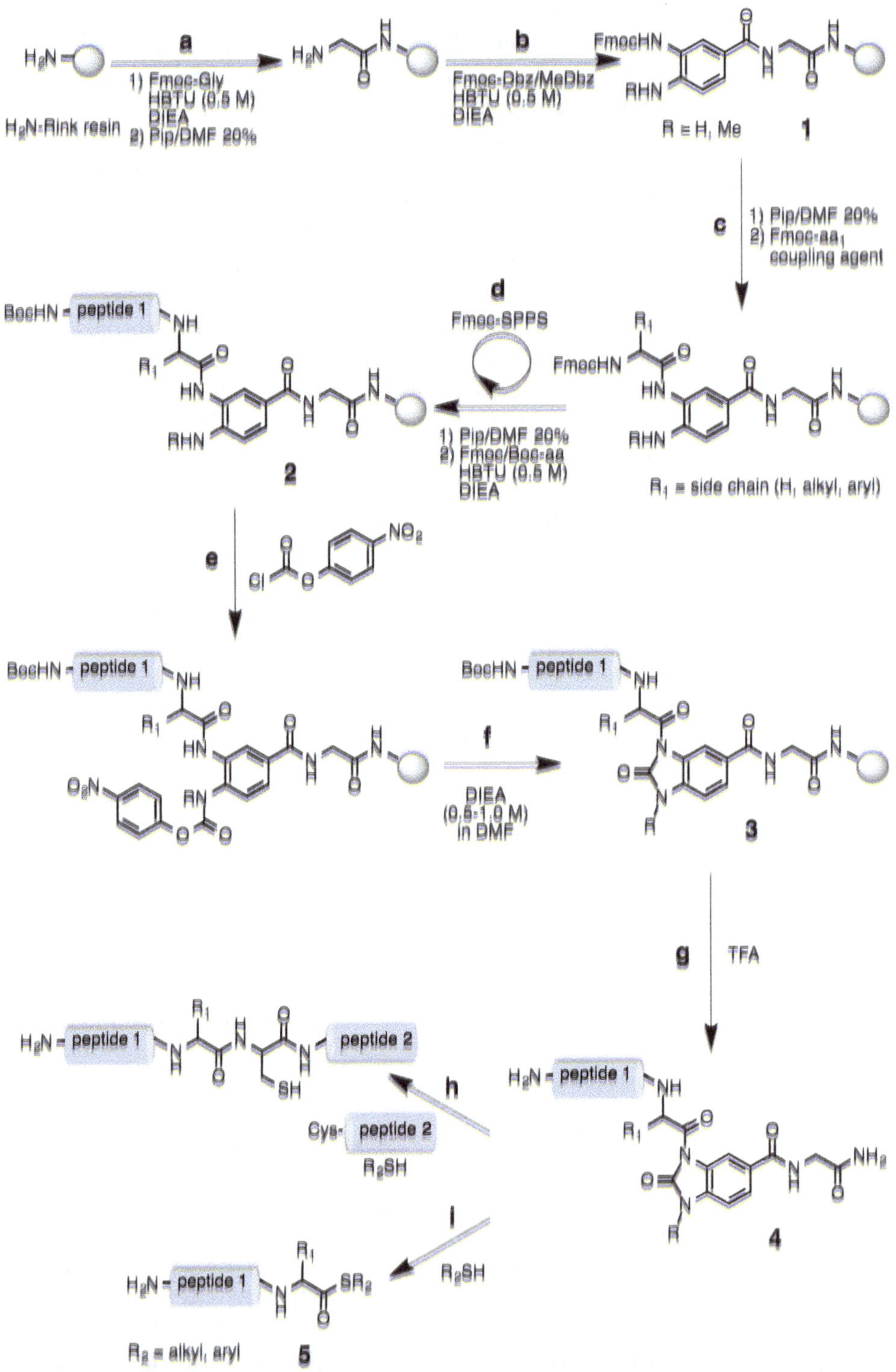

Fig. 2 Synthesis of α-thioester peptides by Fmoc chemistry using the Nbz/MeNbz strategy: (**a**) coupling of Gly, (**b**) coupling of the Fmoc-Dbz/MeDbz linker, (**c**) coupling of the C-terminal residue (Fmoc-aa$_1$), (**d**) chain elongation, (**e**) activation with *p*-nitrophenylchloroformate, (**f**) base induced cyclization to render the resin bound *N*-acylurea peptide, (**g**) acidolytic cleavage, (**h**) NCL in the presence of N-terminal Cys peptide, (**i**) transthioesterification in the presence of R_2SH thiols (R_2 = alkyl or aryl)

Table 1
C-terminal aa coupling conditions for Dbz resins

		Coupling agent (equiv.)					
Amino acid[a]	**equiv.**	**HBTU**	**HATU**	**DIC:HOOBt**	**DIEA (equiv.)**	**Conc[b] (mmol/mL)**	**Time/min**
Ala	5	5			5	0.4	30
Ile Thr Val	5		5		10	0.4	60
Pro	5		5		5.5	0.4	60
Cys[c] His[c]	5			5:5		0.4	30 60
Gly	4	4			4	0.2	30
Glu Phe Met Tyr Trp	3		3		3	0.2	30
Arg Gln Leu Lys Asn Ser	3		3		3	0.3	30

[a]NCL at Asp results in the formation of two regioisomeric products due to reaction of the Cys peptide with either the α and ß-thioesters. Thus, side chain protection of Asp during NCL is necessary in order to avoid the ß-linear isomer
[b]Final concentration of the amino acid in the coupling solution
[c]Preactivate for 3 min before adding to the resin

conditions for Fmoc-SPPS including TFA cleavage. After peptide elongation, the peptide-Dbz (but not the peptide-MeDbz) can be alternatively cleaved from the resin and used to generate benzotriazoles that also undergo thioesterification and NCL [26–28].

The resin-bound *o*-aminoanilide/*o*-(methylamino)anilide peptide **2** is acylated with *p*-nitrophenylchloroformate and, upon short basic treatment with concomitant intramolecular cyclization, the resulting peptide-Nbz/MeNbz **3** is released from the resin through TFA-mediated cleavage to yield the unprotected peptide-Nbz/MeNbz **4**. Remarkably, the integrity of the peptide is not compromised during the harsh acidic cleavage conditions.

The peptide-Nbz/MeNbz is captured in the presence of alkyl/aryl thiols, forming peptide-thioesters **5**, which can be isolated or directly ligated to N-terminal Cys-peptides in a one-pot procedure [30, 32, 33]. The rate-limiting step of the reaction is usually determined by the N-to-S intermolecular acyl transfer. The optimal transfer rates are achieved employing thiols with good water solubility (above 100 mM) and a pKa ~ 7, principally aryl thiols such as 4-mercaptophenol (MPOH) and 4-mercaptophenylacetic acid, which minimize hydrolysis and give high conversions (>95%) [34].

The Nbz/MeNbz technology has been successfully implemented in the preparation of glycoproteins, phosphoproteins, lipoproteins, and cyclic proteins [32, 35–44]. In addition, this cleavable

Nbz/MeNbz linker can undergo the attack of other nucleophiles rendering C-terminal modified peptides selectively labeled with different functionalities such as N-alkyl amides, esters, or cyclic peptides [45–47].

Herein, we describe a methodology to synthetize Nbz and MeNbz peptides as thioester precursors. Furthermore, we provide protocols to use them in NCL to afford larger ligated peptides. In addition, to provide general protocols for the use of this new technology, we include methods to apply it for the synthesis of cyclic peptides such as the heterodimeric RTD-1 defensin.

2 Materials

2.1 General

1. Analytical, reversed-phase high-performance liquid chromatography (RP-HPLC) system equipped with a photodiode array detector (PDA) or a UV-visible dual wavelength detector. Column: C-18, 5 μm, 150 × 4.5 mm.
2. Buffer A: 0.045% TFA (v/v) in water.
3. Buffer B: 0.036% TFA (v/v) in CH_3CN.
4. Semipreparative RP-HPLC system hyphenated to a UV-visible dual wavelength detector, equipped with a C-18 column (5 μm, 100 × 21.2 mm).
5. Ultrapure water with a resistivity of 18.2 MΩcm from a Milli-Q purification system.
6. Mass spectrometer, either electrospray ionization (ESI) or matrix-assisted laser desorption ionization (MALDI) with a time of flight (TOF) or similar mass analyzer (e.g., quadrupole or Q-trap).
7. Lyophilizer.

2.2 Fmoc-SPPS of C-Terminal Nbz/MeNbz Peptides

2.2.1 Equipment

1. Reaction vessel (RV) comprising a polypropylene syringe fitted with a frit and a Teflon stopcock. A 3 mL RV has capacity for 50–100 mg of resin, a 6 mL RV for 100–200 mg, a 12 mL for 200–300 mg, a 20 mL RV for 300–500 mg of resin, and a 60 mL RV for synthesis up to 2 g of starting resin.
2. Büchner flask (2 L capacity or larger) or a similar waste container for SPPS. This is placed between the RV and a vacuum trap.
3. Vacuum trap connected to the vacuum source and to the Büchner flask.
4. A vacuum source, ideally a diaphragm pump able to reach a vacuum of 7 mbar. The vacuum pump is connected to the vacuum trap.

5. Teflon rod or a N_2 source. The Teflon rod, or a gentle N_2 stream bubbling through the resin–solvent mixture, provide a mean of mixing reagents and resin during SPPS.
6. Glass or plastic vials (10, 20 mL).
7. Disposable glass test tubes for Kaiser test (1 mL).
8. Polypropylene centrifuge tubes (15, 50 mL capacity).
9. Centrifuge with rotor to spin 50 mL centrifuge tubes at ~1100 rcf.
10. Block heater with block to hold 1 mL test tubes.
11. Rotary evaporator.
12. Micropipettes for either use with organic and aqueous solutions able to deliver with high precision 2–20 μL, 20–200 μL, and 200–1000 μL.
13. Vacuum desiccator.

2.2.2 Reagents

1. 3-(Fmoc-amino)-4-aminobenzoic acid (Fmoc-Dbz) and 3-(Fmoc-amino)-4-(methylamino)benzoic acid (Fmoc-MeDbz).
2. Rink Amide PS-resin (100–200 mesh, loading: 0.5–0.7 mmol/g).
3. Dawson Dbz AM resin (loading 0.49 mmol/g) or resin preloaded with the Fmoc-Dbz/MeDbz linkers obtained from commercial suppliers.
4. Fmoc- and Boc-protected amino acids (*see* **Note 1**).
5. *p*-nitrophenylchloroformate.
6. Dichloromethane (DCM).
7. 1,2-dichloroethane (DCE).
8. *N,N*-dimethylformamide (DMF).
9. Solution C: 20% (v/v) solution of piperidine in DMF.
10. Trifluoroacetic acid (TFA).
11. Diethyl ether (Et_2O) (*see* **Note 2**).
12. *N,N*-diisopropylethylamine (DIEA).
13. Solution D: 0.5 M solution of 2-(1H-Benzotriazol-1-yl)-1,1,3,3-tetramethyluronium hexafluorophosphate (HBTU) in DMF.
14. 2-(7-aza-1H-Benzotriazol-1-yl)-1,1,3,3-tetramethyluronium hexafluorophosphate (HATU).
15. N,N′-diisopropylcarbodiimide (DIC).
16. Hydroxy-3,4-dihydro-4-oxo-1,2,3-benzotriazine (HOOBt).
17. Kaiser test solutions.
18. Triisopropylsilane (TIS).

19. 1,2-ethanedithiol (EDT). Caution, toxic and malodorous (*see* **Note 3**).
20. 2,2′-(ethylenedioxy)diethanethiol.

2.3 NCL of Nbz/MeNbz Peptides

2.3.1 Equipment

1. Safe-lock plastic tubes (1.5 mL or 2 mL capacity).
2. pH meter equipped with a microprobe.
3. 250 mL polystyrene vacuum bottle filter (0.45 μm filter pore size) or a 250 mL glass bottle and 0.45 μm PES membrane filters to store the preligation NCL buffer.
4. 500 mL beaker.
5. 250 mL volumetric flask.

2.3.2 Reagents

1. Buffer E: 0.2 M Na_2HPO_4, 6 M Gdm.HCl (*see* **Note 4**) in H_2O, pH 7.2. Dissolve 7 g of Na_2HPO_4 in 150 mL of MilliQ water in a 500 mL beaker. Then, add 143.3 g of Gdm.HCl and stir until the mixture becomes homogeneous (*see* **Note 5**). Transfer this solution into a volumetric flask and bring it up to 250 mL with Milli Q water. Finally, filter using 0.45 μm filters. The buffer E can be stored at room temperature.
2. Buffer F: 20 mM tris(2-carboxyethyl)phosphine hydrochloride (TCEP.HCl), 100 mM 4-mercaptophenol (MPOH, *see* **Note 6**) in buffer E. Seal with a rubber septum a vial containing 20 mL of buffer E. Degas this solution by bubbling through a stream of N_2 for 15 min. Then, add solid TCEP.HCl (114.7 mg, 20 mM), and MPOH (252.4 mg, 100 mM) by removing the septum briefly. The pH will drop to about 4–5. Degas for further 5 min. Then, stop the N_2 bubbling and keep the vial sealed under N_2 atmosphere. Buffer F can be stored in the dark under N_2 atmosphere at room temperature for 5–7 days.
3. 0.1 M $(NH_4)_2CO_{3(aq)}$.
4. 0.5 M solution of DIEA in dry DMF.
5. 1 M solution of DIEA in dry DMF.
6. 10 M $NaOH_{(aq)}$.
7. 1 M $NaOH_{(aq)}$.

3 Methods

3.1 SPPS of C-Terminal Nbz/MeNbz Peptides

Here we provide a method for the SPPS of Nbz/MeNbz peptides. Synthesis starts from a resin functionalized with a TFA labile linker (Fig. 2) to which the Dbz (or MeDbz) linker is coupled. Alternatively, preloaded Dbz/MeDbz resins could be used.

In order to avoid truncations or double acylations of the C-terminal amino acid attached to the Dbz moiety, some specific amino acid coupling conditions are required and are listed in Table 1. On the other side, the MeDbz derivative displays a greater selectivity which allows the incorporation of the C-terminal amino acid using a broad range of conditions, independently of the residue, without detectable diacylations of the *o*-(methylamino)anilide.

3.1.1 SPPS Synthesis of Nbz/MeNbz Peptides

The synthetic scheme and coupling conditions are shown in Fig. 2 and Table 1, respectively. **Steps 1–7** can be skipped when using commercially available preloaded Fmoc-Dbz/Fmoc-MeDbz resins (Subheading 2.2.2).

1. Weigh and transfer the Rink Amide PS-resin (2.0 mmol) to a 60 mL RV. Swell the resin in DCM (3 × 20 mL, 3 × 30 s), and then in DMF (3 × 20 mL, 3 × 30 s).
2. Removal of Fmoc: add 10 mL of solution C, stir with the Teflon rod (~10 s) and drain. Repeat the step. Add another 20 mL and stir for 10 min. Drain the RV. Next, wash the resin with DMF (5 × 15 mL, 5 × 30 s) and drain.
3. Dissolve Fmoc-Gly (2.97 g, 10.0 mmol) in solution D (20 mL) in a 50 mL polypropylene tube. Add neat DIEA (1.9 mL), shake for 30 s, and then add the solution to the RV. Stir the mixture (Teflon rod or N_2 bubbling) for 1 h (*see* **Note 7**).
4. Drain the RV and wash the resin with DMF (3 × 15 mL, 3 × 30 s) and DCM (2 × 15 mL, 2 × 30 s). Perform the Kaiser test [48, 49]. If the color is blue (it means there are free uncoupled amines), then repeat the coupling. A pale yellow color indicates that the coupling was completed. Then, proceed to the next step.
5. Proceed to **step 2** for Fmoc removal. After completing **step 2** proceed to **step 6**.
6. Dissolve Fmoc-Dbz/MeDbz (10.0 mmol) in solution D (20 mL). Add neat DIEA (1.9 mL) and shake the solution for 3 min. Add this solution to the RV and stir the mixture for 2 h.
7. Perform **step 4**. Once the Fmoc-Dbz/Fmoc-MeDbz has been incorporated to the resin, the Fmoc-Dbz/MeDbz-Gly resin can be stored dried at 2–8 °C, and then used as a convenience.

   The following steps describe the method to incorporate the different amino acids onto the Dbz/MeDbz and subsequent chain elongation. In Table 1 the optimized conditions to avoid diacylation of Dbz are shown. If MeDbz is used, then use the general conditions described in **step 10**.

8. Transfer the desired amount of Fmoc-Dbz/MeDbz-Gly resin to a RV. Swell the resin using 3 resin-bed-volumes (rv) of DCM (3 × 3 rv, 3 × 30 s), and then DMF (3 × 3 rv, 3 × 30 s).
9. Removal of Fmoc: add 2 rv of solution C, stir with the Teflon rod (~10 s) and drain. Repeat the step. Add other 2 rv of solution C and stir for 10 min. Then, drain the RV and wash the resin with DMF (5 × 5 rv, 5 × 30 s) and drain.
10. Following the conditions in Table 1 dissolve the Fmoc-aa_1 and the coupling agent (HBTU, HATU or DIC/HOOBt) in the specified volume of DMF. Add DIEA (if necessary) and shake the resulting solution for 30 s (or 3 min if the activation is performed with DIC/HOOBt). Add the mixture to the RV and stir for the indicated time. For MeDbz, coupling with HATU (5 equiv.), Fmoc-aa_1 (5 equiv.) and DIEA (5.5 equiv.) for 1 h can be used as a general protocol (*see* **Note 8**). Stir the mixture with the Teflon rod periodically.
11. Drain the RV and wash the resin with DMF (3 × 3 rv, 3 × 30 s) (*see* **Note 9**).
12. Remove Fmoc (*see* **step 9**) from Fmoc-aa_1-Dbz/MeDbz-Gly-resin.
13. Dissolve the next Fmoc-aa_2 (5 equiv.) in solution D (5 equiv.). Add DIEA (5 equiv.) and shake the mixture for 30 s. Add the Fmoc-aa_2 solution to the RV and stir periodically for 30 min (*see* **Note 10**).
14. Drain the RV and wash the resin with DMF (3 × 3 rv, 3 × 30 s) and DCM (2 × 3 rv, 2 × 30 s). Perform the Kaiser test. Proceed to **step 9** if the test result is negative. Otherwise, repeat the coupling (**step 13**).
15. Couple the remaining amino acids following **steps 12–14** (*see* **Notes 1** and **11**).
16. Wash the resin with DCM (3 × 3 rv, 3 × 30 s) and drain the RV.
17. Dissolve *p*-nitrophenylchloroformate (5 equiv.) in DCM (add DCM up to a final concentration around 0.3 M, *see* **Note 12**). Add to the RV and stir (Teflon rod) for 30 min (*see* **Note 13**).
18. Drain the RV and wash the resin with DMF (3 × 3 rv, 3 × 30 s) and DCM (3 × 3 rv, 3 × 30 s).
19. Swell the resin in DMF (2 × 3 rv, 2 × 30 s). Add a solution of 0.5 or 1 M DIEA (in dry DMF, 2 rv) for Nbz and MeNbz, respectively. Stir for 20 min for Nbz and 30–60 min for MeNbz (*see* **Note 14**).
20. Drain the RV and wash the resin with DMF (3 × 3 rv, 3 × 30 s) and DCM (3 × 3 rv, 3 × 30 s).
21. Finally, dry the resin under vacuum in a vacuum desiccator for 2 h.

3.1.2 Cleavage and Concomitant Side Chain Deprotection of Nbz/MeNbz Peptides

1. Place the resin in a round-bottom flask together with a magnetic stir bar.
2. Prepare a cleavage mixture of TFA:H_2O:TIS (95:2.5:2.5, 1 mL/50 mg peptidyl-resin), and chill to ~0 °C (*see* **Note 15**).
3. Add the cleavage cocktail to the flask and stir mildly the resulting mixture at room temperature for 1 h using magnetic stirring (*see* **Note 16**).
4. Filter the resin using the polypropylene syringe and collect the filtrates in a round-bottom flask. Perform additional TFA washes of the resin (3 × 3 rv, 3 × 1 min) and combine them.
5. Remove the TFA by evaporation in a rotary evaporator (caution: concentrate the TFA in a fume hood!) until a minimum volume is left. Alternatively, the TFA solution containing the peptide can be concentrated in a fume hood under a N_2 stream.
6. Add drop-wise the TFA-peptide over cold Et_2O (1 mL TFA/10 mL Et_2O) in a 50 mL conical tube. Centrifuge at ~1100 rcf for 10 min. A pale-white precipitate should appear at the bottom of the tube (*see* **Note 17**). Decant the solution carefully and dry the pellet under a gentle stream of N_2.
7. Dissolve the pellet in a 1:1 mixture of HPLC buffer A and B. Freeze down and lyophilize.
8. Purify, if needed, the Nbz/MeNbz peptide by semipreparative RP-HPLC (*see* **Notes 18** and **19**).

3.2 NCL of Nbz/MeNbz Peptides

Peptides displaying C-terminal Nbz/MeNbz react with N-terminal Cys fragments at low millimolar concentrations (both fragments about 1–5 mM). These ligations achieve conversions greater than 95% in less than 4 h at room temperature, except for Val, Ile and Pro, which need ~20 h (Val, Ile) and 48 h (Pro), to reach similar conversions [30, 33]. The following protocol describes a general procedure for the ligation of Nbz/MeNbz peptides to Cys peptides.

1. Weigh the peptide-Nbz (ca. 1–2 mg) and the N-terminal Cys peptide (1.0–1.2 equiv.) in a safe-lock plastic tube (1.5–2 mL capacity).
2. Use a 10 M $NaOH_{(aq)}$ solution and carefully (add 5 μL increments) adjust the pH of the required volume of buffer F to 7 (monitor with pH meter). When the pH gets close to 7, use a 1 M $NaOH_{(aq)}$ solution to perform a fine adjustment.
3. Dissolve the peptides in the pH adjusted buffer F from **step 2** (final concentration of both peptides should be around 1–5 mM with a slight excess 1.2 equiv., of the Cys peptide).

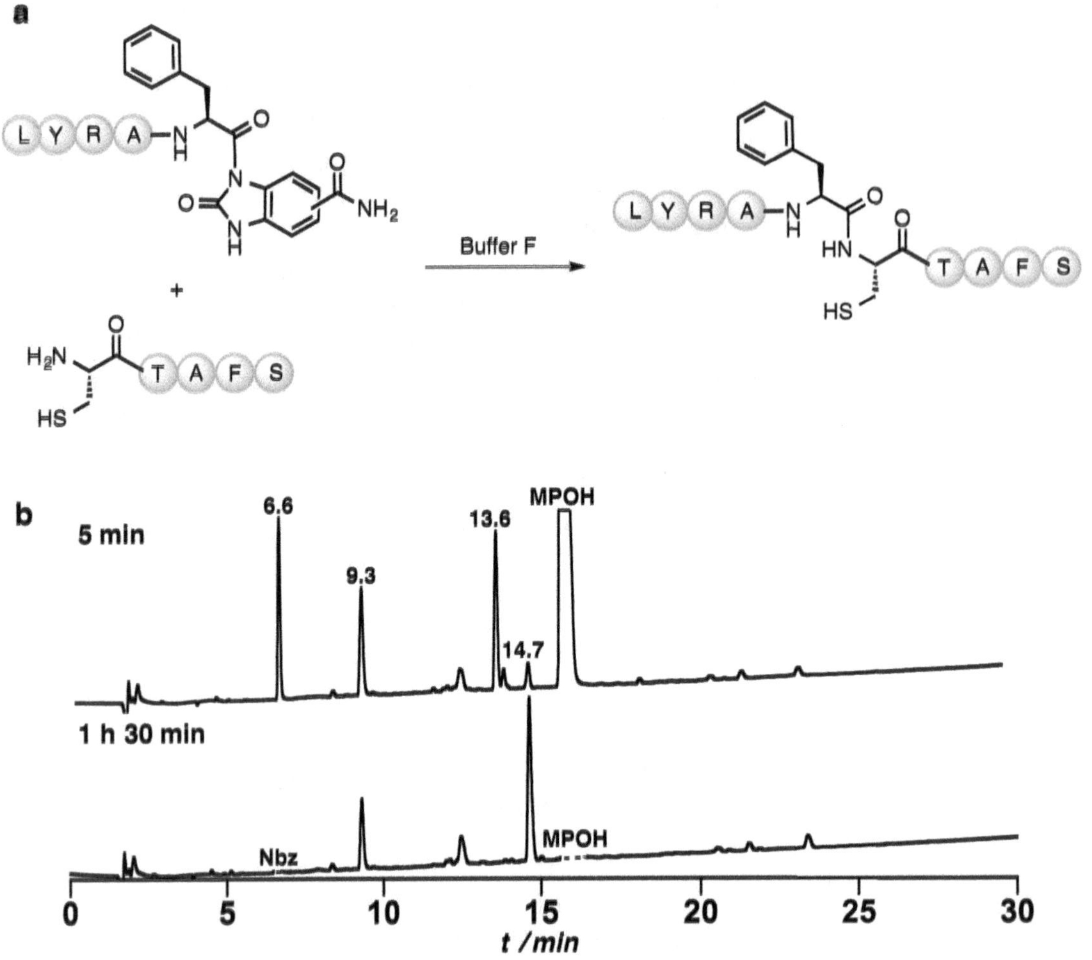

Fig. 3 (**a**) Reaction scheme: LYRAF-Nbz (2.3 mg, 2.7×10^{-3} mmol) and CTAFS (2.3 mg, 4.3×10^{-3} mmol) were dissolved in 1.0 mL of buffer F (6 M Gdm.HCl, 0.2 M Na_2HPO_4, 100 mM MPOH, 20 mM TCEP.HCl, pH = 7.0, room temperature). (**b**) HPLC trace (220 nm) of the reaction at 5 min and 1 h 30 min. Retention time: 6.6 min = Nbz, 9.3 min = CTAFS, 13.6 min = LYRAF-Nbz, 14.7 min = LYRAFCTAFS. Gradient: 0 to 70% B in 30 min. MPOH = 4-mercaptophenol

4. Check the pH with pH meter. If the pH drops below 6.8, then carry out a readjustment using a 1 M $NaOH_{(aq)}$ solution (*see* **Note 20**).
5. Purge the top of the tube with a gentle stream of N_2 and close.
6. As an example of monitoring the reaction (Fig. 3a), we show the ligation of the model peptides LYRAF-Nbz (synthesized on Dawson Dbz AM resin, loading 0.49 mmol/g, Subheading 3.1.1) and CTAFS. The ligation is analyzed using RP-HPLC and mass spectrometry by taking aliquots and diluting them in HPLC buffer (Fig. 3b).

Fig. 4 (**a**) Cyclic structure of the heterodimeric θ-defensin RTD-1 showing the disulfide laddered pattern. (**b**) Linear sequence of the chemically synthesized RTD-1-Nbz

3.3 Synthesis of Heterodimeric θ-Defensin RTD-1

The heterodimeric θ-defensin RTD-1 is a 18-mer antimicrobial peptide component of the innate immune system of some primates. It was first isolated from rhesus macaque leukocytes. RTD-1 and related defensins have a unique folding pattern characterized by a cyclic head to tail structure threaded by three disulfide bonds arranged in a laddered topology that delimits six loops (Fig. 4a) [50, 51].

Synthesis of circular proteins bearing a Cys amino acid can be accomplished through an intramolecular NCL. RTD-1 has 6 Cys, meaning that the linear synthesis can be started at any given amino acid before a Cys residue. In the present approach the cyclization point was chosen between ^{2}Phe and ^{3}Cys and so the Fmoc-SPPS started at ^{2}Phe (Fig. 4b).

3.3.1 Fmoc-SPPS of RTD-1-Nbz

This section describes a protocol for the linear synthesis of RTD-1 on commercially available Dawson Dbz AM resin on a 0.2 mmol scale using a fivefold excess (1.0 mmol) of Fmoc/Boc-aa per coupling cycle (no double couplings).

1. Weigh the resin (408 mg, 0.2 mmol) and transfer it to a 20 mL RV. Swell the resin in DCM (3 × 10 mL, 3 × 30 s) and DMF (3 × 10 mL, 3 × 30 s).
2. Add 5 mL of solution C and stir for 10 s. Drain the RV and repeat the treatment. Drain the RV and add another 10 mL. Stir the mixture for 10 min.
3. Drain the RV and wash the resin with DMF (5 × 10 mL, 5 × 30 s).
4. Dissolve Fmoc-Phe (232 mg, 0.6 mmol) and HATU (228 mg, 0.6 mmol) in DMF (3 mL). Add neat DIEA (0.104 mL, 0.6 mmol), shake the mixture for 30 s and transfer the activated amino acid to the RV. Stir for 30 min.
5. Drain the RV and wash the resin with DMF (3 × 10 mL, 3 × 30 s).
6. Proceed to **step 2** for Fmoc removal.
7. Dissolve Fmoc-Gly (297 mg, 1.0 mmol) in solution D (2 mL) and add DIEA (0.191 mL, 1.1 mmol). Shake the mixture for 30 s and transfer it to the RV. Stir for 40 min.

8. Drain the RV and wash the resin with DMF (3 × 10 mL, 3 × 30 s).
9. Repeat **steps 6–8** to couple the remaining amino acids (*see* **Note 21**).
10. Wash the resin with DCM (3 × 10 mL, 3 × 30 s) and drain. Dissolve *p*-nitrophenylchloroformate (201 mg, 1.0 mmol) in DCM (3 mL) and add to the resin. Stir the resulting suspension occasionally (Teflon rod) for 1 h.
11. Drain the RV and wash the resin with DMF (3 × 10 mL, 3 × 30 s) and DCM (3 × 10 mL, 3 × 30 s).
12. Swell the resin in DMF (2 × 10 mL, 2 × 30 s). Add 0.5 M DIEA in dry DMF (10 mL) and stir for 20 min.
13. Drain the RV and wash the resin with DMF (3 × 10 mL, 3 × 30 s) and DCM (3 × 10 mL, 3 × 30 s).
14. Dry the resin under vacuum in a vacuum desiccator for 2 h.

3.3.2 Cleavage

1. Prepare a cleavage cocktail solution containing TFA:TIS:H_2O:EDT (92.5:2.5:2.5:2.5, 20 mL, *see* **Note 3**). Chill the solution to 0 °C.
2. Transfer the resin (1.16 g) to a 100 mL round bottom flask equipped with a magnetic stir bar. Add the cleavage cocktail and stir mildly at room temperature for 2 h.
3. Filter the mixture using the polypropylene syringe and collect the TFA solution containing the peptide in a 250 mL round bottom flask.
4. Wash the resin with TFA (3 × 5 mL, 3 × 1 min). Combine the TFA filtrates and remove the TFA until reaching a final volume of ca. 5 mL.
5. Add (drop-wise) the TFA mixture over cold Et_2O in two conical tubes containing 40 mL each.
6. Spin out in a centrifuge at ~1100 rcf for 10 min. Discard the supernatant and dry the pellet under a gentle stream of N_2.
7. Dissolve the resulting peptide in 25 mL of HPLC buffer (1:1, A and B) in each tube (use sonication if needed). Freeze down and lyophilize.
8. From 1.16 g of cleaved resin were obtained 450 mg of crude peptide after lyophilization (95% isolated yield) and an analytical purity of 59% (calculated by integration of the HPLC trace at 220 nm).
9. The HPLC trace of the synthesis after cleavage from the solid support and the ESI-MS of the product corresponding to the main peak are shown in Fig. 5.

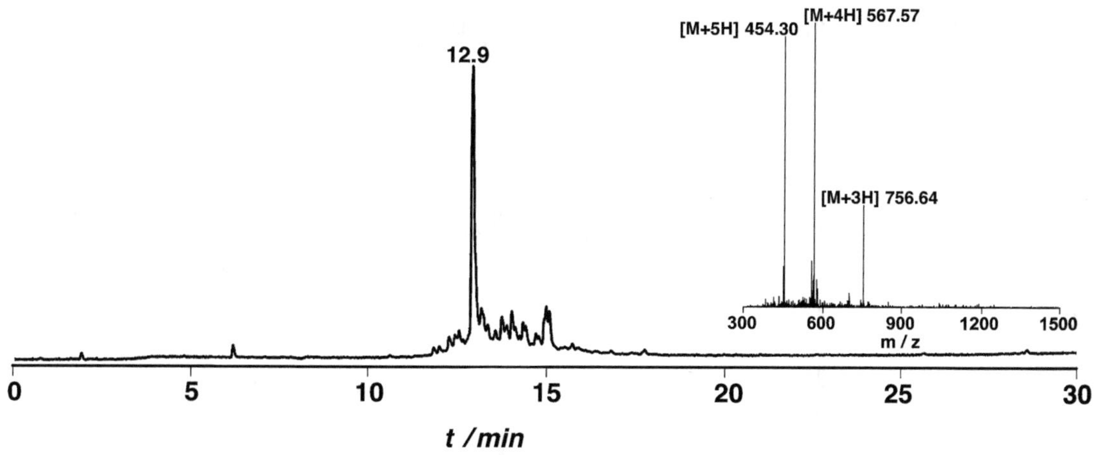

Fig. 5 Analytical RP-HPLC trace (220 nm) of the crude reaction following cleavage of the crude linear RTD-1-Nbz resin. The mass (ESI) of the peak at 12.9 min is shown in the inset and corresponds to the expected peptide. ESI-MS ($C_{90}H_{151}N_{36}O_{21}S_6$, $[M + H]^+$): calcd. 2265.79; found 2266.54. Gradient: 0 to 70% B in 30 min

3.3.3 Cyclization of Linear RTD-1-Nbz

1. Weigh and transfer 53 mg of crude linear RTD-1-Nbz peptide (~1.8×10^{-2} mmol) to a 50 mL falcon tube.
2. Add 20 mL of ligation buffer F (pH 7). Purge the top of the solution with N_2 and close the falcon tube.
3. Take aliquots at 5 min and 2 h and monitor the reaction by directly running analytical RP-HPLC and mass spectrometry (Fig. 6a).
4. At 2 h, acidify the reaction with 10% $HCl_{(aq)}$ until pH 2.
5. Add 20 mL of Et_2O. Shake the mixture vigorously and let the phases to separate. Remove the Et_2O with a pipette.
6. Repeat the extraction (2×20 mL). After three rounds of Et_2O extraction, the MPOH is removed from the aqueous phase containing the cyclic peptide (Fig. 6b).
7. Purify the cyclic-RTD-1 by semipreparative RP-HPLC. Collect and pool the fractions containing the cyclic peptide, combine and lyophilize.
8. Isolated yield: 17 mg (34% from crude linear RTD-1-Nbz).

3.3.4 Oxidative Folding of Cyclic-RTD-1

1. Weigh 1 mg (3.6×10^{-4} mmol) of cyclic-RTD-1 in a safe-lock plastic tube (1.5 mL).
2. Dissolve the cyclic-RTD-1 in 0.75 mL of a $(NH_4)_2CO_{3(aq)}$ buffer (0.1 M, pH 8.7), close the safe-lock plastic tube and stir occasionally.
3. Monitor the progress of the reaction by LC-MS. The mass of the oxidized RTD-1 is 6 Da units smaller than the cyclic-RTD-1 due to the formation of the three disulfide bonds (Fig. 7a

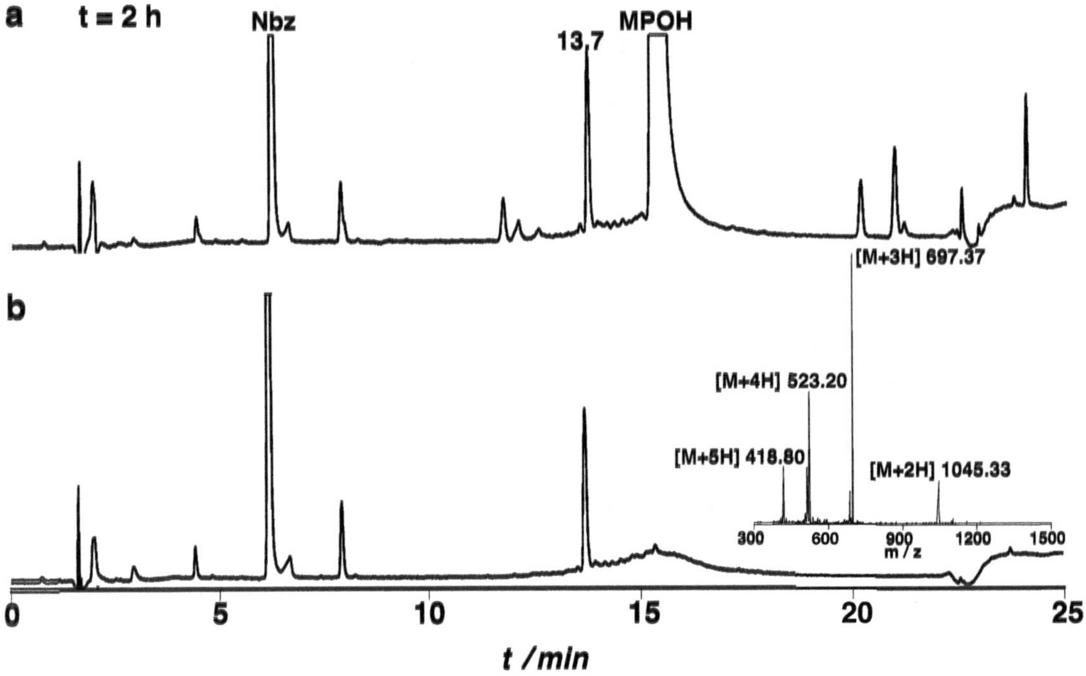

Fig. 6 HPLC profile of the intramolecular cyclization reaction of RTD-1-Nbz crude peptide to yield cyclic-RTD-1: (**a**) HPLC chromatogram of the reaction at 2 h. (**b**) HPLC trace after Et_2O extraction. Inset, ESI-MS of the peak at 13.7 min corresponding to the expected cyclic-RTD-1. ESI-MS ($C_{82}H_{144}N_{33}O_{19}S_6$, $[M + H]^+$): calcd. 2088.63; found 2088.87. Gradient: 0 to 46% B in 20 min, then 100% B for 5 min

inset). Moreover, the oxidized RTD-1 has a shorter HPLC retention time than the reduced protein. Complete oxidative folding of RTD-1 is achieved in about 15 h (Fig. 7b).

4. Purify by semipreparative RP-HPLC the folded RTD-1. Collect and pool the fractions containing the product, combine and lyophilize.

4 Notes

1. The amino acid at the N-terminal position is usually introduced with Boc protection on the α-amino group to release the free amine upon TFA cleavage. Those Boc-amino acids that require side chain protection must have a protecting group compatible with standard TFA cleavage conditions.
2. Et_2O must be peroxide free to prevent the oxidation of sensitive residues such as Met.
3. It is conceivable that EDT may induce hydrolysis of Nbz/MeNbz peptides. This hydrolysis might be sequence dependent and it could take place during the TFA acidolytic cleavage or even during resuspension of the pellet in HPLC or NCL

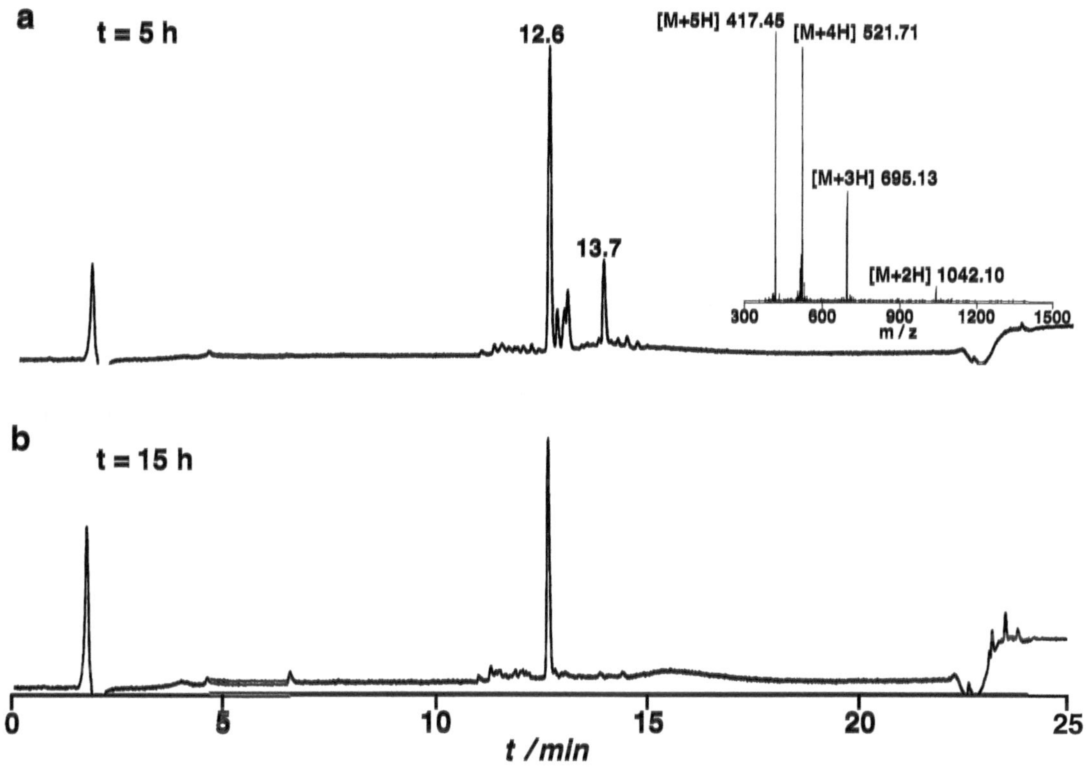

Fig. 7 Oxidative folding of cyclic-RTD-1: (**a**) 5 h reaction. (**b**) HPLC trace after 15 h. Inset is shown the ESI-MS of the peak at 12.6 min corresponding to the folded RTD-1. Peak at 13.7 min corresponds to cyclic-RTD-1. ESI-MS ($C_{82}H_{138}N_{33}O_{19}S_6$, $[M + H]^+$): calc. 2082.58; found 2082.39. Gradient: 0 to 46% B in 20 min, then 100% B for 5 min

ligation buffer due to the presence of traces of EDT after Et_2O precipitation. Hence, we recommend the use of 2,2-′-(ethylenedioxy)diethanethiol as an alternative to EDT. 2,2-′-(ethylenedioxy)diethanethiol is almost odorless and it has similar scavenging properties than EDT. Despite its caveats, and as it is described in this chapter (Subheading 3.3.2), we have used EDT in the cleavage mixture of RTD-1-Nbz with no detectable hydrolysis (Fig. 5). If thiol scavengers need to be used in the cleavage reaction, we suggest to perform microtest cleavages to identify the optimal thiol scavenger.

4. For NCL experiments it is used Gdm.HCl of high purity (>99.5%), which can be obtained from regular suppliers.
5. Warming the suspension in a hot plate improves the solubility of Gdm.HCl.
6. 4-mercaptophenylacetic acid (MPAA) can also be used as thiol additive for NCL.
7. This step is optional. However, the incorporation of a Gly spacer facilitates the next coupling of Fmoc-Dbz/MeDbz.

8. The number of equiv. makes reference to the initial mmol of resin used in the synthesis.
9. The Kaiser test is not sensitive to aryl amines. An alternative method to determine the loading of the C-terminal amino acid on the Dbz/MeDbz is the quantification of the piperidine-dibenzofulvene adduct by removal of the Fmoc group from an aliquot of the resin and reading the absorbance at 289.8 and 301.0 nm [52].
10. Adding the 5 equiv. of HOBt minimizes the risk of *o*-aminoanilide acylation with Gly residues during chain elongation.
11. If desired, acetylation of the N-terminal amino acid can be carried out using fivefold excess of N-acetyl-N-[2-ispropyl-4-oxo-3(4H)-quinazolinyl]acetamide during 30 min without capping the *o*-aminoanilide group [32].
12. Working at large scale synthesis (>1 g resin), the carbamoylation reaction runs cleaner when carried out in DCE likely due to the better swelling properties of this solvent.

    For resins derivatized with highly acid-labile linkers (i.e., Trityl), it is needed to add 2,4,6-trimethylpyridine in an equal amount of equiv. to *p*-nitrophenylchloroformate to avoid premature cleavage.
13. For larger peptides where reaction times may be sequence dependent, it is recommended to extend the activation with *p*-nitrophenylchloroformate for 2 h.
14. The solution should turn yellow due to the *p*-nitrophenolate anion.
15. Add 2,2′-(ethylenedioxy)diethanethiol, 2.5% v/v to the TFA/H_2O/TIS mixture in Cys-containing peptides.
16. The duration of the cleavage will depend in a great extension on the $(Arg(Pbf))_n$ content of the peptide. For those sequences with more than three Arg(Pbf) residues cleavage times around 2–2.5 h are required for fully Arg deprotection.
17. If the peptide does not precipitate out of the solution, then remove all the solvent, including TFA and Et_2O, dissolve the resulting residue in HPLC buffer, freeze down and lyophilize.
18. Nbz and MeNbz peptides are stable and can be purified by HPLC. However, we notice that under typical acidic conditions of HPLC buffers, they also tend to hydrolyze. In order to avoid so, we recommend to lyophilize immediately after RP-HPLC purification.
19. Nbz peptides typically give two regioisomeric products resulting from the attachment of the peptide to 3 or 4-position of the Dbz. These regioisomers, which have the same mass, can be

clearly observed in short peptide sequences as they have different HPLC retention time.

20. Add 5 μL increments of 1 M $NaOH_{(aq)}$ solution to the inside part of the plastic tube cap and vortex or shake. Check the pH after each addition. By adding the $NaOH_{(aq)}$ solution to the cap and not directly to the reaction mixture it is prevented a localized hydrolysis of the thioester peptide and peptide-Nbz/MeNbz due to the high basicity of the $NaOH_{(aq)}$ solution.
21. The N-terminal Cys is introduced as Boc-Cys(Trt)-OH.

Acknowledgments

This work was supported by the Spanish Ministerio de Economía y Competitividad (grants CTQ2012-31197 and RYC-2011-09001). J.P.-P. acknowledges an FPI scholarship (BES-2013-065237).

References

1. Dawson PE, Muir TW, Clark-Lewis I, Kent SB (1994) Synthesis of proteins by native chemical ligation. Science 266:776–779
2. Dawson PE, Kent SB (2000) Synthesis of native proteins by chemical ligation. Annu Rev Biochem 69:923–960
3. TCJr E, Benner J, Xu MQ (1998) Semisynthesis of cytotoxic proteins using a modified protein splicing element. Protein Sci 7:2256–2264
4. Muir TW, Sondhi D, Cole PA (1998) Expressed protein ligation: a general method for protein engineering. Proc Natl Acad Sci U S A 95:6705–6710
5. Muir TW (2003) Semisynthesis of proteins by expressed protein ligation. Annu Rev Biochem 72:249–289
6. Vila-Perello M, Muir TW (2010) Biological applications of protein splicing. Cell 143:191–200
7. Yan LZ, Dawson PE (2001) Synthesis of peptides and proteins without cysteine residues by native chemical ligation combined with desulfurization. J Am Chem Soc 123:526–533
8. Malins LR, Payne RJ (2014) Recent extensions to native chemical ligation for the chemical synthesis of peptides and proteins. Curr Opin Chem Biol 22:70–78
9. Canne LE, Bark SJ, Kent SBH (1996) Extending the applicability of native chemical ligation. J Am Chem Soc 118:5891–5896
10. Offer J, Boddy CN, Dawson PE (2002) Extending synthetic access to proteins with a removable acyl transfer auxiliary. J Am Chem Soc 124:4642–4646
11. Ma J, Zeng J, Wan Q (2015) Postligation-desulfurization: a general approach for chemical protein synthesis. Top Curr Chem 363:57–101
12. Wan Q, Danishefsky SJ (2007) Free-radical-based, specific desulfurization of cysteine: a powerful advance in the synthesis of polypeptides and glycopolypeptides. Angew Chem Int Ed 46:9248–9252
13. Rohde H, Seitz O (2010) Ligation-desulfurization: a powerful combination in the synthesis of peptides and glycopeptides. Biopolymers 94:551–559
14. Shin Y, Winans KA, Backes BJ, Kent SBH, Ellman JA, Bertozzi CR (1999) Fmoc-based synthesis of peptide-thioesters: application to the total chemical synthesis of a glycoprotein by Native Chemical Ligation. J Am Chem Soc 121:11684–11689
15. Ingenito R, Bianchi E, Fattori D, Pessi A (1999) Solid phase synthesis of peptide C-terminal thioester by Fmoc/*t*-Bu chemistry. J Am Chem Soc 121:11369–11374
16. Futaki S, Sogawa K, Maruyama J, Asahara T, Niwa M, Hojo H (1997) Preparation of peptide thioesters using Fmoc-solid-phase peptide synthesis and its application to the construction of a Template-Assembled Synthetic Protein (TASP). Tet Lett 38:6237–6240
17. Alsina J, Yokum TS, Albericio F, Barany G (1999) Backbone Amide Linker (BAL) strategy

for N(alpha)-9-Fluorenylmethoxycarbonyl (Fmoc) solid-phase synthesis of unprotected peptide p-Nitroanilides and Thioesters (1). J Org Chem 64:8761–8769

18. Ficht S, Payne RJ, Guy RT, Wong CH (2008) Solid-phase synthesis of peptide and glycopeptide thioesters through side-chain-anchoring strategies. Chemistry 14:3620–3629
19. Ishii A, Hojo H, Nakahara Y, Ito Y, Nakahara Y (2002) Silyl linker-based approach to the solid-phase synthesis of Fmoc glycopeptide thioesters. Biosci Biotechnol Biochem 66:225–232
20. Tulla-Puche J, Barany G (2004) On-resin native chemical ligation for cyclic peptide synthesis. J Org Chem 69:4101–4107
21. Ollivier N, Dheur J, Mhidia R, Blanpain A, Melnyk O (2010) Bis(2-sulfanylethyl)amino native peptide ligation. Org Lett 12:5238–5241
22. Botti P, Villain M, Manganiello S, Gaertner H (2004) Native chemical ligation through in situ O to S acyl shift. Org Lett 6:4861–4864
23. Sato K, Shigenaga A, Tsuji K, Tsuda S, Sumikawa Y, Sakamoto K, Otaka A (2011) N-sulfanylethylanilide peptide as a crypto-thioester peptide. Chembiochem 12:1840–1844
24. Fang GM, Li YM, Shen F, Huang YC, Li JB, Lin Y, Cui HK, Liu L (2011) Protein chemical synthesis by ligation of peptide hydrazides. Angew Chem Int Ed 50:7645–7649
25. Wang P, Layfield R, Landon M, Mayer RJ, Ramage R (1998) Transfer active ester condensation: a novel technique for peptide segment coupling. Tet Lett 39:8711–8714
26. Katritzky AR, Shestopalov AA, Suzuki K (2004) A new convenient preparation of thiol esters utilizing *N*-acylbenzotriazoles. Synthesis:1806–1813
27. Wang JX, Fang GM, He Y, Qu DL, Yu M, Hong ZY, Liu L (2015) Peptide *o*-aminoanilides as crypto-thioesters for protein chemical synthesis. Angew Chem Int Ed 54:2194–2198
28. Weidmann J, Dimitrijevic E, Hoheisel JD, Dawson PE (2016) Boc-SPPS: compatible linker for the synthesis of peptide *o*-aminoanilides. Org Lett 18:164–167
29. Nagalingam AC, Radford SE, Warriner SL (2007) Avoidance of epimerization in the synthesis of peptide thioesters using Fmoc protection. Synlett:2517–2520
30. Blanco-Canosa JB, Dawson PE (2008) An efficient Fmoc-SPPS approach for the generation of thioester peptide precursors for use in native chemical ligation. Angew Chem Int Ed 47:6851–6855
31. Pascal R, Chauvey D, Sola R (1994) Carboxyl-protecting groups convertible into activating groups. Carbamates of *o*-aminoanilides are precursors of reactive *N*-acylureas. Tet Lett 35:6291–6294
32. Tiefenbrunn TK, Blanco-Canosa J, Dawson PE (2010) Alternative chemistries for the synthesis of thrombospondin-1 type 1 repeats. Biopolymers 94:405–413
33. Blanco-Canosa JB, Nardone B, Albericio F, Dawson PE (2015) Chemical protein synthesis using a second-generation *N*-acylurea linker for the preparation of peptide-thioester precursors. J Am Chem Soc 137:7197–7209
34. Johnson EC, Kent SB (2006) Insights into the mechanism and catalysis of the native chemical ligation reaction. J Am Chem Soc 128:7140–7141
35. Hemantha HP, Brik A (2013) Non-enzymatic synthesis of ubiquitin chains: where chemistry makes a difference. Bioorg Med Chem 21:3411–3420
36. Hemantha HP, Bavikar SN, Herman-Bachinsky Y, Haj-Yahya N, Bondalapati S, Ciechanover A, Brik A (2014) Nonenzymatic polyubiquitination of expressed proteins. J Am Chem Soc 136:2665–2673
37. Fauvet B, Butterfield SM, Fuks J, Brik A, Lashuel HA (2013) One-pot total chemical synthesis of human alpha-synuclein. Chem Commun 49:9254–9256
38. Lindgren J, Ekblad C, Abrahmsen L, Eriksson Karlstrom A (2012) A native chemical ligation approach for combinatorial assembly of affibody molecules. Chembiochem 13:1024–1031
39. Harris PWR, Brimble MA (2012) The chemical synthesis of the collagenous domain of the hormone adiponectin. Int J Pept Res Ther 18:63–70
40. Pentelute BL, Barker AP, Janowiak BE, Kent SB, Collier RJ (2010) A semisynthesis platform for investigating structure-function relationships in the N-terminal domain of the anthrax lethal factor. ACS Chem Biol 5:359–364
41. Gunasekera S, Aboye TL, Madian WA, El-Seedi HR, Goransson U (2013) Making ends meet: microwave-accelerated synthesis of cyclic and disulfide rich proteins via in situ thioesterification and native chemical ligation. Int J Pept Res Ther 19:43–54
42. Zhan C, Varney K, Yuan W, Zhao L, Lu W (2012) Interrogation of MDM2 phosphorylation in p53 activation using native chemical ligation: the functional role of Ser17 phosphorylation in MDM2 reexamined. J Am Chem Soc 134:6855–6864

43. Okamoto R, Mandal K, Ling M, Luster AD, Kajihara Y, Kent SB (2014) Total chemical synthesis and biological activities of glycosylated and non-glycosylated forms of the chemokines CCL1 and Ser-CCL1. Angew Chem Int Ed 53:5188–5193
44. Chiang KP, Jensen MS, McGinty RK, Muir TW (2009) A semisynthetic strategy to generate phosphorylated and acetylated histone H2B. Chembiochem 10:2182–2187
45. Arbour CA, Saraha HY, McMillan TF, Stockdill JL (2017) Exploiting the MeDbz linker to generate protected or unprotected C-terminally modified peptides. Chem Eur J 23:12484–12488
46. Monaim S, Acosta GA, Royo M, El-Faham A, de la Torre BG, Albericio F (2018) Solid-phase synthesis of homodetic cyclic peptides from Fmoc-MeDbz-resin. Tet Lett 59:1779–1782
47. Arbour CA, Kondasinghe TD, Saraha HY, Vorlicek TL, Stockdill JL (2018) Epimerization-free access to C-terminal cysteine peptide acids, carboxamides, secondary amides, and esters via complimentary strategies. Chem Sci 9:350–355
48. Coin I, Beyermann M, Bienert M (2007) Monitoring solid phase peptide synthesis. Protoc Exchange. https://doi.org/10.1038/nprot.2007.461
49. Sarin VK, Kent SB, Tam JP, Merrifield RB (1981) Quantitative monitoring of solid-phase peptide synthesis by the ninhydrin reaction. Anal Biochem 117:147–157
50. Tran D, Tran PA, Tang YQ, Yuan J, Cole T, Selsted ME (2002) Homodimeric theta-defensins from rhesus macaque leukocytes: isolation, synthesis, antimicrobial activities, and bacterial binding properties of the cyclic peptides. J Biol Chem 277:3079–3084
51. Tang YQ, Yuan J, Osapay G, Osapay K, Tran D, Miller CJ, Ouellette AJ, Selsted ME (1999) A cyclic antimicrobial peptide produced in primate leukocytes by the ligation of two truncated alpha-defensins. Science 286:498–502
52. Eissler S, Kley M, Bächle D, Loidl G, Meier T, Samson D (2017) Substitution determination of Fmoc-substituted resins at different wavelengths. J Pept Sci 23:757–762

Chapter 8

Fused Split Inteins: Tools for Introducing Multiple Protein Modifications

Byung Joon Lim, Raymond F. Berkeley, and Galia T. Debelouchina

Abstract

The split inteins from the DnaE cyanobacterial family are efficient and versatile tools for protein engineering and chemical biology applications. Their ultrafast splicing kinetics allow for the efficient production of native proteins from two separate polypeptides both in vitro and in cells. They can also be used to generate proteins with C-terminal thioesters for downstream applications. In this chapter, we describe a method based on a genetically fused version of the DnaE intein Npu for the preparation of doubly modified proteins through recombinant expression. In particular, we provide protocols for the recombinant production of modified ubiquitin through amber suppression where fused Npu is used (1) as a traceless purification tag or (2) as a protein engineering tool to introduce C-terminal modifications for subsequent attachment to other proteins of interest. Our purification protocol allows for quick and facile separation of truncated products and eliminates the need for engineering protease cleavage sites. Our approach can be easily adapted to different proteins and applications where the simultaneous presence of internal and C-terminal modifications is desirable.

Key words Intein, Ubiquitin, Amber suppression, Unnatural amino acid incorporation, Protein modification, Tagless protein purification

1 Introduction

Inteins are a group of autoprocessing domains that can excise themselves from their host polypeptide chain while concurrently building a native peptide bond that connects the flanking host residues. This process, called intein splicing, does not require energy input in the form of ATP and relies instead on the correct folding of the intein and a series of steps that involve strategically positioned cysteine or serine residues [1–4]. There are two types of inteins found in nature: contiguous versions where the intein domain represents one continuous polypeptide chain [5, 6], and split versions where the intein domain is separated into two polypeptides flanked by the N- or C-terminal segments of the host protein (also called extein) [7–9]. Both the contiguous and split

Miquel Vila-Perelló (ed.), *Expressed Protein Ligation: Methods and Protocols*, Methods in Molecular Biology, vol. 2133, https://doi.org/10.1007/978-1-0716-0434-2_8,

inteins have become valuable tools for protein engineering including the purification, modification, cyclization, and ligation of proteins and peptides of various lengths [1, 10]. Historically, these applications have been performed with contiguous inteins such as the robust gyrase A intein from *Mycobacterium xenopi* (*Mxe* GyrA) [11–13]. For example, commercial kits for affinity column purification are available where the protein of interest is fused to an engineered version of GyrA and a chitin binding domain [14]. After expression, the construct is bound to a chitin resin column and on column cleavage is performed with a thiol reagent such as dithiothreitol (DTT), resulting in the release of pure full-length protein of interest. Similarly, the GyrA intein has been a reliable and useful tool for the generation of C-terminal thioesters and their downstream use in expressed protein ligation applications [11, 13]. The use of genetically fused split inteins from the cyanobacterial DnaE split intein family in lieu of GyrA, however, offers several advantages [7]. First, Npu and other members of this family exhibit ultrafast kinetics and high splicing efficiency, properties that can increase the yield of the purified or modified protein and reduce the time required to handle aggregation prone samples. Second, engineered versions of Npu can work efficiently at high temperatures and high concentrations of urea enabling the purification and modification of proteins that require denaturing conditions [15, 16].

In this chapter, we describe a generalizable method to harness the beneficial properties of genetically fused Npu and to introduce modifications at multiple sites in a single protein of interest, in this case the 76-residue protein ubiquitin. Our construct starts with ubiquitin, followed by fused Npu, a small linker and a hexahistidine purification tag (Fig. 1). The hexahistidine tag allows the separation of the full-length protein from any truncations, while the intein subsequently removes the histidine tag in a traceless and efficient manner without the need of proteases. In the intein portion of the construct, the last residue of Npu—typically an asparagine residue important for the splicing mechanism [7]—is mutated to alanine (Fig. 2). This mutation biases the intein reactivity toward N-terminal cleavage at the junction between ubiquitin and Npu [1]. Cleavage can be promoted by water molecules at slightly basic pH resulting in the release of the protein of interest, or by appropriate thiol reagents leading to the modification of the C-terminus. Here, we show how the water-mediated cleavage reaction can be exploited to efficiently purify ubiquitin produced by amber suppression [17] and to incorporate the unnatural amino acid (UAA) norbornene lysine at position 6 of the ubiquitin sequence [18]. The thiolysis reaction, on the other hand, enables the introduction of a C-terminal thioester that can be used in protein ligation applications and in the modification of the C-terminus with various reagents. In this chapter, we illustrate this process using the reagent

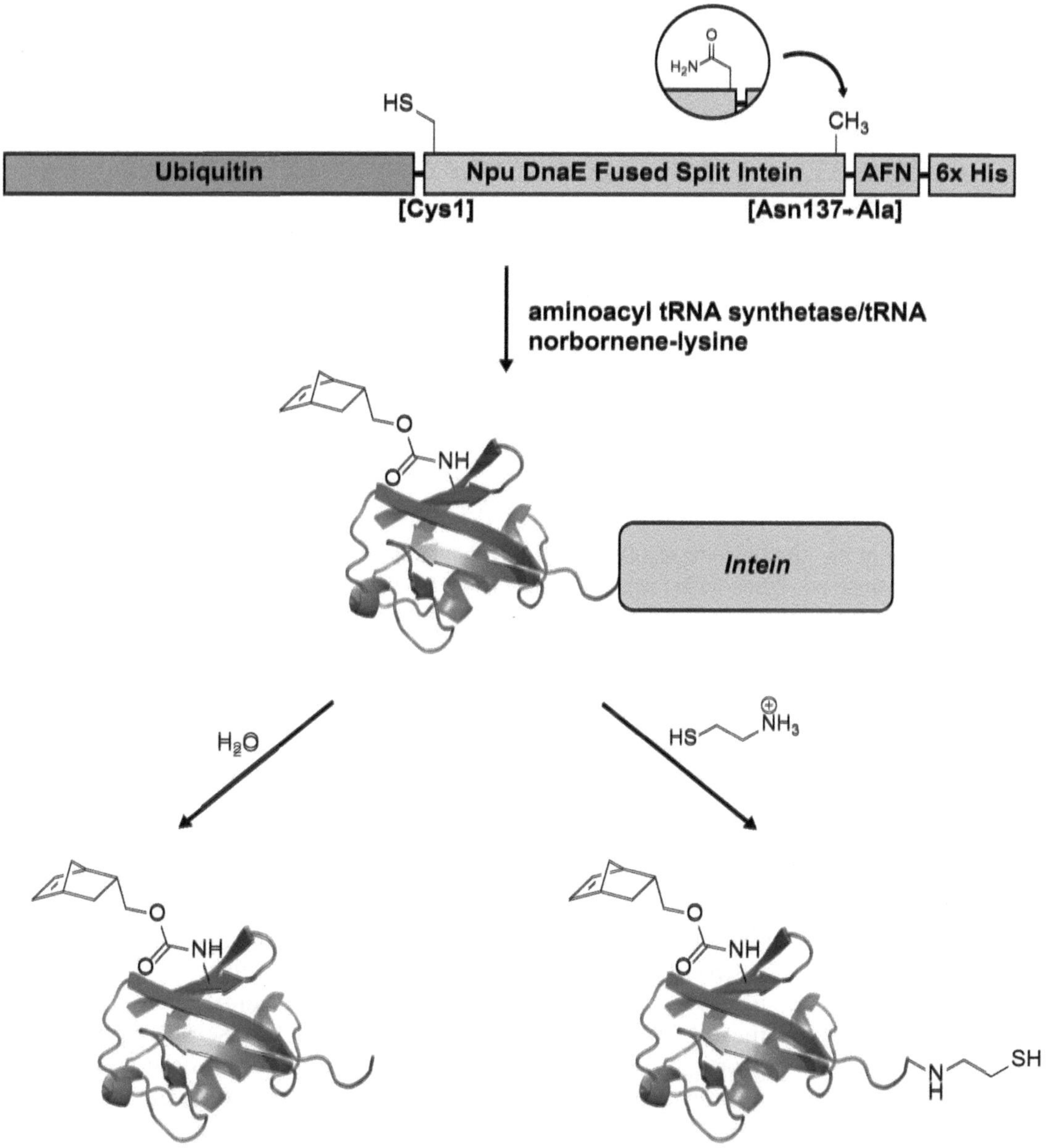

Fig. 1 The fused split intein Npu can be used as a traceless purification tag for the recombinant production of modified proteins through amber suppression and/or as a tool to introduce C-terminal thioesters for downstream protein ligation. In our construct, ubiquitin is genetically fused to Npu, followed by a small linker (AFN) and a hexahistidine tag for affinity purification. The last residue of Npu, N137, is mutated to alanine to promote intein cleavage at the ubiquitin–Npu junction

aminoethanethiol. This generates a ubiquitin construct that can be attached to other proteins of interest through an asymmetric disulfide linkage [19]. More importantly, the two strategies can be combined to produce ubiquitin containing both an internal and a C-terminal modification [20, 21]. We have also successfully used

1 10	11 20	21 30
MQIFVXTLTG	KTITLEVEPS	DTIENVKAKI
31 40	41 50	51 60
QDKEGIPPDQ	QRLIFAGKQL	EDGRTLSDYN
61 70	71 80	81 90
IQKESTLHLV	LRLRGG**CLSY**	**ETEILTVEYG**
91 100	101 110	111 120
LLPIGKIVEK	**RIECTVYSVD**	**NNGNIYTQPV**
121 130	131 140	141 150
AQWHDRGEQE	**VFEYCLEDGS**	**LIRATKDHKF**
151 160	161 170	171 180
MTVDGQMLPI	**DEIFERELDL**	**MRVDNLPNIK**
181 190	191 200	201 210
IATRKYLGKQ	**NVYDIGVERD**	**HNFALKNGFI**
211 221		
ASA*AFNHHHHHH*		

Fig. 2 Sequence of the K6X-Ubiquitin-NpuDnaE-AFN-His6 protein used as an example in this protocol. The position of the norbornene lysine amino acid (X) is underlined, the ubiquitin sequence is shown in black, the NpuDnaE intein is in bold characters, while the AFN-His6 sequence at the C-terminus is shown in italic font

this strategy to produce isotopically labeled and modified ubiquitin proteins [20], further illustrating the utility and efficiency of the protocol described below.

Our protocol starts with the optimization of protein expression and the incorporation of the unnatural amino acid norbornene lysine through amber suppression on a small scale. Following a scale-up of the protein expression, cell lysis, and affinity purification, we split the protocol into two parts. The first part describes the intein hydrolysis protocol used to generate the native ubiquitin C-terminus, while the second part describes the intein thiolysis steps necessary to modify the protein with C-terminal aminoethanethiol moiety. We also present a preparative-scale reverse-phase purification protocol applicable to both intein cleavage reactions and end with a refolding protocol for the modified ubiquitin construct.

2 Materials

All solutions and media are prepared in ultrapure deionized water (purified to a resistivity of 18.2 MΩ·cm at 25 °C) using analytical grade reagents, unless indicated otherwise in the text. All materials used in the preparation of gels and purification columns are reagent grade, unless otherwise stated.

2.1 Plasmid Constructs

1. The pTXB1-Ubiquitin-NpuDnaE-AFN-His6 plasmid used herein was provided as a generous gift by the lab of Dr. Tom Muir [7]. The plasmid contains an ampicillin resistance marker

and expression of the ubiquitin-fused split intein construct Ubiquitin-NpuDnaE-AFN-H6 (Fig. 2) is placed under control of the T7 promoter. Store at −20 °C until use.

2. Primers to replace the codon for lysine 6 in the ubiquitin sequence with the amber codon (TAG).
3. The pEVOL-tRNA^{Pyl}/pylRS^{Y306A,Y384F} plasmid used herein was provided as a generous gift by the lab of Dr. Carsten Schultz [18]. The plasmid contains a chloramphenicol resistance marker and expression of the aminoacyl tRNA synthetase–tRNA pair is placed under control of the pBAD promoter. The encoded aminoacyl tRNA synthetase can efficiently incorporate lysine variants carrying large head groups such as the norbornene lysine UAA used here. Store at −20 °C until use.

2.2 Protein Expression

1. Ampicillin stock solution: 100 mg in 1 mL water. Filter with a 0.22 μm syringe-tip filter. Store at −20 °C until use.
2. Chloramphenicol stock solution: 35 mg in 1 mL ethanol. Filter with a 0.22 μm syringe-tip filter. Store at −20 °C until use.
3. IPTG stock solution: 500 mM in water. Filter with a 0.22 μm syringe-tip filter. Store at −20 °C until use.
4. Arabinose stock solution: 20% w/v in water. Filter with a 0.22 μm syringe-tip filter. Store at −20 °C until use.
5. BL21 (DE3) competent cells. Store as received at −80 °C until use.
6. LB Agar Plates: 40 g Miller LB Agar (*see* **Note 1**) in 1 L water. Autoclave and when the temperature of the media has dropped below 65 °C, add 1 mL of ampicillin stock solution for a final concentration of 100 μg/mL and 1 mL of chloramphenicol stock solution for a final concentration of 35 μg/mL. Pour media into sterile petri dishes and allow them to cool to room temperature before use. Store unused plates at 4 °C.
7. LB media: 25 g Miller LB (*see* **Note 2**) in 1 L water. Autoclave and allow media to cool to room temperature before use. Store unused media at 4 °C.
8. Norbornene lysine: 0.1 M in 0.5 M sodium hydroxide. See synthesis protocol below.
9. 12% Tris–glycine gels for SDS-PAGE.
10. Thiol-free SDS-PAGE loading buffer: 2% w/v sodium dodecyl sulfate (SDS), 10% v/v glycerol, 12.5 mM ethylenediaminetetraacetic acid (EDTA), and 0.02% w/v bromophenol blue in 50 mM Tris–HCl, pH 6.8.

2.3 Protein Purification

2.3.1 Ni-NTA Affinity Purification

1. 1 M tris(2-carboxyethyl)phosphine (TCEP): 280 mg in 530 μL of 5 M sodium hydroxide and 470 μL of 1 M unbuffered Tris base.
2. Lysis buffer: 50 mM sodium phosphate (monobasic), 300 mM sodium chloride, and 5 mM imidazole (*see* **Note 3**). Adjust to a final pH of 7.5. Add 1 protease inhibitor tablet per 50 mL buffer immediately before use (*see* **Note 4**).
3. Wash buffer 1: 50 mM sodium phosphate (monobasic), 300 mM sodium chloride, and 20 mM imidazole. Adjust to a final pH of 7.5.
4. Wash buffer 2: 50 mM sodium phosphate (monobasic), 300 mM sodium chloride, and 50 mM imidazole. Adjust to a final pH of 7.5.
5. Elution buffer: 50 mM sodium phosphate (monobasic), 300 mM sodium chloride, and 250 mM imidazole. Adjust to a final pH of 7.5.
6. Nickel (II) nitrilotriacetic acid (Ni-NTA) resin (*see* **Note 5**).
7. 150 mL aluminum beaker.
8. Sonicator: 1/8″ diameter probe tip, power 125 W, frequency 20 kHz.

2.3.2 High-Performance Liquid Chromatography (HPLC) Purification

1. XBridge Peptide BEH C18 300 Å, 5 μm, 4.6 mm × 150 mm analytical HPLC column (Waters).
2. XBridge Peptide BEH C18 300 Å, 5 μm, 10 mm × 250 mm preparative-scale HPLC column (Waters).
3. HPLC solvent A: 0.1% TFA in ultrapure or HPLC grade water.
4. HPLC solvent B: 0.1% TFA in HPLC grade acetonitrile.
5. Trifluoroacetic acid (TFA).

2.3.3 Modification of the C-Terminus

1. 2-aminoethanethiol stock solution: 1 M in 50 mM sodium phosphate and 300 mM sodium chloride. Adjust to a final pH of 7.5.

2.4 Synthesis of Norbornene Lysine

1. Triphosgene: 4.8 g (1 equivalent), CAS 32315-10-9.
2. 5-Norbornene-2-methanol: 2 g (1 equivalent), CAS 95-12-5.
3. Boc-L-Lysine-OH: 4.8 g (1.2 equivalent), CAS 13734-28-6.
4. Anhydrous tetrahydrofuran (THF): 40 mL.
5. 1.0 M NaOH/THF: 2 to 1 v/v. Combine 60 mL of 1.0 M sodium hydroxide (NaOH) and 30 mL of anhydrous THF.
6. Ethyl acetate.
7. Hexanes.
8. *N*,*N*-dimethyl formamide (DMF): 90 mL.

9. Anhydrous sodium sulfate.
10. 1 M HCl solution: 83.3 mL of 12 N HCl and deionized water up to 1 L total volume.
11. 50 mM HCl solution: 10 mL of 1 M HCl solution to 190 mL of deionized water.
12. Saturated NaCl solution.
13. 60% formic acid in chloroform: 6 to 4 v/v. Combine 15 mL of formic acid and 10 mL of chloroform.
14. Schlenk line with argon gas and high vacuum valve.
15. Magnetic stirrer and stir bars.
16. Rotary evaporator.
17. Separatory funnel.
18. Glass column for flash chromatography.
19. Silica gel powder: CAS 7631-86-9.

2.5 Ubiquitin Refolding

1. Denaturing buffer: 6 M guanidium chloride, 10 mM Tris–HCl, and 10 mM KCl. Adjust to a final pH of 7.5.
2. Refolding buffer: 10 mM Tris–HCl and 10 mM KCl. Adjust to a final pH of 7.5.
3. Dialysis cassette or dialysis membrane, 3500 MWCO, 3 mL capacity.

3 Methods

Our methodology section is organized in the following way. First, in Subheading 3.1 we outline the synthesis of norbornene lysine. While we use this as an example modification, the subsequent steps in our protocol are compatible with the incorporation of other UAAs provided that the appropriate matching aminoacyl tRNA synthetase plasmid is used. Then, in Subheading 3.2 we provide protocols for the expression and purification of the Ub-NpuDnaE-His6 construct on a small and large scale, respectively. We also provide details on how to optimize the incorporation of the UAA. Subheading 3.3 outlines the protocols for intein cleavage through hydrolysis and thiolysis, while Subheading 3.4 describes the preparative HPLC purification of the generated constructs. We end with Subheading 3.5 which presents the refolding protocol for the doubly modified final ubiquitin construct. We should note that Subheadings 3.2, 3.3, and 3.4 are general and applicable to any other protein construct of interest.

3.1 Synthesis of Norbornene Lysine

Since to our knowledge norbornene lysine is not commercially available, we provide our synthesis protocol, based on the published literature [22]. All synthetic steps should be performed in a hood equipped with standard synthesis equipment (*see* Subheading 2.4). Care should be taken when handling all reagents. All applicable material safety data sheets should be consulted and followed before proceeding with the protocol.

1. Dissolve triphosgene (4.8 g) in dry THF (0.5 M, 35 mL) while stirring and add 5-norbornene-2-methanol (2 g) dropwise for 1 h to the solution at 0 °C under argon atmosphere.
2. Stir the reaction mixture for 6 h.
3. Remove all volatile components under reduced pressure using a rotary evaporator and dry under high vacuum for an additional 1 h.
4. Add dry THF to the oily intermediate to prepare a 5 mL solution at 3.5 M.
5. Dissolve Boc-L-Lys-OH (4.8 g) in 1.0 M NaOH/THF (2:1 v/v, 90 mL, 0.3 M) and slowly add the intermediate solution from the previous step to the solution at 0 °C.
6. Warm the reaction mixture up to room temperature and stir for 16 h.
7. Add ethyl acetate (100 mL) to the mixture and acidify the aqueous layer to pH 4 with 1 M HCl solution.
8. Collect the organic layer and extract the aqueous layer with additional ethyl acetate (100 mL each) three times.
9. Combine all organic layer fractions and wash them with saturated NaCl solution (400 mL).
10. Dry the organic solution with anhydrous sodium sulfate, and remove it by filtration.
11. Evaporate all volatile components under reduced pressure in a rotary evaporator.
12. Purify the crude product by silica gel flash column chromatography (ethyl acetate/hexane, gradient: 1:2 to 2:1).
13. Remove all volatile components by rotary evaporation and dissolve the residue in 60% formic acid in chloroform (6:4 v/v, 90 mL, 0.2 M).
14. Stir the reaction mixture for 24 h at room temperature.
15. Add DMF (90 mL) to the reaction mixture and remove all volatile components under reduced pressure.
16. Add 50 mM HCl (200 mL) to the residue and lyophilize the solution to obtain the pure hydrochloride salt of norbornene lysine.

3.2 Ub-NpuDnaE-His6 Expression

This section describes the expression and purification protocol for the ubiquitin–intein construct and the incorporation of norbornene lysine at position 6 of the ubiquitin sequence. We refer to this construct as K6X-Ubiquitin-NpuDnaE-AFN-His6. We describe cloning and transformation steps (Subheading 3.2.1), a protocol for the optimization of protein expression on a small scale (Subheading 3.2.2) and the large-scale expression and purification of the modified construct (Subheading 3.2.3).

3.2.1 Cloning and Transformation

The first steps of the protocol involve the cloning of the UAA incorporation site into the plasmid of interest. This is followed by the double transformation of the ubiquitin–intein and aminoacyl tRNA synthetase plasmids into competent cells for protein expression.

1. Pick the desired site for unnatural amino acid incorporation in the protein plasmid sequence. This site will be replaced with the TAG amber stop codon. In our case, we choose to replace lysine 6 in ubiquitin with the unnatural amino acid norbornene lysine.
2. Using the template plasmid (pTXB1-Ubiquitin-NpuDnaE-AFN-His6) and appropriate primers, introduce the TAG mutation by site-directed mutagenesis (*see* **Note 6**). We denote the modified plasmid as pTXB1-K6X-Ubiquitin-NpuDnaE-AFN-His6.
3. Sequence the modified plasmid and store at −20 °C.
4. Retrieve a 50 μL aliquot of the BL21 (DE3) competent cells, the pTXB1-K6X-Ubiquitin-NpuDnaE-AFN-His6 plasmid stock, and the pEVOL-tRNA^{Pyl}/pylRS$^{Y306A,\ Y384F}$ plasmid stock and thaw them on ice.
5. Using sterile technique, add 1.5 μL (50–100 ng of DNA) of each plasmid stock to the BL21 (DE3) competent cells (*see* **Note 7**).
6. Incubate the cells and plasmids on ice for 20 min.
7. Heat-shock the cells at 42 °C for 40 s.
8. Incubate the cells on ice for 5 min.
9. Using sterile technique, add 150 μL of antibiotic-free, sterile LB media to the transformation mixture. Gently mix the cells by pipetting. Do not vortex.
10. Incubate the cells at 37 °C for 1.5 h in an incubator at 220 rpm.
11. Plate 200 μL of the transformation mixture on LB agar containing chloramphenicol and ampicillin.
12. Incubate the plate at 37 °C. Colonies are generally observed after 10–15 h (*see* **Note 8**). Once colonies are visible, proceed to **step 1** in Subheading 3.2.2. Colonies should remain viable for a week if stored at 4 °C but are best used immediately.

3.2.2 Optimization of Protein Expression

Here, we provide a protocol to optimize protein expression when the incorporation of unnatural amino acids by amber suppression is desired. This protocol is general and can be adapted for the optimization of UAA incorporation for other proteins or for other UAAs. The optimal expression conditions (temperature, time, antibiotic) may differ for other systems.

1. Pick a single colony from the plate prepared in Subheading 3.2.1 and inoculate 5 mL of autoclaved LB media containing 5 μL of the stock solutions of ampicillin (final 100 μg/mL) and chloramphenicol (final 35 μg/mL) in a 15 mL culture tube under sterile conditions.
2. Incubate the preculture for 12–16 h at 37 °C while shaking at 220 rpm. If desired, cell stocks can be prepared after incubation (*see* **Note 9**).
3. Prepare two 15 mL culture tubes (one for norbornene lysine incorporation and one for a negative control) containing 5 mL of autoclaved LB and antibiotic as described in **step 1** above. Inoculate each tube with 50 μL of the overnight preculture.
4. Incubate the cultures at 37 °C while shaking at 220 rpm until the OD_{600} reaches 0.2–0.4. This should take 2.5–4 h.
5. Add 50 μL of the 0.1 M norbornene lysine stock solution to one of the cultures for a final concentration of 1 mM norbornene lysine. Add 50 μL of 0.5 M NaOH to the other tube to serve as a control.
6. Incubate the cultures for 30 min to 1 h until an OD_{600} of 0.4–0.6 has been reached.
7. Take a 250 μL aliquot from each culture for analysis by SDS-PAGE. To prepare each SDS-PAGE sample, measure the OD_{600} of each culture, pellet the cells by centrifugation at 4 °C and 3000 × *g*, and decant the supernatant. Resuspend the cells in 10 μL of SDS loading buffer per 0.1 units of OD_{600}. This aliquot will serve as the "Time 0" timepoint for characterizing the expression of the ubiquitin-fused split intein construct. Store on ice or at −20 °C until analysis by SDS-PAGE.
8. Add 5 μL of IPTG stock and 5 μL of arabinose stock to each culture tube for a final concentration of 0.5 mM and 0.02% (w/v), respectively. The addition of IPTG and arabinose will induce the expression of the ubiquitin-fused split intein construct and the aminoacyl tRNA synthetase–tRNA pair.
9. Take aliquots from both cultures at defined time points and prepare them for SDS-PAGE analysis as described in **step 7** above. For the K6X-Ubiquitin-NpuDnaE-AFN-His6 construct, we typically take time points at 0, 1, 2, 3, and 4 h after the addition of IPTG. After all time points have been taken, the cultures can be discarded as biohazardous waste.

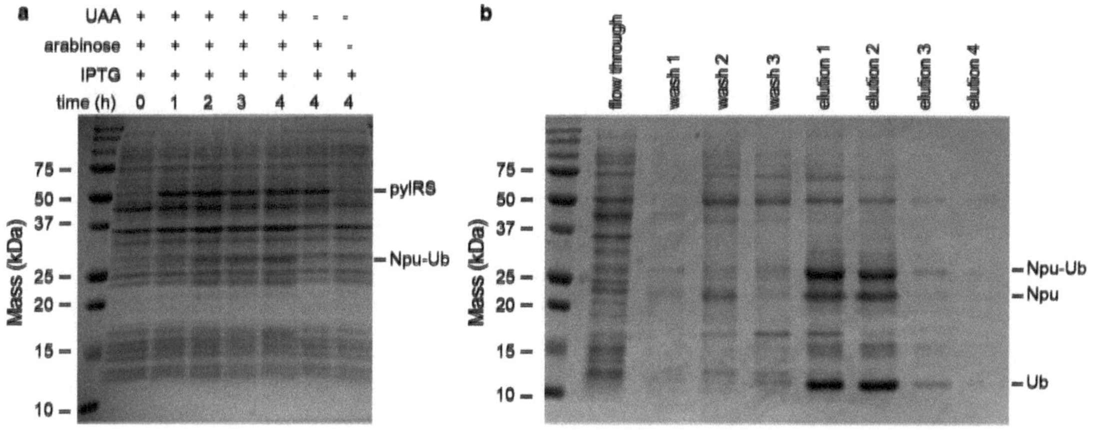

Fig. 3 (**a**) Test expression of ubiquitin-Npu–containing norbornene lysine at position 6 of the ubiquitin sequence, a modification introduced genetically through amber suppression. Expression of the construct is expected only when the unnatural amino acid (UAA), arabinose and IPTG are added to the culture media. In our case, the optimal expression conditions are 4 h and 37 °C. (**b**) Ni-NTA column purification of the ubiquitin–Npu construct analyzed by SDS-PAGE gel electrophoresis. Intein hydrolysis should not be visible in the wash fractions and should only be initiated upon elution. In this gel image, some hydrolysis has already taken place in the elution fractions

10. Heat the SDS-PAGE samples at 85 °C for 5 min in thiol-free SDS PAGE loading buffer (*see* **Note 10**).
11. Analyze the samples on a 12% SDS-PAGE Tris–glycine gel.
12. Choose the best time course for scaling up protein production (Fig. 3a). The aminoacyl tRNA synthetase bands should appear in both samples, but the desired ubiquitin-fused split intein band should only be present in the samples containing norbornene lysine. For optimal expression, determine the time point when the ubiquitin-fused split intein band is maximized and in vivo hydrolysis is minimized (*see* **Note 11**).

3.2.3 Large-Scale Protein Expression and Purification

Once protein expression has been optimized on a small scale (Subheading 3.2.2), large scale cultures (1 L or more) can be prepared. In this section, we provide a method to scale the culture growth and the purification steps that need to be performed before intein cleavage. This includes cell lysis and affinity column purification of the His-tagged ubiquitin–intein construct. These steps remove truncation products that may be generated due to premature translation termination at the stop TAG codon site. It is important to perform these steps as fast as possible while keeping the protein sample at 4 °C to avoid premature intein cleavage during purification.

1. Start with a frozen cell stock or a freshly transformed plate as described in Subheading 3.2.1 (*see* **Note 12**). Inoculate 10 mL of autoclaved LB media containing chloramphenicol and

ampicillin in a sterile 50 mL culture tube and grow cells overnight (12–16 h) at 37 °C while shaking at 220 rpm.

2. Add 10 mL of saturated preculture to 1 L of sterile LB media containing chloramphenicol and ampicillin in a 4 L Erlenmeyer flask.
3. Incubate the cultures at 37 °C while shaking at 220 rpm until OD_{600} reaches 0.2–0.4. This should take 2.5–4 h.
4. Add 10 mL of the 0.1 M norbornene lysine stock solution to the culture for a final concentration of 1 mM norbornene lysine.
5. Incubate the cultures for 30 min to 1 h until an OD_{600} of 0.4–0.6 has been reached.
6. Take a 250 μL aliquot from each culture for analysis by SDS-PAGE as described in **step** 7 of Subheading 3.2.2.
7. Add 1 mL of 0.5 M of IPTG (for a final concentration of 0.5 mM) and 20% w/v arabinose (for a final concentration of 0.02% w/v) to the culture.
8. Allow the protein to express under the conditions and time determined in Subheading 3.2.2. In our example, we express the protein at 37 °C for 4 h and shaking at 220 rpm.
9. Pour the culture into a 1 L Nalgene centrifuge bottle and pellet the cells by centrifugation at 10,000 × *g* for 30 min.
10. Decant the supernatant and carefully transfer the pellet to a 50 mL centrifuge tube using a plastic scraper. If desired, the pellet can be flash frozen at this point and stored at −80 °C until purification can be resumed.
11. Add 20 mL of cold lysis buffer and 20 μL of 1 M TCEP stock solution to the pellet. While keeping the sample on ice, resuspend the pellet until the solution becomes homogeneous. If necessary, pass the solution through an 18-gauge syringe needle to break any leftover clumps.
12. Transfer the suspension into a 150 mL aluminum beaker and place the beaker on ice. Lyse the cells by sonication. We use a cycle set at 65% amplitude with 30 s on/off intervals for 6 min (*see* **Note 13**).
13. Clarify the lysate by centrifugation at 30,000 × *g* for 1 h at 4 °C.
14. Prepare the Ni-NTA column by pouring 2 mL of Ni-NTA agarose resin (4 mL of stock solution slurry) into a glass or disposable plastic column outfitted with a filter frit and a flow adaptor.
15. The Ni-NTA resin is supplied as a suspension in 20% ethanol. Drain the ethanol from the resin and wash with 10 mL (5 column volumes, CV) of water, followed by 10 mL (5 CV) of lysis

buffer. To avoid premature intein hydrolysis, perform the Ni-NTA purification in a 4–6 °C cold room as quickly as possible.

16. Pour the supernatant of the centrifuged lysate into the column.
17. Mix the resin and lysate by pipetting up and down. Transfer the suspension into a 50 mL centrifuge tube and incubate on a rotator for 1 h at 4 °C to allow the protein to bind to the resin.
18. Return the suspension back to the column to perform the wash and elution steps. An aliquot from each sample should be collected for SDS-PAGE analysis. Drain and collect the flow-through.
19. Perform the following washes:
    (a) 10 mL (5 CV) of lysis buffer—collect the eluent as *Wash 1.*
    (b) 10 mL (5 CV) of wash buffer 1—collect the eluent as *Wash 2.*
    (c) 5 mL (2.5 CV) of wash buffer 2—collect the eluent as *Wash 3.*
20. Elute the protein with 3 mL of elution buffer (1.5 CV). Repeat four times and collect all fractions as *Elution 1*, *2*, *3*, and *4* (*see* **Note 14**).
21. Analyze each fraction by SDS-PAGE and combine the fractions that contain the desired protein (Fig. 3b).

3.3 Intein Cleavage

The following portion of the protocol is split into two discrete sections. If the generation of the native C-terminal carboxylic acid is required, follow the instructions in Subheading 3.3.1. In this case, the intein is used as a purification tag. When the generation of C-terminal thioesters is desired, follow the protocol in Subheading 3.3.2. In our case, we modify the C-terminus of ubiquitin with an aminoethanethiol moiety that can be used to attach ubiquitin to other proteins of interest. In both cases, ubiquitin also contains the unnatural amino acid norbornene lysine at position 6 as introduced above.

3.3.1 Generation of the Native C-Terminus via Hydrolysis

1. Add 1 M TCEP stock to the pooled fractions for a final TCEP concentration of 10 mM. Verify that the pH is ~7.5 using 2–3 μL of the pooled sample and a pH indicator strip.
2. Incubate the sample overnight (12–16 h) at room temperature to allow intein hydrolysis to occur (*see* **Note 15**).
3. Prepare a small amount of a 1:1 solution of the hydrolysis reaction and HPLC solvent A (50–100 μL total volume). Centrifuge to remove any precipitate and load the sample onto an analytical HPLC column. For the ubiquitin sample described here, we use the analytical column described in Subheading

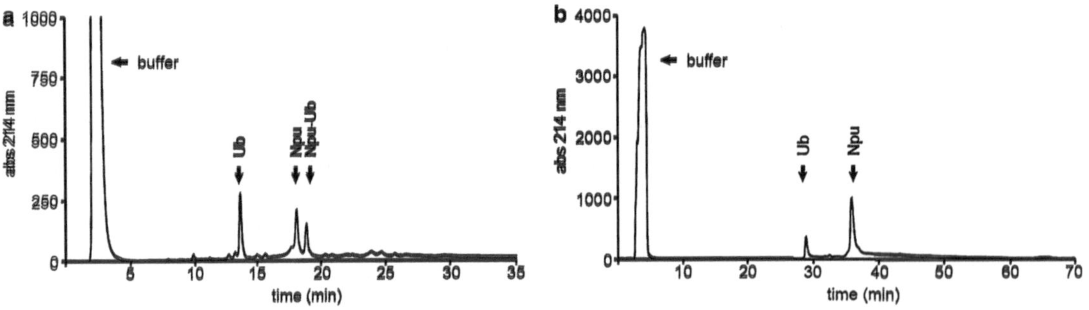

Fig. 4 Analysis and purification of the intein cleavage by reverse-phase HPLC. (**a**) Analytical HPLC chromatogram of an intermediate stage where a fraction of the ubiquitin–intein construct is still intact, and another fraction has hydrolyzed to form ubiquitin and Npu. The chromatogram was collected using a 30–70% HPLC Solvent B gradient over 30 min with a flow rate of 1 mL/min. (**b**) Preparative scale reverse-phase HPLC purification of the fully hydrolyzed ubiquitin–intein construct. Here only two peaks are visible, one corresponding to ubiquitin and the other one to the intein. Fractions containing ubiquitin were collected and lyophilized. Purification was performed with a 5–70% HPLC Solvent B gradient over 60 min with a flow rate of 4 mL/min

2.3.2 with a flow rate of 1 mL/min, a solvent gradient of 30–70% B for 30 min, and UV absorbance detection at 214 nm.

4. The hydrolysis is complete if the ubiquitin–intein peak has disappeared and two separate peaks are visible, one corresponding to the hydrolyzed ubiquitin and the other to the intein (Fig. 4). Confirm the identity of the protein of interest by electrospray mass spectrometry (ESI-MS).

3.3.2 C-Terminal Modification via Thiolysis

1. Add 1 M 2-aminoethanethiol solution and 1 M TCEP solution to the combined Ni-NTA fractions to a final concentration of 100 mM and 10 mM, respectively. Verify that the pH is ~7.5 using 2–3 μL of the pooled sample and a pH indicator strip.
2. Incubate the sample overnight (12–16 h) at room temperature to allow the reaction to complete (*see* **Note 15**).
3. Prepare a small amount of 1:1 solution of the hydrolysis reaction and HPLC solvent A (50–100 μL total volume). Centrifuge to remove any precipitate and load the sample onto an analytical HPLC column. For the ubiquitin sample described here, we use the analytical column described in Subheading 2.3.2 with a flow rate of 1 mL/min, a solvent gradient of 30–70% B for 30 min, and UV absorbance detection at 214 nm.
4. The reaction is complete if the ubiquitin–intein peak has disappeared and two separate peaks are visible, one corresponding to the modified ubiquitin and the other to the intein (Fig. 4). Confirm the identity of the protein of interest by electrospray mass spectrometry (ESI-MS).

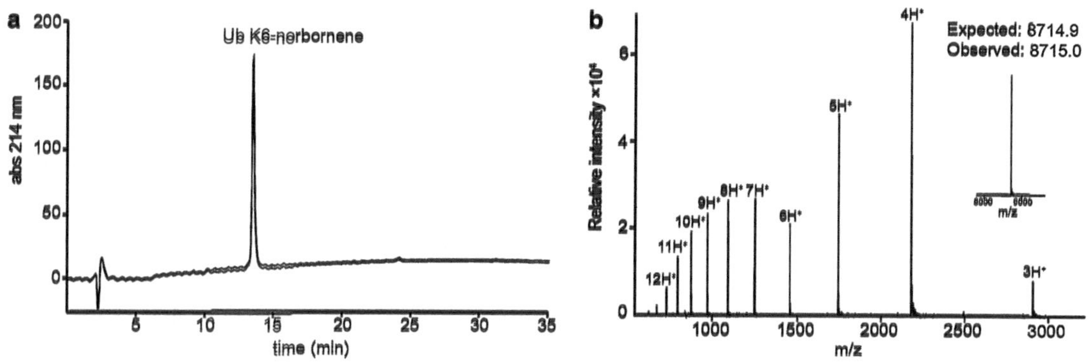

Fig. 5 Analysis of the purified ubiquitin construct containing norbornene lysine at position 6 of the sequence. (**a**) Analytical reverse-phase HPLC chromatogram depicting a single ubiquitin peak. The chromatogram was collected with a 30–70% HPLC Solvent B gradient over 30 min with a flow rate of 1 mL/min. (**b**) ESI-MS analysis of the purified protein. Calculated expected mass is 8714.9, experimentally determined mass is 8715.0

3.4 Preparative-Scale HPLC Purification

The reaction products from the intein cleavage can be easily separated by reverse-phase C18 HPLC chromatography as described below. This purification protocol is compatible with any protein construct that can be lyophilized and refolded.

1. Take the reaction mixture from Subheading 3.3.1 or 3.3.2 and adjust the pH to ~4 by the dropwise addition of concentrated HPLC grade TFA (*see* **Note 16**).
2. Centrifuge the sample to remove any precipitate (*see* **Note 17**), filter through a 0.45 μL HPLC certified syringe-tip filter (e.g., GHP Acrodisc from Pall Corporation) and load the sample onto a preparative-scale HPLC column. For the ubiquitin samples described here, we use the preparative column described in Subheading 2.3.2 with a flow rate of 4 mL/min, a solvent gradient of 30–70% B for 60 min, and UV absorbance detection at 214 nm. Under these conditions, ubiquitin elutes at ~25 min and the intein elutes at ~30 min (Fig. 4b).
3. Collect the desired fractions and verify purity by analytical HPLC and ESI-MS (Fig. 5). Combine the pure fractions.
4. Flash-freeze the sample by dipping the tube into liquid nitrogen and lyophilize. Lyophilized samples can be stored at −80 °C indefinitely. Typical yields for the doubly modified ubiquitin construct described here are ~5 mg/1 L of culture (*see* **Notes 18** and **19**).

3.5 Protein Refolding

Ubiquitin can be easily refolded by dialysis. We have found that the final ubiquitin fold is not perturbed by the presence of the UAA or the C-terminal modification [20]. If working with other proteins, follow their respective refolding steps and, if possible, confirm that the final structure is not perturbed, for example, by circular dichroism (CD), nuclear magnetic resonance (NMR) spectroscopy, X-ray crystallography, or other suitable biophysical methods.

1. Dissolve the lyophilized ubiquitin into 6 M guanidium chloride denaturing buffer and adjust the concentration to 1 mg/mL. In this case, we use 1 mL of protein solution.
2. Centrifuge the sample to remove any precipitation and load the supernatant into a dialysis cassette.
3. Place the cassette into 1 L of refolding buffer at 4 °C and gently stir the solution with a magnetic stir bar for 4–6 h. Replace with fresh refolding buffer and continue dialysis overnight.
4. Carefully remove the solution from the dialysis cassette, centrifuge to remove precipitation, and measure the absorbance at A280 to determine the concentration of protein.

4 Notes

1. Any common LB agar recipe will likely work in this case. The Miller LB agar recipe that we use contains 10 g tryptone, 5 g yeast extract, 10 g sodium chloride, and 15 g agar per liter.
2. Any common LB recipe will likely work in this case. The Miller LB agar recipe that we use contains 10 g tryptone, 5 g yeast extract, and 10 g sodium chloride per liter.
3. A stock solution of 4× buffer will be useful in preparing the lysis buffer, wash buffer 1, wash buffer 2, and elution buffer. To make the 4× stock, prepare a solution containing 200 mM sodium phosphate (monobasic) and 1.2 M sodium chloride. Adjust to a final pH of 7.5.
4. We use cOmplete™ Protease Inhibitor Cocktail EDTA-free tablets (Roche) since they contain a combination of multiple protease inhibitors for serine and cysteine proteases but do not inhibit metalloproteases or aspartic proteases.
5. We use HisPur™ Ni-NTA resin (Thermo Fisher Scientific) and we follow the purification protocols suggested by the manufacturer. If another vendor is used, the protocol should be adapted according to the manufacturer's instructions.
6. Any site-directed mutagenesis protocol or vendor kit should work. Follow the instructions provided by the vendor.
7. Other competent cell lines can be used for expression provided that they do not carry chloramphenicol resistant plasmids.
8. The appearance of colonies and cell culture growth may be slower than usual due to the use of two antibiotics.

9. Working near a flame, take 250 μL of cell culture and add 250 μL of autoclaved 50% glycerol solution. Flash-freeze in liquid nitrogen and store at −80 °C for future protein expression.
10. If SDS-PAGE gel loading buffer with thiols (dithiothreitol or β-mercaptoethanol) is used, the gel analysis may show misleading premature intein cleavage.
11. If substantial in vivo hydrolysis is observed, protein expression may be performed at lower temperature. In our case, we found that the aminoacyl tRNA synthetase does not express well below 30 °C, which affects the efficiency of UAA incorporation. To improve expression of the aminoacyl tRNA synthetase, the final concentration of arabinose may also be increased up to a final concentration of 0.2% w/v.
12. To inoculate from a transformed plate, use a sterilized glass pipette to pick a colony from the plate and dip the pipette into the autoclaved LB media. To inoculate from a frozen cell stock, thaw the stock on ice and use a sterile pipette tip to transfer 2–5 μL of cell stock to the LB media.
13. If premature hydrolysis is an issue, a French press may be used for lysis.
14. If C-terminal modification is desired, we recommend moving to **step 1** in Subheading 3.3.2 as soon as possible. We collect the Ni-NTA elution fractions quickly while working at 4 °C, take a small aliquot for gel analysis and add 2-aminoethanthiol and TCEP immediately. As soon as the thiol is added to the elution fractions, they can be moved to room temperature. This is necessary to prevent premature hydrolysis of the protein before the thiolysis reaction can take place.
15. Since ubiquitin is a stable protein, we routinely incubate overnight. However, the incubation times can be shortened if more unstable proteins are being purified. In this case, we recommend following the hydrolysis or thiolysis reaction by SDS-PAGE or analytical HPLC in real time. If the protein of interest requires denaturants, we also recommend using the fused Cfa intein instead of Npu [15]. On the other hand, if the reaction is not complete after 12–16 h of incubation time, we recommend adding fresh TCEP solution (10 mM) and EDTA to a final concentration of 0.2 mM. To speed up hydrolysis, the pH can also be raised to 8.0 and the sample can be incubated at 37 °C for a few hours.
16. If the protein of interest contains cysteines or C-terminal thioesters, incubate the sample with freshly added TCEP (10 mM) for 30 min at room temperature before the addition of TFA. This prevents undesired disulfide formation and ensures that only monomeric protein is loaded on the column.

17. It is common to observe a significant amount of precipitation at this step. We found that the precipitate contains the intein while ubiquitin remains in solution. Should the protein of interest also precipitate at low pH, the precipitate can be redissolved in 6 M unbuffered guanidinium hydrochloride and loaded onto the preparative HPLC column directly.
18. Based on our experience purifying and modifying different proteins with the protocol described here, several factors can affect the final yield. The most significant factor is the efficiency of the UAA incorporation. For example, higher yields can be obtained with the orthogonal tyrosyl tRNA–tRNA synthetase pair that can incorporate *p*-benzoyl-L-phenylalanine by amber suppression [20, 23]. Second, the expression temperature can be an important factor in preventing in vivo hydrolysis. If possible, we recommend expressing the protein of interest-split intein fusions at 18 °C overnight. Third, ubiquitin contains two glycine residues at the C-terminus and the flexibility imparted by those residues should aid the hydrolysis reaction. Thus, ubiquitin might be more prone to in vivo hydrolysis compared to most proteins.
19. Finally, we note that reverse-phase HPLC purification and refolding are not required for this protocol. If native conditions are to be maintained, the removal of the intein can be performed by reverse Ni-NTA purification, ion exchange, or size exclusion. The removal of leftover reagents such as imidazole and 2-aminoethanethiol can also be performed by dialysis.

References

1. Shah NH, Muir TW (2014) Inteins: nature's gift to protein chemists. Chem Sci 5 (2):446–461. https://doi.org/10.1039/c3sc52951g
2. Volkmann G, Mootz HD (2013) Recent progress in intein research: from mechanism to directed evolution and applications. Cell Mol Life Sci 70(7):1185–1206. https://doi.org/10.1007/s00018-012-1120-4
3. Klabunde T, Sharma S, Telenti A, Jacobs WR, Sacchettini JC (1998) Crystal structure of GyrA intein from Mycobacterium xenopi reveals structural basis of protein splicing. Nat Struct Biol 5(1):31–36. https://doi.org/10.1038/nsb0198-31
4. Liu ZH, Frutos S, Bick MJ, Vila-Perello M, Debelouchina GT, Darst SA, Muir TW (2014) Structure of the branched intermediate in protein splicing. Proc Natl Acad Sci USA 111(23):8422–8427. https://doi.org/10.1073/pnas.1402942111
5. Hirata R, Ohsumi Y, Nakano A, Kawasaki H, Suzuki K, Anraku Y (1990) Molecular-structure of a gene, Vma1, encoding the catalytic subunit of H^+-translocating adenosine-triphosphatase from vacuolar membranes of saccharomyces-cerevisiae. J Biol Chem 265 (12):6726–6733
6. Kane PM, Yamashiro CT, Wolczyk DF, Neff N, Goebl M, Stevens TH (1990) Protein splicing converts the yeast Tfp1 gene-product to the 69-Kd subunit of the vacuolar H^+-adenosine triphosphatase. Science 250(4981):651–657. https://doi.org/10.1126/science.2146742
7. Shah NH, Dann GP, Vila-Perello M, Liu ZH, Muir TW (2012) Ultrafast protein splicing is common among cyanobacterial split inteins: implications for protein engineering. J Am

Chem Soc 134(28):11338–11341. https://doi.org/10.1021/ja303226x

8. Wu H, Hu ZM, Liu XQ (1998) Protein trans-splicing by a split intein encoded in a split DnaE gene of Synechocystis sp. PCC6803. Proc Natl Acad Sci USA 95(16):9226–9231. https://doi.org/10.1073/pnas.95.16.9226
9. Carvajal-Vallejos P, Pallisse R, Mootz HD, Schmidt SR (2012) Unprecedented rates and efficiencies revealed for new natural split inteins from metagenomic sources. J Biol Chem 287 (34):28686–28696. https://doi.org/10.1074/jbc.M112.372680
10. Debelouchina GT, Muir TW (2017) A molecular engineering toolbox for the structural biologist. Q Rev Biophys 50:e7. https://doi.org/10.1017/S0033583517000051
11. Southworth MW, Amaya K, Evans TC, Xu MQ, Perler FB (1999) Purification of proteins fused to either the amino or carboxy terminus of the Mycobacterium xenopi gyrase A intein. Biotechniques 27(1):110–114. https://doi.org/10.2144/99271st04
12. Batjargal S, Walters CR, Petersson EJ (2015) Inteins as traceless purification tags for unnatural amino acid proteins. J Am Chem Soc 137 (5):1734–1737. https://doi.org/10.1021/ja5103019
13. Frutos S, Goger M, Giovani B, Cowburn D, Muir TW (2010) Branched intermediate formation stimulates peptide bond cleavage in protein splicing. Nat Chem Biol 6 (7):527–533. https://doi.org/10.1038/Nchembio.371
14. Chong SR, Mersha FB, Comb DG, Scott ME, Landry D, Vence LM, Perler FB, Benner J, Kucera RB, Hirvonen CA, Pelletier JJ, Paulus H, Xu MQ (1997) Single-column purification of free recombinant proteins using a self-cleavable affinity tag derived from a protein splicing element. Gene 192(2):271–281. https://doi.org/10.1016/S0378-1119(97)00105-4
15. Stevens AJ, Brown ZZ, Shah NH, Sekar G, Cowburn D, Muir TW (2016) Design of a split intein with exceptional protein splicing activity. J Am Chem Soc 138(7):2162–2165. https://doi.org/10.1021/jacs.5b13528
16. Stevens AJ, Sekar G, Shah NH, Mostafavi AZ, Cowburn D, Muir TW (2017) A promiscuous split intein with expanded protein engineering applications. Proc Natl Acad Sci USA 114 (32):8538–8543. https://doi.org/10.1073/pnas.1701083114
17. Wang L, Brock A, Herberich B, Schultz PG (2001) Expanding the genetic code of Escherichia coli. Science 292(5516):498–500. https://doi.org/10.1126/science.1060077
18. Hoffmann JE, Plass T, Nikic I, Aramburu IV, Koehler C, Gillandt H, Lemke EA, Schultz C (2015) Highly stable trans-cyclooctene amino acids for live-cell labeling. Chem-Eur J 21 (35):12266–12270. https://doi.org/10.1002/chem.201501647
19. Chatterjee C, McGinty RK, Fierz B, Muir TW (2010) Disulfide-directed histone ubiquitylation reveals plasticity in hDot1L activation. Nat Chem Biol 6(4):267–269. https://doi.org/10.1038/Nchembio.315
20. Debelouchina GT, Gerecht K, Muir TW (2017) Ubiquitin utilizes an acidic surface patch to alter chromatin structure. Nat Chem Biol 13(1):105–110. https://doi.org/10.1038/nchembio.2235
21. Lim BJ, Ackermann BE, Debelouchina GT (2019) Targetable tetrazine-based dynamic nuclear polarization agents for biological systems. Chembiochem https://doi.org/10.1002/cbic.201900609
22. Lang K, Davis L, Torres-Kolbus J, Chou CJ, Deiters A, Chin JW (2012) Genetically encoded norbornene directs site-specific cellular protein labelling via a rapid bioorthogonal reaction. Nat Chem 4(4):298–304. https://doi.org/10.1038/nchem.1250
23. Chin JW, Martin AB, King DS, Wang L, Schultz PG (2002) Addition of a photocrosslinking amino acid to the genetic code of Escherichia coli. Proc Natl Acad Sci USA 99 (17):11020–11024. https://doi.org/10.1073/pnas.172226299

Chapter 9

Streamlined Expressed Protein Ligation: Site-Specific Antibody–Drug Conjugate

Silvia Frutos

Abstract

Protein semisynthesis is a powerful tool for studying proteins and has contributed to a better understanding of protein structure and function and also driven innovations in protein science. Expressed protein ligation (EPL) is a widely used method to generate chemically modified proteins. However, EPL has some limitations, particularly relevant to modify challenging proteins such as antibodies. The method termed streamlined expressed protein ligation (SEPL) overcomes some of the problems of EPL, and other methods of protein semisynthesis, to generate challenging modified proteins such as antibody–drug conjugates (ADCs). ADCs targeting highly cytotoxic molecules to cancer cells, offer an attractive strategy to selectively eliminate tumor cells with improved therapeutic index than the antibodies or cytotoxic molecules themselves. Despite the potential of ADCs, the development of such complex molecules is challenging. We provide here protocols to prepare site-specifically modified ADCs by streamlined expressed protein ligation (SEPL), which does not require the incorporation of unnatural modifications into the antibody. Therefore, fully native antibodies, with only the desired cytotoxic molecules attached, can be generated.

Key words Site-specific conjugation, SEPL, Antibody–drug conjugate, protein modification, Expressed protein ligation

1 Introduction

Expressed protein ligation (EPL) methods, based on the use of inteins to generate protein thioester intermediates, have been used to modify several types of proteins including transmembrane receptors [1], enzymes [2], transcription factors [3], ion channels [4], and antibodies [5]. However, despite the successes of EPL, the realization of its full potential is limited by the properties of the inteins used to generate the key protein thioester intermediate. Some of these limitations include slow reaction kinetics of thioester formation, premature cleavage reactions, and dependence on extein sequence.

To address the main caveats of the intein thiolysis process in standard EPL protocols, a technology called streamlined expressed

Miquel Vila-Perelló (ed.), *Expressed Protein Ligation: Methods and Protocols*, Methods in Molecular Biology, vol. 2133, https://doi.org/10.1007/978-1-0716-0434-2_9, © Springer Science+Business Media, LLC, part of Springer Nature 2020

Fig. 1 Reaction scheme for the synthesis of ADCs via SEPL

protein ligation (SEPL) [6], based on the use of ultrafast split inteins, was developed (Fig. 1). This strategy relies on the use of split inteins to generate the required protein thioester for EPL. Split inteins consist of two polypeptides, termed InteinN (IntN) and InteinC (IntC), which upon binding, catalyze protein splicing in *trans* [7], in contrast to commonly used inteins in standard EPL, which are contiguous polypeptides that catalyze protein splicing in *cis*. The trans-splicing reaction of naturally split inteins is initiated by a very tight protein-protein interaction, with very fast on-rates. Ultrafast split inteins have very fast reaction kinetics and high splicing efficiency which overcome problems of contiguous inteins traditionally used in standard EPL [8, 9].

The use of split inteins as thioester surrogates also solves one of the major limitations of standard EPL, which is premature cleavage during recombinant expression and initial purification of intein fusions. Such premature cleavage can be a significant side reaction, particularly for expression systems and proteins, such as antibodies, that require long incubations (days) to obtain the required intein fusions. In the case of a split intein, premature cleavage cannot take place because the intein fragment (IntN), fused to the protein of interest, is completely inactive, and so it can be stably produced without any risk of undesired side reactions [10, 11]. This makes ultrafast split inteins and therefore SEPL an attractive tool for protein semisynthesis of challenging substrates such as ADCs.

ADCs combine the properties of antibodies (selective binding to tumor, long half-lives, low toxicity in circulation and induced cell toxicity upon receptor binding) with the lethality of cancer-cell killing agents (cytotoxic drugs) to specifically kill cancer cells [12, 13]. Despite all the advantages of ADCs, the development of these molecules is challenging because of their molecular complexity. Importantly, the lack of chemical control and the heterogeneity of ADCs obtained via traditional bioconjugation strategies has tremendously impacted their development [13]. Several methods of site-specific modification of proteins have been developed over the last years and applied to ADCs. Most of them are based on the incorporation of an orthogonal chemical handle on the antibody for its subsequent modification, such as genetic incorporation of unnatural amino acids [14, 15], enzymatic modification of side chains [16–18], manipulation of the glycans [19], transpeptidase based approaches [20] or engineering of surface exposed Cys residues [21]. Although these strategies do provide access to homogeneous ADCs, they all have their specific limitations: premature release of cytotoxic payloads observed for cysteine substituted ADCs, need for additional enzymatic processing steps for enzymatic methods, and low expression yields of the antibody for the genetically incorporated unnatural amino acids methods [22].

SEPL overcomes problems and limitations of existing methods for the development of therapeutic ADCs and others protein conjugates. SEPL does not require the coexpression of any other enzymes or cofactors neither the reengineering of the protein synthesis machinery of the expressing cell line. SEPL has been successfully used to conjugate different types of cargos to antibodies and antibody fragments, as well as for the generation of homogeneous and highly potent ADCs [23, 24].

Here we provide an SEPL protocol to generate a site-specific ADC with DAR 2. The protocol describes the synthesis of an anti-Her2-Auristatin F conjugate, as an example and can be easily adapted to conjugate any other tumor targeting antibody to a cytotoxic molecule. The chapter includes protocols for the expression of the IntC fragment in *E. coli*, cloning and expression of antibody–intein fusions, synthesis of the cytotoxic cargo, and conjugation reaction. Additionally, protocols for the purification of the resulting ADC and its analytical characterization are also provided.

2 Materials

All solutions and reagents are prepared using analytical grade reagents and high-purity salts, buffers, HPLC grade solvents, and ultrapure water.

2.1 InteinC Fragment Production (IntC-OMe)

1. IntC-GyrA-His6 expression plasmid (*see* **Note 1**).
2. BL21(DE3) *E. coli* cells.
3. LB medium.
4. 1 M IPTG (Isopropyl β-D-1-thiogalactopyranoside): 1 M IPTG in water.
5. 1 M TCEP (tris(2-carboxyethyl)phosphine): 1 M TCEP in 1 M Tris in water, pH 8.
6. Nickel resin for immobilized metal affinity chromatography.
7. Lysis buffer: 50 mM sodium phosphate, 300 mM NaCl, 5 mM imidazole, 1 mM TCEP in water and pH 7.5.
8. Wash buffer: 50 mM sodium phosphate, 300 mM NaCl, 20 mM imidazole in water and pH 8.0.
9. Elution Buffer: 50 mM sodium phosphate, 300 mM NaCl, 250 mM imidazole in water and pH 8.0.

10. L-Cysteine methyl ester hydrochloride (Cys-OMe): 1M Cys-OMe in water.
11. MES running buffer: 50 mM MES, 50 mM Tris-Base, 0.1% SDS (Sodium dodecyl sulfate), 1 mM EDTA (ethylenediaminetetraacetic acid) in water and pH 7.3.
12. Coomassie staining solution: 1.2 mM Brilliant Blue R, 45% MeOH, 10% acetic acid in water.
13. Destaining solution: 15% MeOH, 10% mL acetic acid in water.
14. SDS-PAGE loading dye: 5% β-Mercaptoethanol, 0.02% bromophenol blue, 30% glycerol, 10% SDS, 250 mM Tris–HCl pH 6.8 in water.
15. Mobile phase A: 0.1% trifluoroacetic (TFA) acid in water.
16. Mobile phase B: 0.1% trifluoroacetic acid (TFA) in acetonitrile.
17. HPLC instrument (*see* **Note 2**).
18. Sonicator: 13 mm probe, 40% amplitude.
19. Empty plastic column.
20. Rotary shaker.

2.2 Cloning of InteinN

1. pFUSE-IgG heavy-chain expression plasmids (*see* **Note 3**).
2. DNA plasmid encoding the N-terminal fragment of the intein (IntN) (*see* **Note 4**).
3. Forward and reverse primers to introduce the DNA encoding the IntN at the heavy-chain expression plasmid (*see* **Note 5**).
4. Phusion high-fidelity DNA polymerase (2 U/mL).
5. DpnI restriction enzyme (10 U/μL).
6. Gel extraction kit.
7. Subcloning grade *E. coli* competent cells (*see* **Note 6**).
8. Miniprep kit.
9. 100 mg/mL Zeocin.
10. Autoclaved low salt LB medium.
11. Low salt LB-agar plates containing 25 μg/mL of zeocin antibiotic.
12. TAE buffer: 40 mM Tris base, 20 mM acetic acid, and 1 mM EDTA in water.

2.3 Fused Protein Production

1. Light-chain expression plasmids and IgG-IntN-heavy chain expression plasmids (*see* **Note 7**).
2. ExpiFectamine 293 expression and transfection kit (*see* **Note 8**).
3. Protein G-Sepharose beads.

4. Phosphate buffered saline (PBS): 137 mM NaCl, 2.7 mM KCl, 10 mM Na_2HPO4, 1.8 mM KH_2PO4 in water. Adjust pH to 7.4 with 6 N HCl.
5. 1 M EDTA: 1 M EDTA in water.
6. Elution buffer: 0.1 M glycine in water and pH 2.7.
7. Neutralization buffer: 1 M Tris–HCl in water and pH 8.0.
8. 8% Bis-tris SDS-PAGE gels.
9. MES running buffer: 50 mM MES, 50 mM Tris base, 0.1% SDS, 1 mM EDTA in water and pH 7.3
10. Coomassie staining solution: 1.2 mM Brilliant Blue R, 45% MeOH, 10% acetic acid in water.
11. Destaining solution: 15% MeOH, 10% mL acetic acid in water.
12. SDS-PAGE loading dye: 5% β-mercaptoethanol, 0.02% bromophenol blue, 30% glycerol, 10% SDS (Sodium dodecyl sulfate), 250 mM Tris–HCl pH 6.8 in water.
13. Dialysis bags.
14. Orbital shaker platform.
15. Hemocytometer.

2.4 Cargo Synthesis

1. HBTU (2-(1H-benzotriazol-1-yl)-1,1,3,3-tetramethyluronium hexafluorophosphate).
2. HOBt (1-Hydroxybenzotriazole hydrate).
3. HOAt (1-Hydroxy-7-azabenzotriazole).
4. EDC (*N*-(3-Dimethylaminopropyl)-*N′*-ethylcarbodiimide hydrochloride).
5. DIEA (*N*,*N*-Diisopropylethylamine).
6. Pyridine.
7. 2-Chlorotrityl resin 0.25 mmol/g (2Cl-Trt).
8. Diaminoethylene glycol.
9. Boc-Cys(Trt)-OH.
10. Reversed-phase column: cartridge of 2 g sorbent and 80 μm particle size.
11. Auristatin F (*see* **Note 9**).
12. Deprotection mixture: 40% TFA, 2.5% of triisopropylsilane (TIS) in dichloromethane (DCM).
13. C_{18} analytical (4.6 mm × 150 mm, 5 μm) and preparative (1 cm × 250 mm, 5 μm) reversed-phase column.
14. Mobile phase A: 0.1% trifluoroacetic (TFA) acid in water.
15. Mobile phase B: 0.1% trifluoroacetic acid (TFA) in acetonitrile.

2.5 Conjugation

1. Conjugation buffer: PBS pH 7.4 and 1 mM EDTA.
2. Superdex 200 Increase 10/300 GL column.
3. Elution buffer: PBS (137 mM NaCl, 2.7 mM KCl, 10 mM Na_2HPO4, 1.8 mM KH_2PO4 in water. Adjust pH to 7.4 with 6 N HCl).
4. InteinC stock: 500 μM IntC-OMe in water (obtained in Subheading 3.1).
5. IgG-InteinN stock: 1 mg/mL in PBS (obtained in Subheading 3.3)
6. Cys-Auristatin F stock: 10 mM Cys-AF (obtained in Subheading 3.4) in water.
7. TCEP stock: 1 M in Tris in water pH 8.
8. 8% Bis-tris SDS-PAGE gels.
9. MES running buffer: 50 mM MES, 50 mM Tris-Base, 0.1% SDS, 1 mM EDTA in water and pH 7.3.
10. Coomassie staining solution: 1.2 mM Brilliant Blue R, 45% MeOH, 10% acetic acid in water.
11. Destaining solution: 15% MeOH, 10% mL acetic acid in water.
12. SDS-PAGE loading dye: 5% β-Mercaptoethanol, 0.02% Bromophenol blue, 30% Glycerol, 10% SDS (Sodium dodecyl sulfate), 250 mM Tris–HCl pH 6.8 in water.
13. Nonreduced loading dye: 0.02% Bromophenol blue, 30% Glycerol, 10% SDS (Sodium dodecyl sulfate), 250 mM Tris–HCl pH 6.8 in water.

2.6 RP-HPLC and Mass Spectrometric (LC-MS) Analysis

1. Zorbax 300SB C8 column (4.6 mm × 100 mm, 3.5 μm).
2. HPLC instrument (*see* **Note 2**).
3. Mobile phase A: 0.25% formic acid and 0.02% TFA in water.
4. Mobile phase B: 90% isopropanol in water with 0.25% formic acid and 0.02 TFA.
5. Deglycosylation enzyme: PNGase F (500 U/μL).
6. 6 M GuHCl (Guanidine hydrochloride): 6H GuHCl in PBS.
7. 1 M DTT (dithiothreitol): 1 M DTT in water.
8. ESI-MS mass detector.
9. C_{18} column (75 mm × 10 mm, 2.0 μm).
10. Deconvolution software.
11. SEPL conjugate (IgG-Auristatin F) (obtained in Subheading 3.5).

2.7 Hydrophobic Interaction Chromatography (HIC) analysis

1. TSK-GEL® Butyl-NPR column (4.6 mm × 3.5 cm) (Tosoh Bioscience).
2. HIC buffer A: 50 mM sodium potassium phosphate pH 7.0 and 1.5 M ammonium sulfate.
3. HIC buffer B: 50 mM sodium potassium phosphate pH 7.0 and 20% isopropanol.
4. HPLC instrument (*see* **Note 2**).
5. Site-specific SEPL conjugate (IgG-Auristatin F) (obtained in Subheading 3.5).
6. Nonspecific ADC (e.g., T-DM1, an anti-Her2 antibody conjugate to emtansine) (*see* **Note 10**).

3 Methods

3.1 InteinC Fragment Production (IntC-OMe)

The inteinC fragment is used as a reagent in the conjugation reaction to reconstitute the active intein domain and generate the thioester in situ. In this section a protocol to obtain the IntC fragment is provided, based on the expression and purification of a GyrA intein fusion, cleavage of the desired IntC and purification by RP-HPLC.

1. Transform BL21(DE3) *E. coli* cells with the IntC-GyrA-His6 plasmid. Add 1 μL of the plasmid to 50 μL *E. coli* competent cells and incubate the cells on ice for 20 min. Incubate for 45 s at 42 °C and transfer reaction tube back to ice for 2 min. Add 900 μL of LB-medium to the tube and keep at 37 °C incubator shaker for 1 h. Plate *E. coli* cells in a LB plate containing 100 μg/mL of ampicillin. Incubate the plate at 37 °C for 18 h.
2. Inoculate 10 mL of LB containing 100 μg/mL of ampicillin with a colony from the LB plate and incubate the culture at 37 °C for 18 h.
3. Add the 10 mL preculture from **step 2** to 1 L of autoclaved LB containing 100 μg/mL of ampicillin. The cells are then grown at 37 °C and 200 rpm until optical density, measured at 600 nm, is 0.6 ($OD_{600} = 0.6$).
4. Induce the expression of the target protein by addition of 0.5 mM IPTG from 1 M IPTG stock for 3 h at 37 °C. Harvest the cells by centrifugation at 4000 × *g*, 4 °C for 20 min. Discard supernatant and keep cell pellets. Cell pellets can also be stored at −20 °C.
5. Resuspend the cell pellet by adding 20 mL of lysis buffer at 4 °C.
6. Lyse cells by sonication (40% amplitude, 8 × 20 s pulses separated by 30 s on ice) and recover the soluble fraction by

centrifugation at 28,000 × *g* and 4 °C for 20 min. Discard the pellet and keep the supernatant for **step 7**.

7. Add 8 mL of slurry nickel resin to an empty plastic column. Wash the beads with 10 CV of water and 10 CV with lysis buffer. Mix the supernatant (soluble fraction) with the beads. Close the column using two caps and mix gently the suspension by shaking at 200 rpm on a rotary shaker, at 4 °C for 60 min.
8. Remove the column caps and collect the flow-through. Wash the column with 5 CV of lysis buffer and 5 CV of wash buffer. Collect the washes fractions in 15 mL tubes. Elute the fused protein (IntC-GyrA-H6) with elution buffer in four 1.5 CV elution fractions. Collect them in 15 mL tubes.
9. Analyze fractions by 8% Bis-tris SDS-PAGE gel. Mix 5 μL of each fraction with 5 μL of water and 2 μL of loading dye. Incubate samples at 95 °C for 5 min and load them into the gel. The SDS-PAGE is run at 165 V for 35 min using the MES running buffer. Stain the gel for 10 min with the staining solution, and destain it for 4 h at room temperature with the destaining solution.
10. Pool fractions containing the protein into a 50 mL tube. Carry out in situ intein cleavage and ligation to Cys-OMe by adding 100 mM Cys-OMe from 1 M stock and 5 mM TCEP from 1 M TCEP stock. Adjust the pH at 7.5.
11. Monitor the reaction by RP-HPLC. Mix 20 μL of reaction, 40 μL 6 M GnHCl +0.1% TFA and 40 μL buffer A. Inject 10 μL and analyze over a 0–73% gradient in C_{18} reversed-phase column.
12. Purify the product by injecting the reaction mixture to a C_{18} preparative RP-HPLC with a linear gradient of 25–40% mobile phase B over 60 min.
13. Combine pure fractions and lyophilize them to obtain IntC-OMe fragment.

3.2 Cloning of IntN (IgG-IntN)

The inteinN fragment should be cloned at the C-terminus of the protein of interest. In this case, the protein of interest is an antibody and the InteinN is cloned at the C-terminus of the heavy chain, in order to obtain the antibody–intein fusion. The inteinN fragment is cloned into the target vector using a restriction enzyme free cloning strategy [25].

1. Set up 50 μL PCR reaction containing Phusion DNA polymerase buffer, 200 μM dNTP mixture, 0.5 μM of each forward and reverse primers, 10 ng of intein (IntN) encoding plasmid, and 0.5 μL of Phusion DNA polymerase (0.2 U/mL).
2. Carry out PCR reaction using the following cycle: 30 s denaturation at 95 °C, followed by 25 cycles of 5 s denaturation at

95 °C, 10 s annealing at 50 °C, and 15 s extension at 72 °C, followed by a final incubation of 5 min at 72 °C.

3. Purify the PCR product by electrophoresis. Add 10 μL of loading dye to the PCR product from **step 2**. Load the mixture into a 1.5% agarose gel and run it using TAE buffer at 90 V for 30 min.
4. Cut and purify the DNA band using a gel extraction kit to obtain the PCR product that will be used in the subsequent insertion reaction as a megaprimer.
5. Set up 25 μL PCR reaction containing Phusion DNA polymerase buffer, 200 μM dNTP mixture, 300 ng of the megaprimer from **step 4**, 10 ng of IgG Heavy chain plasmid, and 0.5 μL of Phusion DNA polymerase (2 U/mL) (*see* **Note 11**).
6. Carry out PCR reactions using the following cycle: 2 min of denaturation step at 95 °C, followed by 25 cycles of 30 s denaturation at 98 °C, 1 min annealing at 52 °C, and 9 min extension at 72 °C, followed by a cycle of 10 min at 72 °C (*see* **Note 12**).
7. Add 1 μL of DpnI to each PCR sample and incubate at 37 °C for 3 h to digests the methylated parental plasmid.
8. Add 10 μL of above DpnI-treated sample to 50 μL *E. coli* competent cells and incubate the cells on ice for 30 min. Incubate the cells for 45 s at 42 °C and transfer reaction tube back to ice for 2 min. Add 900 μL of LB-medium to the tube and keep at 37 °C incubator shaker for 1 h at 200 rpm.
9. Plate *E. coli* cells on low salt LB-agar plates containing 25 μg/mL of zeocin. Incubate the plate at 37 °C for 18 h.
10. Inoculate a single colony in 5 mL low salt-LB medium containing 25 μg/mL Zeocin and grow the culture for 18 h at 37 °C and 200 rpm.
11. Isolate the plasmid DNA using a miniprep isolation kit and perform DNA sequencing to verify desired inteinN sequence has been cloned at the right site.

3.3 Fused Protein Production

The fused antibody–inteinN (IgG-IntN) is produced using a mammalian transient expression system and purified by affinity chromatography. The antibody–intein fusion will be used as a substrate at the subsequent SEPL conjugation reaction.

1. Seed 2.5 × 10^6 viable Expi293 cells/mL and allow them to grow overnight at 37 °C and 8% CO_2 on an orbital shaker platform.
2. Determine viable cell density with a hemocytometer. The cells should have reached a density of approximately 4.5–5.5 × 10^6 cells/mL.

3. Dilute the cells into 25.5 mL fresh Expi293 expression medium (125 mL shaker flask) prewarmed to 37 °C to a final density of 3×10^6 viable cells/mL.
4. Dilute 10 μg of IgG-light chain and 20 μg of IgG-IntN heavy chain into a total of 1.5 mL of Opti-MEM medium (*see* **Note 13**).
5. Mix gently the transfection reagent. Dilute 80 μL of transfection reagent into 1.4 mL of Opti-MEM medium. Incubate at room temperature for 5 min.
6. Add the transfection reagent to the DNA and mix gently 3 times pipetting up and down slowly.
7. Incubate transfection reagent/DNA complexes at room temperature for 20 min.
8. Slowly transfer the solution from **step** 7 to the 125 mL shaker flask (prepared in **step 3**), swirling the flask gently during addition. Incubate the cells in a 37 °C incubator with 8% CO_2 in air on an orbital shaker for 20 h.
9. Add 150 μL of transfection enhancer 1 and 1.5 mL of transfection enhancer 2 to the flask and returned the flask to the incubator for 5 days.
10. Spin down the cells at 2000 ×*g* for 20 min at 4 °C and transfer the supernatant to a dialysis bag (10 KDa cutoff). Dialyze the sample against 1 L of PBS for 8 h at 4 °C.
11. Transfer the dialyzed sample to 50 mL tube and add 0.2 mL of Protein G-sepharose beads, previously equilibrated with PBS, and incubate the tube on a rotator at 4 °C overnight.
12. Transfer the beads to a small column and collect the flow-through in a 50 mL tube. Wash the beads with 15 CV (3.75 mL) of PBS and collect the fraction in 15 mL tube. Elute the antibody with 10 CV of elution buffer. Collect each CV in an Eppendorf tube containing 25 μL of 1 M Tris–HCl pH 8 to neutralize the sample.
13. Analyze by SDSPAGE. Mix 10 μL of each fraction with 2 μL of loading dye. Incubate samples at 95 °C for 10 min and load them into the gel. The SDS-PAGE is run at 165 V for 35 min using the MES running buffer. Stain the gel for 10 min with the staining solution and destain it for 4 h at room temperature with destaining solution.
14. Pool fractions containing the fused antibody and dialyze the resulting sample into PBS and 1 mM EDTA at 4 °C overnight.
15. Concentrate the sample to a final concentration of 1 mg/mL with an ultrafiltration concentrator. The sample can be stored at −80 °C.

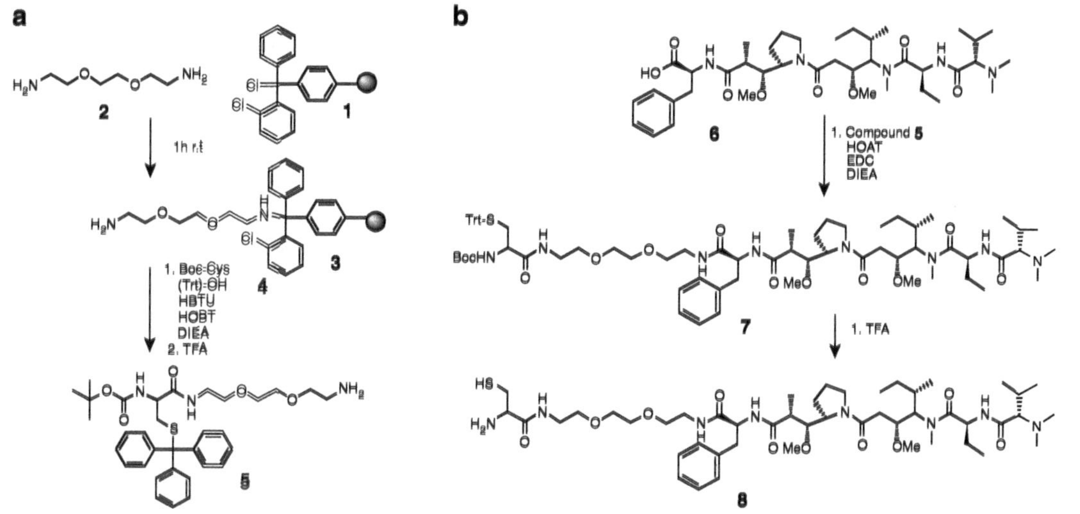

Fig. 2 (**a**) Scheme for the synthesis of Boc-Cys(Trt)-NH-diethylene glycol-NH_2. (**b**) Scheme for the synthesis of Cys-diethylene glycol-Auristatin F (Cys-AF)

3.4 Cargo Synthesis

In order to conjugate Auristatin F to an antibody using SEPL it needs to be modified to introduce an N-terminal Cys residue. This section provides a protocol for the modification of Auristatin F to incorporate an N-terminal Cys and a noncleavable polyethylene glycol linker (Fig. 2).

3.4.1 Synthesis of Boc-Cys(Trt)-NH-Diethylene Glycol-NH_2

1. Place 0.25 mmol (200 mg) of 2Cl-Trt resin (**1**) on a glass reaction vessel. Wash the resin with 5 mL of DCM three times and 5 mL of DMF three times.
2. Add 40 eq. of diaminoethylene glycol (**2**) (1.6 mL). Let the reaction go for 1 h to obtain the intermediate (**3**).
3. Block un-reacted sites by washing resin 3 times with 3 mL of DCM/MeOH/DIEA (17:2:1 v/v/v), 2 times with 5 mL of DCM and 3 times with 5 mL of DMF.
4. Couple the amino acid Boc-Cys(Trt)-OH (**4**) dissolving 5 eq. amino acid (579.5 mg), 4.5 eq. HBTU (426.3 mg), 5 eq. HOBt (201 mg) and 10 eq. DIEA (436 μL) in 1 mL of DMF. Incubate the mixture 5 min at room temperature and add the mix to the resin. Let the reaction at room temperature for 1 h with occasional agitation. Wash the resin 3 times with 5 mL of DMF and 5 times with 5 mL of DCM.
5. Carry out the cleavage from the resin by using 7.5 mL of 1% TFA/DCM for 2 min, 10 times. Collect each filtrate in 50 mL tubes containing 75 μL of TIS. Immediately add 25 mL of MeOH and 100 μL pyridine to each tube.

6. Transfer the collected fractions to a round bottom flask and rota-evaporate the MeOH. Dilute the residue formed with 20 mL acetonitrile. Transfer the solution to a 50 mL tube and lyophilize the product.
7. Purify the Boc-Cys(Trt)-NH-diethylene glycol-NH_2 **(5)** using a cartridge of reversed-phase resin (*see* **Note 14**).

3.4.2 Synthesis of Cys-Diethylene Glycol-Auristatin F (Cys-AF)

1. Dissolve 5 mg of Auristatin F **(6)** (6.7 μmols) in 130 μL of DMF. Add 1.58 mg of HOAt, 1.49 μL of EDC, and 2.96 μL of DIEA.
2. Mix and add the entire solution to the lyophilized Boc-Cys (Trt)-linker **(5)** (7 mg). Incubated at room temperature for 1 h.
3. Once the reaction is complete, add 20 mL of ethyl acetate (EtOAc). Transfer the resulting solution to a separatory funnel for a liquid-liquid extraction. Remove the excess of water-soluble reagents by two extractions with 20 mL of 100 mM HCl in water. The organic fraction containing the desired product **(7)** is collected.
4. Rota-evaporate the organic fraction and dissolve the residue formed in 50% water/acetonitrile and lyophilize.
5. Add 1 mL of 40% TFA/DCM containing 2.5% TIS to the lyophilized Boc-Cys(Trt)-linker-AF **(7)** to remove the protecting groups. The reaction proceeds for 1 h at room temperature with occasional agitation.
6. Evaporate the TFA and dilute the residue formed with 20 mL acetonitrile. Transfer the solution to a 50 mL tube and lyophilize the product.
7. Purify the Cys-AF product **(8)** by injecting the reaction crude to a C_{18} preparative RP-HPLC with a linear gradient of 20–70% mobile phase B over 60 min.

3.5 Conjugation

In this section, the protocol to conjugate the Auristatin F (Cys-AF) from Subheading 3.4 to the antibody intein (IgG-IntN) from Subheading 3.3, is provided.

1. Prepare a solution of all the reagents required for the reaction in ligation buffer (PBS + 1 mM EDTA) except the antibody. For a reaction mixture of 1 mL the solution is prepared by mixing 549 μL of H_2O, 100 μL of PBS-EDTA 10×, 2 equivalents of IntC from a 500 μM stock (50 μL), 1 mM Cys-AF from a 10 mM stock (100 μL), 200 mM of MESNa from 1 M stock (200 μL) and 1 mM TCEP from 1 M stock (1 μL).
2. Mix the solution from **step 1** with the same volume of antibody obtained in Subheading 3.3. Adjust pH to 7.5 with 0.5 M NaOH and incubate the reaction in the dark at room temperature for 24 h.

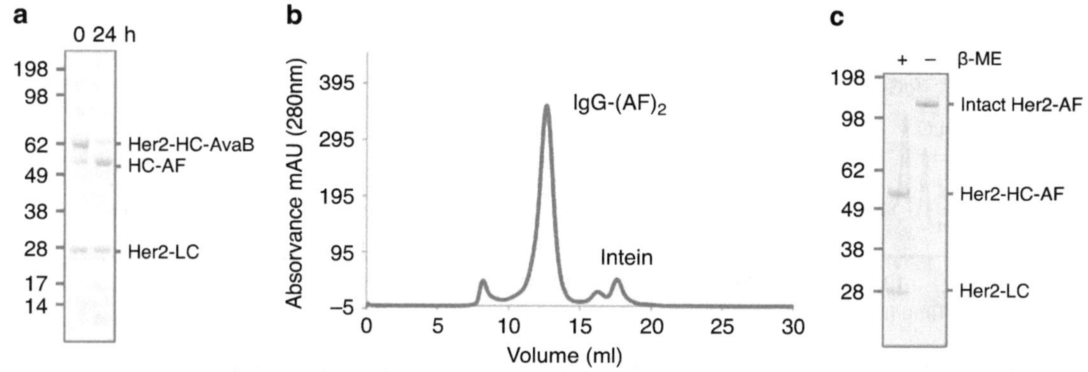

Fig. 3 (**a**) Coomassie stained SDS-PAGE analysis, under reducing conditions, of one-pot thiolysis/ligation reaction for 24 h at rt. (**b**) Purification of the Her2-Auristatin F ADC by size exclusion chromatography (SEC). (**c**) Coomassie-stained SDS-PAGE analysis of purified Her2-AF under reducing (+) and nonreducing (−) conditions

3. Monitor the reaction by 8% Bis-tris SDS-PAGE gel. Mix 3 μL of each time points with 7 μL of water and 2 μL of loading dye. Incubate samples at 95 °C for 10 min and load them into the gel. Run the gel at 165 V for 35 min using the MES running buffer. The gel is stained for 10 min with the Coomassie staining solution and it is destained for 4 h at room temperature with destaining solution (Fig. 3a).
4. Remove excess of reagents such as MESNa and TCEP, and cargo by dialyzing the sample into PBS pH 7.4 at 4 °C for 16 h.
5. Purify the resulting ADC by size-exclusion chromatography. Load the sample on a Superdex 200 10/300 chromatography column equilibrated with PBS buffer. Elute the sample with PBS buffer (*see* **Note 15**). Monitor the protein peak using UV280 absorption (Fig. 3b).
6. Pool the pure fractions and concentrate down to 3 mg/mL. Quantify the amount of conjugate obtained by absorbance at 280 nm (*see* **Note 16**). Store the sample at −80 °C.
7. Analyze the final ADC by 8% Bis-tris SDS-PAGE gel under reducing and nonreducing conditions as is illustrated at Fig. 3c. Mix 1 μL of ADC with 9 μL of water and either 2 μL of loading dye or 2 μL of nonreducing loading dye. Incubate samples at 95 °C for 10 min and load them into the gel. The SDS-PAGE is run at 165 V for 35 min using the MES running buffer. Stain the gel for 10 min with the staining solution, and destain it for 4 h at room temperature with destaining solution.

3.6 RP-HPLC and Mass Spectrometric (LC-MS) Analysis

Protocol for the analysis of the site-specifically modified ADC by revered-phase HPLC and mass spectrometry is provided. LC-ESI-MS method is used to determine the yield of the SEPL reaction.

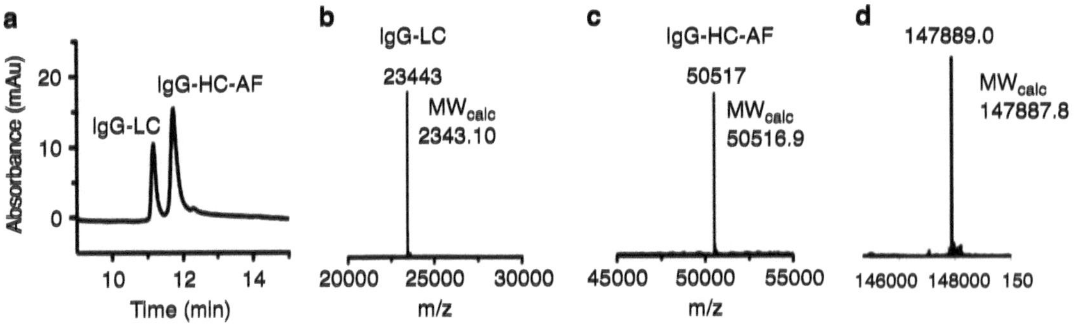

Fig. 4 (**a**) RP-HPLC analysis (detection at 280 nm) of one-pot thiolysis/ligation reaction after 24 h at rt. (**b**) ESI-MS analysis of Her2-Auristatin F fully reduced and deglycosylated sample. (**c**) ESI-MS analysis of the Her2-Auristatin F ADC obtained by SEPL and analyzed under nonreducing conditions

1. Add 1 μL of PNGase F enzyme to 100 μg of ADC (30–35 μL) from Subheading 3.5 and incubate at 37 °C for 16 h.
2. Denature and reduce the sample by adding 472 μL of 6 M GuHCl in PBS pH 7.4 and 10 mM DTT. Incubate the mixture at 37 °C for 1 h.
3. Inject the sample on a Zorbax 300SB C8 column.
4. Elute with a 15–70% linear gradient of mobile phase A (0.25% Formic acid and 0.02% TFA in water) and mobile phase B (90% isopropanol in water with 0.25% Formic acid and 0.02 TFA) over 30 min at 1 mL/min flow rate and 70 °C (Fig. 4a).
5. Collect the peaks and analyze them using ESI-MS mass detector. Deconvolute the spectra and determine the masses of the eluted peaks LC and HC (*see* **Note 17**). MS results of fully reduced conjugated antibody are illustrated in Fig. 4b.
6. For the intact mass analysis of nonreduced ADC, add 1 μL of PNGase F enzyme to 100 μg of ADC. Incubate the sample at 37 °C for 16 h. Analyze the sample using ESI-MS mass detector. Deconvolute the spectra and determine the mass of the peak (Fig. 4c).

3.7 HIC Analysis

Hydrophobic interaction chromatography (HIC) separates species based on hydrophobicity and can be used to resolve antibody conjugates with different amounts of drugs conjugated to them. In this section a protocol for the analysis by HIC of ADCs is provided, as well as a method to calculate their DAR.

1. Inject 100 μg of ADC onto a TSK-GEL Butyl-NPR column previously equilibrated with HIC buffer A.
2. Elute with a linear gradient from 0 to 100% of HIC buffer B at 0.8 mL/min over 20 min. Examples of a HIC profile for site-specific SEPL ADC and a conventional ADC (Trastuzumab-mcc-DM1) are illustrated in Fig. 5.

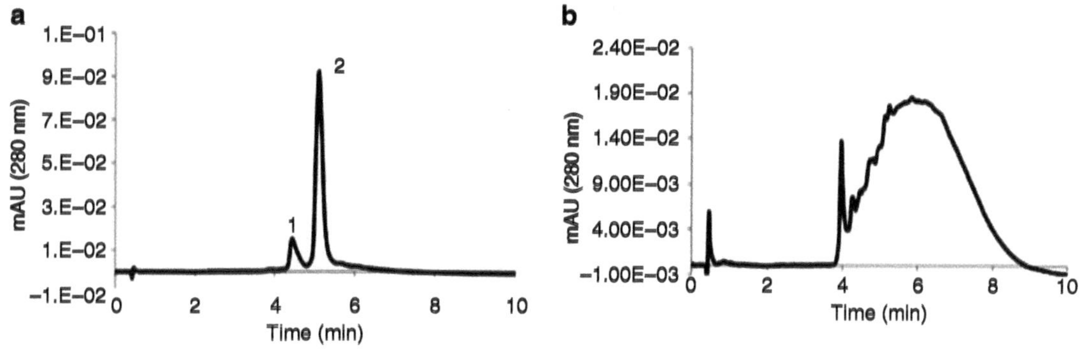

Fig. 5 HIC (Hydrophobic Interaction Chromatography) analysis of ADCs. (**a**) Site-specific ADC (Her2-Auristatin F) prepared by SEPL and (**b**) nonspecific ADC, T-DM1

3. Collect the peaks and analyze them using ESI-MS mass detector to assign the drug load of the different peaks.
4. Calculate the DAR for the SEPL antibody drug conjugate. Using the previously assignment by ESI-MS, quantify antibody species with different ratios of drug per antibody by integrating peaks in the chromatogram. Calculate the peak area in percentage and multiply by the corresponding drug load to obtain the weighted peak area. Finally, calculate the average DAR by summing the weighted peak area and dividing the sum by 100. DAR = (Σ Weighted Peak Area)/100).

4 Notes

1. The IntC gene (NpuC in this case) is cloned into a modified pTXB1 vector from NEB in which a His-tag and stop codon were inserted between MxeGyrA intein and the chitin-binding domain to generate IntC-GyrA-H6 plasmid [6]. The sequence is:

   MIKIATRKYLGKQNVYDIGVERDHNFALKNGFIA-SAAFNSGGCITGDALVALPEGESVRIADIVPGARPNSD-NAIDLKVLDRHGNPVLADRLFHSGEHPVYTVRTVEG LRVTGTANHPLLCLVDVAGVPTLLWKLIDEIKPGDYA-VIQRSAFSVDCAGFARGKPEFAPTTYTVGVPG LVRFLEAHHRDPDAQAIADELTDGRFYYAKVASVTDAG VQPVYSLRVDTADHAFITNGFVSHAHHHHHH.
2. Analytical RP-HPLC and HIC are carried out on a system with quaternary pump and autosampler. Preparative HPLC is performed on a system with a binary pump.
3. The expression plasmid used for the heavy chain contains all the constant domains of the human IgG1 heavy chain (CH1, CH2, and CH3) and the variable heavy chain (HV) of anti-Her2

mAb. Plasmid is under the hEF1-HTLV promoter, and contains zeocin resistance genes for antibiotic selection. For the protocol described in this chapter the pFUSE-CHIg (Invivogen) was used, but other plasmids might be used following manufacturer's instructions.

4. InteinN fragment is encoded in a pET expression plasmid. In this case the intein AvaN is used [6].
5. The forward primer should be designed with 20–25 bases overlapping with the vector sequence 5′ to the desired point of insertion, followed by 20–25 bases corresponding to the 5′ end of the gene of interest. The reverse primer should be designed with 20–25 bases overlapping with the vector sequence 3′ to the point of insertion followed by 20–25 bases corresponding to the 3′ end of the gene of interest.
6. It is important to perform the transformation with ultracompetent *E. coli* cells, which increases the transformation efficiency of large DNA molecules.
7. The expression plasmid used for the light chain contains the constant domain of the human kappa light chain and the variable light chain (LV) of anti-Her2 mAb. Plasmid is under the hEF1-HTLV promoter and contains blasticidin resistance genes for antibiotic selection. For the protocol described in this chapter the pFUSE-CLIg-hK (Invivogen) was used, but other plasmids might be used following manufacturer's instructions.
8. The ExpiFectamine expression and transfection kit from Life Technology was used. The protocol described in this chapter was optimized using this expression system. Other eukaryotic expression systems might be used, following manufacturer's protocols and recommendations.
9. Auristatin F is obtained from a CRO specialized in synthesis of cytotoxics.
10. T-DM1 (trastuzumab emtansine, nonspecific conjugate) is obtained from a CRO specialized in antibody–drug conjugates.
11. It is important to use a molar ratio between 1:3 and 1:4 of vector to insert.
12. In general, using 52 °C as an annealing temperature for the PCR should produce the desired result, but it could be reduced to 40 °C for problematic reactions. The elongation is performed at 68 °C and 2 min per kb of new construct.
13. Antibody production depends greatly on the ratio of heavy chain and light chain expression. In this case, light-chain plasmids and IgG-IntN-heavy chain plasmids ratio of 2:1 is used to co-transfect mammalian cells.

14. Load the sample by gravity into a cartridge of 2 g reversed-phase resin. Load by gravity one column volume of 5% of acetonitrile in water. Repeat this step increasing the amount of acetonitrile by a 10% until 80% of acetonitrile is reached.
15. The elution volume is in good agreement with the estimated MW for a tetrameric IgG.
16. The typical overall yield is 40%, including thiolysis, ligation, and purification.
17. In this case, a BioSuite pPhenyl 1000RPC column (2.0 × 75 mm 10 mm) was used with a 5–80% linear gradient of mobile phase A (1% formic acid in water) and mobile phase B (1% formic acid in acetonitrile).

References

1. Singla N, Erdjument-Bromage H, Himanen JP et al (2011) A semisynthetic Eph receptor tyrosine kinase provides insight into ligand-induced kinase activation. Chem Biol 18:361–371
2. Lu W, Gong D, Bar-Sagi D et al (2001) Site-specific incorporation of a phosphotyrosine mimetic reveals a role for tyrosine phosphorylation of SHP-2 in cell signaling. Mol Cell 8:759–769
3. Wu JW, Hu M, Chai J et al (2001) Crystal structure of a phosphorylated Smad2. Recognition of phosphoserine by the MH2 domain and insights on Smad function in TGF-beta signaling. Mol Cell 8:1277–1289
4. Valiyaveetil FI, Sekedat M, Muir TW et al (2004) Semisynthesis of a functional K+ channel. Angew Chem Int Ed Engl 43:2504–2527
5. Möhlmann S, Bringmann P, Greven S et al (2011) Site-specific modification of ED-B-targeting antibody using intein-fusion technology. BMC Biotechnol 11:76. https://doi.org/10.1186/147267501176
6. Vila-Perelló M, Liu Z, Shah NH et al (2012) Streamlined expressed protein ligation using split inteins. J Am Chem Soc 135:286–292
7. Shah NH, Muir TW (2011) Split inteins: nature's protein ligases. Isr J Chem 51:854–861
8. Zettler J, Schütz V, Mootz HD (2009) The naturally split Npu DnaE intein exhibits an extraordinarily high rate in the protein trans-splicing reaction. FEBS Lett 583:909–914
9. Shah NH, Dann GP, Vila-Perello M et al (2012) Ultrafast protein splicing is common among cyanobacterial split inteins: implications for protein engineering. J Am Chem Soc 134:11338–11341
10. Volkmann G, Sun W, Liu X-Q (2009) Controllable protein cleavages through intein fragment complementation. Protein Sci 18:2393–2402
11. Mootz HD (2009) Split inteins as versatile tools for protein semisynthesis. Chembiochem 10:2579–2589
12. Teicher BA, Chari RVJ (2011) Antibody conjugate therapeutics: challenges and potential. Clin Cancer Res 17:6389–6397
13. Chari RVJ, Miller ML, Widdison WC (2014) Antibody–drug conjugates: an emerging concept in cancer therapy. Angew Chem Int Ed 53:3796–3827
14. Axup JY, Bajjuri KM, Ritland M et al (2012) Synthesis of site-specific antibody-drug conjugates using unnatural amino acids. Proc Natl Acad Sci U S A 109:16101–16106
15. Zimmerman ES, Heibeck TH, Gill A et al (2014) Production of site- specific antibody–drug conjugates using optimized non-natural amino acids in a cell-free expression system. Bioconjug Chem 25:351–361
16. Lhospice F, Delphine B, Belmant C et al (2015) Site-specific conjugation of monomethyl auristatin E to anti-CD30 antibodies improves their pharmacokinetics and therapeutic index in rodent models. Mol Pharm 12:1863–1871
17. Strop P, Liu S-H, Dorywalska M et al (2013) Location matters: site of conjugation modulates stability and pharmacokinetics of antibody drug conjugates. Chem Biol 20:161–167
18. Agarwal P, Van der Weijden J, Sletten EM et al (2013) A Pictet-Spengler ligation for protein chemical modification. Proc Natl Acad Sci U S A 110:46–51
19. Van Geel R, Wijdeven MA, Heesbeen R et al (2015) Chemoenzymatic conjugation of toxic

payloads to the globally conserved N-glycan of native mAbs provides homogeneous and highly efficacious antibody-drug conjugates. Bioconjug Chem 26:2233–2242

20. Swee LK, Guimaraes CP, Sehrawat S et al (2013) Sortase-mediated modification of αDEC205 affords optimization of antigen presentation and immunization against a set of viral epitopes. Proc Natl Acad Sci U S A 110:1428–1433

21. Jackson D, Atkinson J, Guevara CI et al (2014) In vitro and in vivo evaluation of cysteine and site specific conjugated herceptin antibody-drug conjugates. PLoS One 9:e83865. https://doi.org/10.1371/journal.pone.0083865

22. Tian F, Lu Y, Manibusan A et al (2014) A general approach to site-specific antibody drug conjugates. Proc Natl Acad Sci U S A 111:1766–1771

23. Frutos S, Jordan JB, Bio MM et al (2016) Access to site-specific Fc-cRGD peptide conjugates through streamlined expressed protein ligation. Org Biomol Chem 14:9549–9553

24. Frutos S, Hernández JL, Otero A et al (2018) Site-specific antibody drug conjugate using streamlined expressed protein ligation. Bioconjug Chem 29:3503–3508

25. Geiser M, Cèbe R, Drewello D, Schmitz R (2001) Integration of PCR fragments at any specific site within cloning vectors without the use of restriction enzymes and DNA ligase. BioTechniques 92:88–90

Chapter 10

In Vivo Histone Labeling Using Ultrafast *trans*-Splicing Inteins

Nicholas A. Prescott and Yael David

Abstract

The development of expressed protein ligation (EPL) widened the scope of questions that could be addressed by mechanistic biochemistry. Protein *trans*-splicing (PTS) relies on the same basic chemical principles, but utilizes split inteins to tracelessly ligate distinct peptide or polypeptide fragments together with native peptide bonds. Here we present a method to adapt PTS methodologies for their use in live cells, in order to deliver synthetic or native histone modifications. As an example, we provide a protocol to incorporate a small molecule fluorophore into chromatinized histones. The protocol should be easily adaptable to incorporate other modifications to chromatin in vivo.

Key words Split inteins, In cellulo labeling, Synthetic biology, Chromatin

1 Introduction

Expressed protein ligation (EPL) has proven to be an invaluable technique for in vitro biochemical studies of post-translation modifications and their role in regulating protein structure and function [1, 2]. However, due to its inherent limitations, this expanded toolbox had not yet managed to bridge the gap between mechanistic studies and the systems-level correlative information gained from in vivo and in cellulo experiments. In recent years, the principles of EPL have been applied to develop effective protein *trans*-splicing (PTS) methods in cultured cells, providing the opportunity to perform experiments that may distinguish causation from correlation in a temporally controlled and chemically defined manner.

Classical protein splicing utilizes intact inteins which are able to autoexcise themselves from a single polypeptide chain. Such inteins have been identified in species across all domains of life [3]. It was later found that whole inteins could be artificially split into separate polypeptide fragments that could facilitate protein *trans*-splicing [4]. Naturally occurring split inteins were subsequently identified in cyanobacteria. In these species, the catalytic subunit of DNA

Miquel Vila-Perelló (ed.), *Expressed Protein Ligation: Methods and Protocols*, Methods in Molecular Biology, vol. 2133, https://doi.org/10.1007/978-1-0716-0434-2_10,

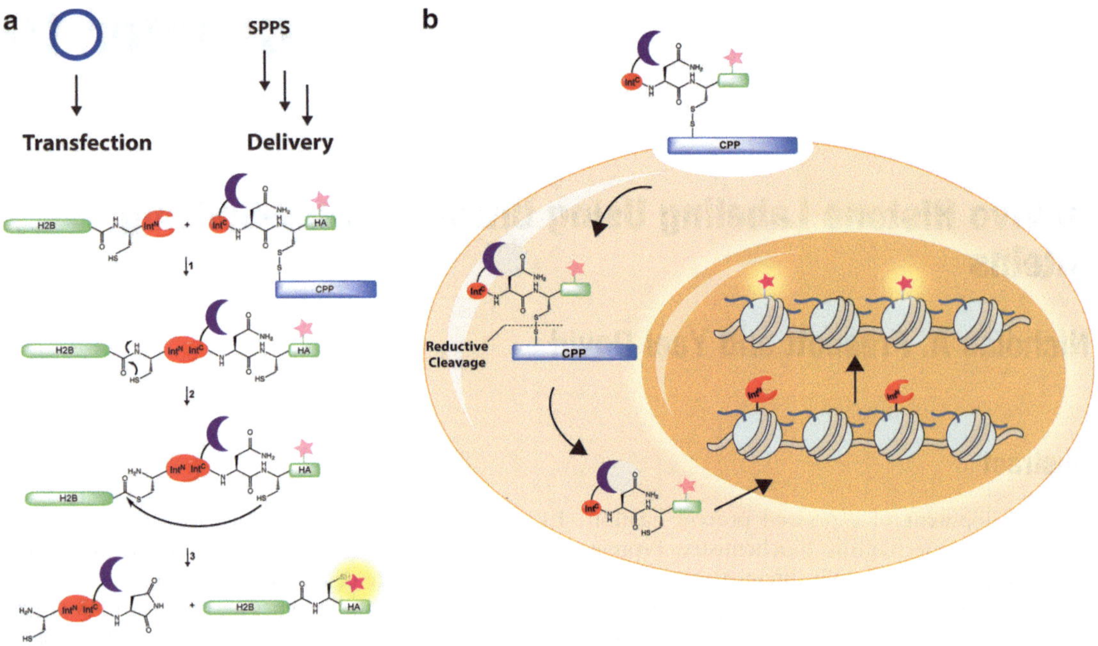

Fig. 1 Synthetic peptide delivery to chromatinized histones. (**a**) Schematic of intein splicing reaction showing [1] initial split intein recognition, [2] the intein-mediated N-to-S acyl shift, and [3] transthioesterification followed by splicing resolution to yield spliced exteins, in this case histone H2B and a peptide containing an HA epitope and small molecule fluorophore, with a native peptide bond. (**b**) Delivery of CPP-cargo peptide into cells containing histone-Int^N fusion proteins. Upon entry to the cytosol, reductive cleavage separates the CPP from the cargo peptide, which proceeds to diffuse into the nucleus, where ultrafast *trans*-splicing occurs after recognition of Int^N by Int^C. Upon splicing, the dark fluorescence quencher is separated from the fluorophore, allowing fluorescence to be detected. CPP, cell-penetrating peptide

Polymerase III, DnaE, is genetically encoded as two so-called extein fragments, which are spliced together by N- and C-inteins, generally termed Int^N and Int^C, respectively [5]. The DnaE class of split inteins have since served as the canonical example of split inteins, though split inteins have been identified in other organisms. For all split inteins, when Int^N and Int^C bind one another, they excise themselves and tracelessly ligate their exteins (Fig. 1). Expressed protein ligation techniques were quickly adapted to capitalize on the accessibility of these split inteins, and protein *trans*-splicing became yet another powerful tool for in vitro biochemical experiments, including isotopic labeling for NMR structural studies, regioselective side chain modification, and mechanistic studies of protein splicing itself [6–8].

Methods to label proteins in live cells have long been sought after. These have historically included genetic fusions to effector domains such as GFP to study subcellular localization, or the FKBP/rapamycin system to chemically induce dimerization of proteins of interest. More recently, techniques utilizing the SNAP- and Halo-tags to efficiently label some proteins with certain synthetic cargoes have been reported [9, 10]. While these developments minimize the remaining artifacts from labeling, they are unable to fully eliminate them.

Efforts to bring protein splicing out of the test tube and into living cells to tracelessly label proteins date back over a decade. Early traceless splicing attempts showed moderate signs of success, owing in part to the slow splicing rates of the split inteins employed [11]. Other attempts used split inteins combined with genetic engineering to conditionally induce protein splicing [12]. Recently, we developed an efficient strategy utilizing ultrafast *trans*-splicing split inteins to label histones in live cells [13]. This method relies on the genetic fusion of a C-terminally truncated histone and an Int^N (in this case we used the N-intein from *Anabaena variabilis*, termed Ava^N) that is expressed in cells and incorporated into chromatin, and a synthetic piece generated by solid-phase peptide synthesis (SPPS), which contains the C-intein fragment (in this case we used the C-intein from *Nostoc punctiforme*, termed Npu^C). The Npu^C-cargo fusion is then conjugated by direct disulfide bonding to the HA2-TAT cell-penetrating peptide (CPP), which is derived from the viral protein transduction domains HA2 and TAT, allowing for its timely delivery into the cells [14].

In this chapter we provide a method for the C-terminal labeling of histone H2B in vivo with a small molecule fluorophore and exogenous epitope tag, including Int^N construct design and transient overexpression; synthesis of synthetic cargo and CPP; delivery of the synthetic cargo to cells; and finally, analysis of splicing efficiency by western blot and live cell imaging. While this methodology was originally applied toward the study of histone proteins incorporated into chromatin, the considerable modularity of this technique makes it generally applicable for many proteins in the nucleus or cytosol. No reports have been published of successful protein *trans*-splicing within mitochondria or the lumen of other organelles, though such a reaction should be possible in principle by appending the proper mitochondrial targeting sequence or signal peptide [15, 16]. In addition, although we originally employed the Ava^N-Npu^C split intein pair for the site-specific manipulation of histones, other intein families can be easily used in combination as many of them are conserved in structure and will readily react with other families. In contrast, orthogonal pairs of split inteins (such as IMDPH/GP41) could in theory be used in tandem to modify multiple histones (or other proteins) simultaneously [4]. Finally, current efforts are aimed toward expanding the methodology from cultured cells to specific tissues in live animals in order to understand the causal role of certain PTMs at the systems or organismal level.

2 Materials

18.2 MΩ-cm ultrapure water or equivalent should be used in the preparation of all aqueous solutions.

2.1 Solid Phase Peptide Synthesis

1. Dimethylformamide (DMF).
2. Dichloromethane (DCM).
3. Activation solution: 0.5 M HBTU (O-(Benzotriazol-1-yl)-*N*,*N*,*N*′,*N*′- tetramethyluronium hexafluorophosphate), 0.5 M HOBt (Hydroxybenzotriazole) in DMF.
4. ChemMatrix Resin™ with Rink Amide linker, 0.5 mmol/g loading capacity (*see* **Note 1**).
5. 20 mL polypropylene gravity flow reaction columns.
6. 20% Piperidine: 20% piperidine solution (v/v) in DMF.
7. Standard Fmoc-SPPS amino acids, with Fmoc protecting groups on the alpha amine and the following side-chain protecting groups: Arg(Pbf), Asn(Trt), Asp(OtBu), Cys(Trt), Glu (OtBu), Gln(Trt), His(Trt), Lys(Boc), Ser(tBu), Thr(tBu), Trp(Boc), and Tyr(tBu), as well as Lys(ivDde) and Cys(StBu).
8. *N*-terminally Boc-protected amino acids (for final *N*-terminal residue of each peptide), including Boc-Cys(Npys)-OH.
9. Acetic anhydride.
10. Diisopropylethylamine (DIPEA).
11. 2% Hydrazine: 2% hydrazine (v/v) in DMF.
12. Tetramethylrhodamine *N*-hydroxysuccinimidyl ester (TMR-SE).
13. Cleavage cocktail: 95% trifluoroacetic acid (TFA), 2.5% triisopropylsilane (TIS), 2.5% water (v/v).
14. Diethyl ether.
15. QSY™ 9 C_5-Maleimide (*see* **Note 2**).
16. Maleimide reaction buffer: 6 M guanidine hydrochloride (GnHCl), 1× PBS, 1 mM EDTA, pH 7.0, in water.
17. L-Cysteine.
18. Tris(2-carboxyethyl)phosphine hydrochloride salt (TCEP).
19. Liquid chromatography–mass spectrometry (LC-MS) system with C18 column.
20. LCMS solvent A: 0.1% formic acid (v/v) in water.
21. LCMS solvent B: 90% acetonitrile (v/v) and 0.1% formic acid (v/v) in water.
22. Reverse Phase-High Performance Liquid Chromatography (RP-HPLC) system with C18 column.
23. HPLC solvent A: 0.1% TFA (v/v) in water.

24. HPLC solvent B: 90% acetonitrile (v/v) and 0.1% TFA (v/v) in water.
25. *N*,*N'*-Diisopropylcarbodiimide (DIC).

2.2 Directed Disulfide Conjugation of CPP to Cargo Peptide

1. Resuspension solution: 6 M GnHCl in water.
2. Conjugation Buffer: 6 M GnHCl, 100 mM phosphate, pH 6.0 in water.
3. Liquid chromatography–mass spectrometry (LC-MS) system with C18 column.
4. LCMS solvent A: 0.1% formic acid (v/v) in water.
5. LCMS solvent B: 90% acetonitrile (v/v) and 0.1% formic acid (v/v) in water.
6. Reverse Phase-High Performance Liquid Chromatography (RP-HPLC) system with C18 column.
7. HPLC solvent A: 0.1% TFA (v/v) in water.
8. HPLC solvent B: 90% acetonitrile (v/v) and 0.1% TFA (v/v) in water.

2.3 Transfection of Cells and Peptide Delivery

1. 6-well, treated tissue culture treated polystyrene cell culture plate.
2. Tissue culture CO_2 incubator.
3. HEK 293T cells.
4. Sterile, ventilated tissue culture hood.
5. Full cell medium: Dulbecco's Modified Eagle Medium (DMEM), supplemented with 10% fetal bovine serum (FBS), 2 mM glutamine, 500 units/mL penicillin/streptomycin antibiotic.
6. 1.7 mL Eppendorf tube.
7. Opti-MEM Reduced Serum Media.
8. Lipofectamine 2000.
9. Opti-MEM Reduced Serum Media.
10. Dimethylsulfoxide (DMSO).

2.4 Analysis of Manipulated Cells

1. 1.7 mL Eppendorf tube.
2. PBS: phosphate buffered saline in water.
3. Iodoacetamide solution: 80 mM iodoacetamide in water.
4. Hypotonic lysis buffer: 10 mM Tris, 15 mM NaCl, 1.5 mM $MgCl_2$, protease inhibitors, pH 7.6 in water.
5. Dounce homogenizer with tight and loos pestles.
6. Triton X-100.

7. SDS RIPA lysis buffer: 25 mM Tris pH 8.0, 150 mM NaCl, 1% (w/v) sodium dodecyl sulfate, 1% (v/v) NP-40, 1% (w/v) deoxycholate, pH 7.5 in water
8. 4× sample buffer (SB): 200 mM Tris–HCl, 280 mM sodium dodecyl sulfate, 6 mM bromophenol blue, 40% v/v glycerol, 20% v/v 2-mercaptoethanol in water.
9. 12% polyacrylamide Bis-Tris gel and gel running apparatus.
10. MOPS running buffer: 40 mM (*N*-morpholino) propanesulfonic acid (MOPS), 10 mM sodium acetate, 1 mM EDTA, pH 7.0 in water.
11. Methanol.
12. Transfer buffer: 25 mM Tris base, 192 mM glycine, 20% (v/v) methanol in water.
13. PVDF membrane and membrane transfer apparatus.
14. Bovine Serum Albumin (BSA).
15. PBST: phosphate buffered saline (PBS) supplemented with 0.1% (v/v) Tween-20 in water.
16. Antibodies: rabbit-αHA, mouse-αH2B, α-rabbit800 and α-mouse680 (*see* **Note 3** about secondary antibodies).
17. 35 mm glass bottom cell culture dish.
18. Opti-MEM media.
19. Leibovitz's L-15 media.
20. Confocal microscope.

3 Methods

3.1 Splicing Construct Design and Cloning

The use of split inteins for in vivo protein semisynthesis provides great flexibility and utility, but also adds complexity in construct engineering. In designing the construct for eventual use, one must first identify a site where the protein of interest (in our case the core histone H2B) can be split into two fragments—one to be expressed within cells, and one to be chemically synthesized. These are to be joined together by preparing each piece as a fusion to one half of a split intein pair, which will rapidly and tracelessly self-excise and ligate the fragments together upon interacting within the cell (Fig. 1). Several key considerations must be made when designing the exact splicing site for PTS, including proximity to the protein terminus and the extein dependence of the selected split intein pair [17]. In this section, we describe the process to identify an adequate splicing site within the protein of interest, which will determine the Int^N-fusion protein to be recombinantly expressed and the Int^C-cargo peptide to be chemically synthesized. This design process was shown to allow for the delivery of a multitude of native and synthetic cargoes, including monoubiquitinated H2B within intact chromatin. However, for clarity, we later illustrate the method by appending a small molecule fluorophore and epitope tag to the C-terminus of full-length histone H2B.

1. Protein *trans*-splicing is most efficient when performed on a protein region that is unstructured, since the truncation and fusion of the Int^N fragment to the C-terminus of the protein may prevent proper folding of an ordered polypeptide. Identify an unstructured or flexible region within the protein of interest. For histone H2B, the C-terminal region spanning residues Ala^{110}-Lys^{125} is adequate (Fig. 1a).
2. Within the identified region, any residue that can be mutated to cysteine without disrupting protein structure or function could plausibly serve as the *N*-terminal cysteine residue of the delivered Int^C-cargo peptide. Determine which residues within the region are good candidates to divide the two splicing fragments. For the region of H2B identified, residues Ala^{110}, Ala^{117}, Ser^{123}, and Ala^{124} meet these criteria (*see* **Note 4**).
3. Once candidate residues have been identified that could be mutated to Cys, the exact splicing site must be determined. Depending on the split intein pair used, the chosen pair may exhibit a dependence of splicing kinetics on the immediately neighboring N- or C-extein residues (*see* **Note 5**) [4]. To circumvent this, choose a splice site from the list of candidates that most closely reflects the optimal extein sequences. Alternately, if no such option is available, it may be advisable to insert an extein linker between the protein of interest and the N-intein (*see* **Note 6**, Table 1 for constructs used herein).
4. With the splice site now designed, use conventional cloning methods to generate a construct for mammalian cell expression of the desired protein-Int^N fusion (*see* **Notes 7** and **8**).

Table 1
Amino acid sequence of transfected protein and synthetic peptides

Component	Amino acid sequence
H2B-Ava^N-Flag fusion	PEPAKSAPAPKKGSKKAVTKAQKKGGKKRKRSRKESYSIYVYKVLKQV HPDTGISSKAMGIMNSFVNDIFERIAGEASRLAHYNKRSTITSREIQTAVR LLLPGELKHAVSEGTKAITKYTSSKGGKFAEYCLSYDTEVLTVEYGFVPI GEIVDKGIECSVFSIDSNGIVYTQPIAQWHHRGKQEVFEYCLEDGSIIKAT KDHKFMTQDGKMLPIDEIFEQELDLLQVKGLPEGDYKDDDDK
Npu^C-cargo peptide	IKIATRKYLGKQNVYDIGVERC$^{(QSY)}$HNFALKNGFIASNCFNK$^{(TMR)}$YPY DVPDYA
HA2-TAT CPP	C$^{(Npys)}$GLFEAIAEFIENGWEGLIEGWYGGRKKRRQRRR

3.2 Synthesis of Int^C-Cargo Peptide

In this section we provide a protocol for the synthesis of the Int^C-cargo peptide using SPPS, which includes the peptide synthesis, conjugation of small molecules, and purification of the final peptide. The construct here is designed to enable conjugation of the fluorescence quencher (QSY-maleimide) to the Int^C fragment and the fluorophore (TMR-SE) to the extein sequence (*see* **Note 9**). This design ensures that fluorescence signal should only be detectable upon completed intein splicing. Standard Fmoc-based SPPS is used for the synthesis of the construct that will be delivered. All peptides are synthesized with Rink Amide linkers to generate C-terminal carboxamide that will result in a near-native (but more stable) C-terminal end on the target protein. *N*-terminally Boc-protected amino acids are incorporated for the final *N*-terminal residue of each peptide. Peptides are synthesized using manual addition of the reagents (using a stream of dry N_2 to agitate the mixture). This protocol describes synthesis on a 0.1 mmol scale using resin with a 0.5 mmol/g capacity, but it can be scaled up as needed (*see* Table 1 for peptide sequence).

1. Before beginning synthesis, prepare 1 mL activation solution per coupling to be performed.
2. Weigh 200 mg of Rink Amide ChemMatrix Resin™ into reaction column, and swell in 5 mL DMF for 30 min. with gentle agitation before draining.
3. Add 3–4 mL of 20% piperidine solution (*see* **Note 10**) to resin and incubate for 1 min to preequilibrate resin. Drain and repeat, incubating this time for 20 min. to allow complete deprotection to take place.
4. Drain and wash the resin under continuous flow, first with 5 column volumes (CV, relative to volume of swelled resin) of DCM and then with 5 CV DMF.
5. Dissolve 5 molar eq. Fmoc-protected amino acid (relative to the resin) in 1 mL activation solution. Add 10 eq. DIPEA to amino acid solution. For residues Lys^{39} and Cys^{36}, the orthogonal side-chain protecting groups ivDde and StBu must be used, respectively, instead of the standard Boc sand Trt groups.
6. Add amino acid solution to the *N*-terminally deprotected resin, and allow to react for 30 min at room temperature, with gentle agitation.
7. Wash resin with 3 CV DMF under continuous flow.
8. Add 10 eq. acetic anhydride and 20 eq. DIPEA into 1 mL DMF, and incubate with the resin for 20 min. with agitation. Drain and repeat once.
9. Wash resin with 3 CV DMF under continuous flow.

10. Add an excess of 20% (v/v) piperidine in DMF to remove Fmoc.
11. Drain and wash resin thoroughly with DMF to remove trace piperidine, as in **step 9**.
12. Repeat **steps 5–11** for each amino acid coupling until the desired peptide has been completed.
13. Upon completed peptide chain assembly, deprotect the Lys^{39} side chain by adding 5 mL of 2% hydrazine solution and incubating for 3 min. Drain, then wash resin as in **step 4**. Repeat entire **step 5** times to ensure full deprotection of the side chain.
14. Dissolve 2 eq. of TMR-SE in approximately 4 mL DMF, then add 10 eq. DIPEA into the solution. Add fluorophore solution to resin, and allow to react for 12 h or overnight at room temperature with agitation. Cover sides of reaction column with foil to keep resin protected from light if necessary.
15. Drain and wash the resin thoroughly, as in **step 4**.
16. To cleave peptide from resin, add 20 μL cleavage cocktail per mg of peptidyl-resin into the reaction column and allow the cleavage reaction to take place for 1.5–2 h at room temperature with agitation.
17. Drain the peptide-containing filtrate into a 15 mL conical tube and evaporate the solution by gently blowing nitrogen over sample for 30 min. The column and resulting cleaved resin can then be disposed of.
18. Add cold diethyl ether to the evaporated peptide solution such that any remaining cleavage cocktail is less than 10% of the total volume in order to precipitate the peptide. Centrifuge at 4000 rcf for 10 min at 4 °C. Carefully decant the supernatant, and repeat twice more by adding 10 mL of ether each time.
19. Allow the resulting peptide pellet to dry by leaving it at room temperature overnight in a chemical hood to remove all ether, protecting from light to avoid possible photobleaching of the fluorophore.
20. To alkylate the free thiol group of Cys^{22} with the QSY quencher, first resuspend the Int^C-cargo peptide to a concentration of 4 mM in maleimide reaction buffer (*see* **Note 11**).
21. Separately, dissolve 2 eq. QSY-maleimide in an equal volume of maleimide reaction buffer (i.e., to make an 8 mM solution). Combine peptide and maleimide solutions, and allow reaction to take place at room temperature with agitation for several hours (*see* **Notes 12** and **13**).

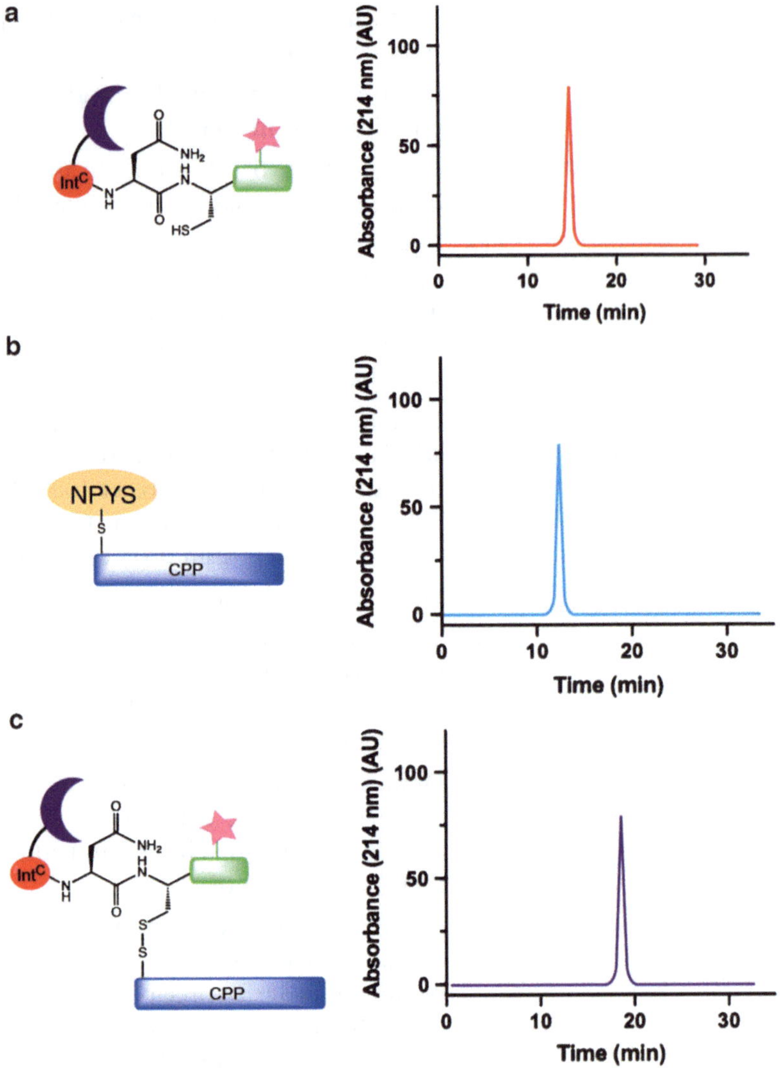

Fig. 2 Representative analytical RP-HPLC chromatograms for synthetic cargo peptide (**a**), HA2-TAT CPP (**b**), and disulfide-conjugated CPP-cargo peptide (**c**)

22. Quench excess maleimide by dissolving 25 eq. L-cysteine (with respect to initial maleimide concentration) into the reaction solution. Allow reaction to take place at room temperature with agitation for several hours (*see* **Note 14**).
23. Once excess maleimide is quenched, proceed to remove the StBu protecting group from Cys^{36} by weighing out TCEP and adding directly into peptide solution to a final concentration of 90 mM.
24. Monitor deprotection periodically by LCMS and only proceed to purify the peptide when complete.
25. Purify product by semi-prep or preparative RP-HPLC (Fig. 2a) and lyophilize before storing peptide at −80 °C.

3.3 Synthesis of CPP and Directed Disulfide Conjugation to Int^C-Cargo Peptide

In this section we provide a protocol for the synthesis of the CPP (HA2-TAT) by SPPS and subsequent disulfide conjugation of the CPP to the Int^{C}-cargo peptide, as well as the purification of the final product. The CPP-mediated delivery of synthetic cargo into cells provides two great strengths when compared to other methods like viral or liposomal delivery: increased synthetic accessibility and lower levels of cytotoxicity (*see* **Note 15**) [18, 19]. Crucially, the targeted disulfide conjugation of the CPP to the catalytic cysteine residue of the Int^{C} fragment also brings along three key advantages: protection of the catalytic residue from oxidation or other reactions prior to delivery, traceless intracellular removal from the Int^{C}-cargo by reductive cleavage within the cytosol, and greater efficacy than when the CPP is fused C-terminally to the cargo peptide.

3.3.1 Synthesis of HA2-TAT Peptide

The standard SPPS protocol from Subheading 3.2 was used to synthesize the CPP, with the following modifications (*see* Table 1 for peptide sequence):

1. When preparing to couple the Boc-Cys(Npys)-OH after the rest of the HA2-TAT peptide has been produced, instead of following the standard protocol, dissolve 5 molar eq. each of Boc-Cys(Npys)-OH, HOBt, and DIC (with respect to the resin) in 1 mL DMF, and incubate with peptidyl resin for 1 h to couple (*see* **Note 16**).
2. Using the same cleavage cocktail as described previously, cleave peptide from resin for 3 h (*see* **Note 17**).
3. After cleavage, drain CPP solution into a 15 mL conical tube and precipitate crude peptide with cold diethyl ether, as described previously.
4. Resuspend the pelleted peptide in a 50:50 mixture of HPLC solvents A and B and lyophilize the resuspended peptide overnight.
5. Dissolve lyophilized peptide in resuspension buffer.
6. Purify CPP by RP-HPLC (Fig. 2b, *see* **Note 18**).

3.3.2 Conjugation of CPP to Cargo Peptide

1. In degassed conjugation buffer, separately resuspend pure HA2-TAT to a concentration of 4.5 mM and cargo peptide to a concentration of 5 mM in equal volumes (*see* **Note 19**). Check that pH remains at 6.0 for both, and adjust with HCl or NaOH if necessary.
2. Combine Int^{C}-cargo and CPP to final concentrations of 2.25 and 1.5 mM, respectively.
3. Allow components to react at room temperature. Reaction completion can be monitored every 30 min by LCMS, and typically is achieved within 1 h.
4. Purify resulting conjugate from unreacted materials by RP-HPLC (Fig. 2c).

3.4 Expression of Histone-Int^N Fusion and Int^C-Cargo Peptide Delivery

In this section we provide a protocol for the transient transfection and overexpression of the histone-Int^N fusion protein, as well as delivery of the synthetic Int^C-cargo peptide (*see* **Note 20**). Stability and functionality of the Int^N-fusion protein should be tested to uncouple any effects of the fusion protein from the effects of downstream delivery or experimental conditions. While we provide a method based on transient, Lipofectamine-mediated transfection protocol, other reagent-based or even viral transduction methods might be employed to produce a cell line stably expressing the Int^N-fusion.

3.4.1 Transient Transfection of Histone-Int^N Fusion

For a single well in a 6-well flat bottom plate of ~2×10^6 cells, our transfection protocol is as follows (*see* **Note 21**):

1. Grow HEK 293T cells at 37 °C in 5% atmospheric CO_2 in incubator to ~70% confluence in full cell medium (*see* **Note 22**).
2. In tissue culture hood, combine 400 ng of plasmid DNA for mammalian expression of the Int^N-fusion protein, as designed Subheading 3.1, with 80 μL of Opti-MEM media. Separately, combine 3 μL of Lipofectamine with 80 μL DMEM. Incubate at room temperature for 15 min.
3. Combine DNA–media mixture with Lipofectamine–media mixture, pipetting gently to mix, and incubate for 30 min. at room temperature.
4. Aspirate media off of cells and add 1 mL of full media. Add the transfection mixture prepared in the previous step to cells in a dropwise manner. Gently mix the media by moving the plate side to side.
5. After 6–12 h of incubation at 37 °C in a CO_2 controlled incubator, aspirate media and supplement with 2 mL of fresh full media.
6. Allow transfected cells to grow for an additional 24 h before proceeding to delivery of synthetic cargo peptide (Fig. 3, *see* **Notes 23** and **24**).

3.4.2 Int^C-Cargo Peptide Delivery

1. Prepare peptide for delivery by dissolving pure, lyophilized peptide stocks to a concentration of 1 mM in DMSO. This concentrated stock solution can be flash frozen and reused if any excess remains unused (*see* **Note 25**).
2. Dilute stock peptide solutions in Opti-MEM cell medium to a final volume of 500 μL and to the necessary delivery concentration, typically around 1–10 μM (*see* **Notes 26** and **27**).
3. Aspirate media off Int^N-expressing cells, then wash them with 2 mL Opti-MEM. Aspirate media once more.

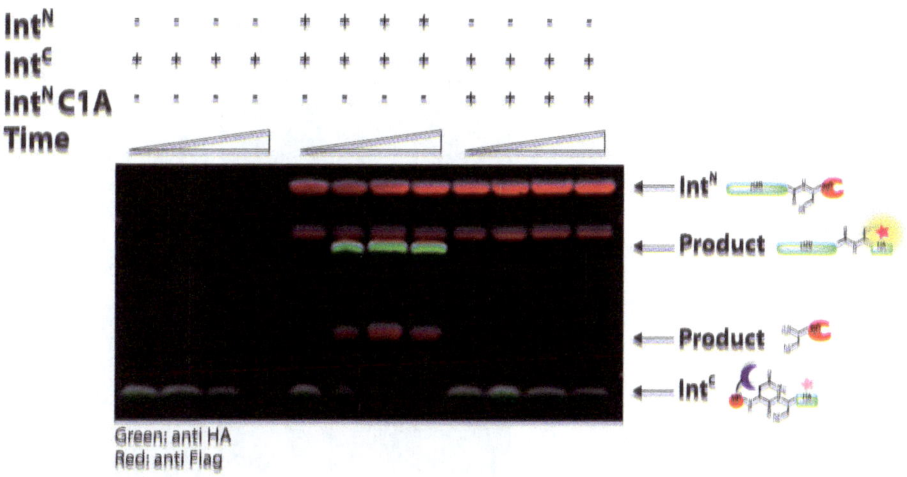

Fig. 3 Western blot analysis showing time course of protein *trans*-splicing. Lanes 1–4: nontransfected sample shows no product formation over time; lanes 5–8: splicing product forms rapidly over course of reaction, concurrent with depletion of free cargo peptide; lanes 9–12: cells transfected with catalytically inactive Int^N fusion protein (C1A) show no formation of splicing product

4. Gently add peptide–Opti-MEM solution to cells by pipetting onto walls of the well and then spread by gently moving plate side to side. Incubate at 37 °C for 1 h.
5. Supplement cells with an equal volume of full cell culture medium and incubate at 37 °C for an additional 1–4 h (Fig. 3).

3.5 Analysis of Manipulated Cells

In this section, we provide two methods to analyze successful protein *trans*-splicing in cells: live-cell confocal microscopy and western immunoblotting analysis (Figs. 3 and 4). Following cargo peptide delivery (Subheading 3.4.2, **step 5**), cells are ready for analysis as described below (*see* **Note 28**), or subsequent experimentation (*see* **Note 29**).

3.5.1 Cell Fractionation and Western Blot Analysis of Splicing

1. Wash cells with 1 mL cold PBS, then aspirate away the buffer.
2. Resuspend cells in 1 mL cold PBS supplemented with 100 μL of iodoacetamide solution (*see* **Note 30**).
3. Transfer cells to a 1.7 mL Eppendorf tube, and pellet by spinning at 400 rcf for 5 min at 4 °C in a tabletop microcentrifuge. Discard the supernatant.
4. Resuspend cell pellet in 1 mL of hypotonic lysis buffer and lyse with 10 strokes of a loose pestle Dounce homogenizer.
5. Centrifuge at 400 rcf for 5 min at 4 °C to pellet nuclei, and remove the supernatant, which contains the cytosolic fraction.
6. Resuspend the nuclear pellet in 1 mL of hypotonic lysis buffer supplemented with 1% (v/v) Triton X-100. Lyse with 10 strokes of a tight pestle Dounce homogenizer.

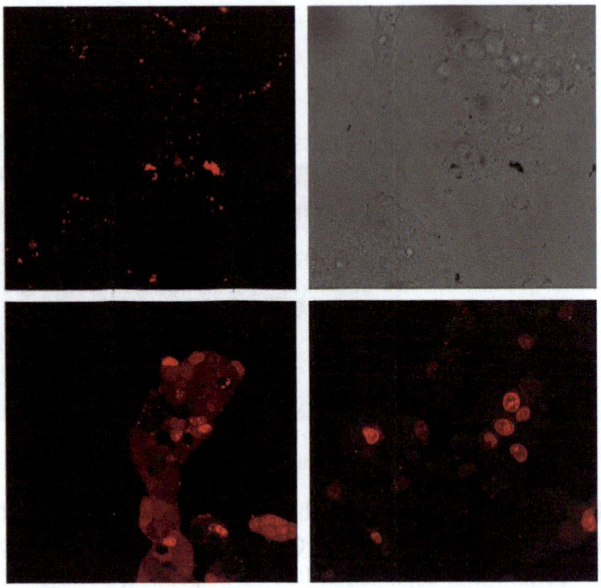

Fig. 4 Live-cell imaging of nontransfected and transfected cells delivered with TMR-carrying CPP-cargo peptide. Top, fluorescent (left) and bright-field (right) images of nontransfected cells after cargo delivery. Bottom, representative fluorescent images of transfected cells after cargo delivery

7. Centrifuge at 10,000 rcf for 10 min at 4 °C. Remove supernatant, which contains the nucleoplasm fraction.
8. Resuspend chromatin pellet in 300 μL of SDS-RIPA lysis buffer and then sonicate for 5 s at 35% amplitude with a rod sonicator.
9. In separate Eppendorf tubes, combine 5 μL of 4× sample buffer with 15 μL of the prepared chromatin fractions from nontransfected, transfected but nondelivered, and transfected and delivered cells. Heat at 95 °C for 5 min.
10. Load samples onto a 12% polyacrylamide bis-tris gel, and run at a constant voltage of 150 V for 1 h in MOPS running buffer.
11. When the gel run is over, transfer protein contents to a PVDF membrane at a constant voltage of 85 V for 75 min at 4 °C using standard gel/membrane transfer apparatus (*see* **Note 31**).
12. When transfer is complete, prepare blocking solution by dissolving 0.75 g BSA in 15 mL PBS. Incubate membrane in this solution for 1 h at room temperature.
13. After blocking, wash the membrane three times in 15 mL PBST for 5 min.

Table 2
Antibodies used for western immunoblotting analysis

Probe order	Host	Epitope	Dilution factor
Primary	Mouse	Anti-Flag	1:1000
Primary	Rabbit	Anti-HA	1:1000
Secondary	Goat	Anti-Mouse-680	1:15,000
Secondary	Goat	Anti-Rabbit-800	1:15,000

14. Blot the membrane with primary antibodies diluted into15 mL PBST according to manufacturer's instructions for 1 h at room temperature or overnight at 4 °C (*see* Table 2 for antibodies).
15. After incubation with the primary antibody, wash the membrane as in **step 13**.
16. Blot the membrane with secondary antibodies diluted in 15 mL PBST according to manufacturer's instructions for 1 h at room temperature (*see* Table 2).
17. Image the blot according to manufacturer's instructions for the secondary antibodies used (Fig. 3).

3.5.2 Live Cell Imaging

1. Rather than growing in a flat-bottom 6-well plate, grow cells on a glass bottom 35 mm dish, carrying out the above-described protocol (Subheading 3.4.2).
2. After cargo delivery, wash cells with Opti-MEM medium.
3. *Optional*: Some microscopes may be sensitive to the color of the cell medium. If so, wash once more with Leibovitz L-15 medium.
4. Immediately after washing, take cells for imaging. We used a confocal microscope to take multiple cross-sectional images per condition. Nontransfected cells should show no TMR fluorescence, as the splicing reaction will be unable to take place in the absence to the Int^N fusion protein. TMR fluorescence within the nucleus would be indicative of successful splicing in transfected and delivered cells (Fig. 4).

4 Notes

1. We typically prefer to use the ChemMatrix Resin™ because it has been shown to work well for long or complicated peptides [20].
2. We have used the QSY™ series of fluorescence quenchers due to their relatively broad absorption spectra, though any compatible fluorophore/dark quencher pair should theoretically work with this procedure.

3. We prefer to use fluorescent secondary antibodies, which allow for multiplexing and imaging of multiple components concurrently, but chemiluminescent antibodies should also work in this context.
4. If an endogenous cysteine residue is present in the protein of interest *N*-terminal to the site of manipulation, this may serve as the ideal splicing site since it would retrosynthetically come from the *N*-terminal Cys of the delivered peptide. If no Cys residue is available, serine, threonine, or alanine residues may be replaced with the necessary Cys of the delivered peptide to best mimic the native protein.
5. Generally, PTS has been performed using naturally occurring split inteins. However, recent work by Muir and colleagues has described engineered split inteins with faster splicing kinetics and significantly lower dependence on extein residues [17, 21]. Application of these engineered species toward PTS may be another way to circumvent the introduction of exogenous sequences to the final spliced product.
6. If none of the identified residues appears better than the others, multiple constructs may need to be tested empirically. If opting to append an extein linker to the native sequence of the protein being studied, we suggest running a series of control experiments to verify that the linker does not disrupt protein function or stability.
7. Any cloning method should be applicable to generate the Int^N-fusion construct.
8. As this DNA construct will be used for transfection into mammalian cells, it will be needed in high purities and at concentrations of ~1 μg/μL.
9. Although we here provide the protocol to couple the QSY-maleimide and TMR-SE, many different cargoes have been incorporated using this strategy, including synthetic amino acids with predefined native or nonhydrolyzable PTMs and larger peptide epitopes for immunoblotting. Delivery of larger cargoes, including an entire ubiquitinated lysine residue, has also been reported, though this was most successful in isolated nuclei rather than whole cells.
10. If the sequence to be synthesized contains Asp or Asn residues followed by Gly, Ala, or Ser, it is possible that the acid or amide side chains may cyclize with the peptide backbone upon treatment with a strong base like piperidine. If such a motif is present, use a solution of 20% piperidine (v/v) and 0.1 M HOBt in DMF to deprotect.

11. We suggest first carrying out an analytical test reaction on a small, 0.1 mg scale, and then scaling up to as large as 10 mg Int^C peptide.
12. Verify that the combined peptide–maleimide solution is still at pH 7.0. If it is not, adjust accordingly by adding either NaOH or HCl solutions.
13. Completion of maleimide alkylation should be monitored hourly by LCMS until all unreacted peptide is consumed. If need be, reaction can be stored overnight at −20 °C and resumed the following morning.
14. As with maleimide alkylation of the Int^C peptide, monitor reaction completion by LCMS. Only proceed when all unreacted maleimide is consumed.
15. Although we use HA2-TAT as our preferred cell-penetrating peptide, other CPPs should work successfully in principle. For a thorough review of common CPPs and their applications, *see* Ref. 15.
16. The coupling of the Boc-Cys(Npys)-OH can be sluggish. We suggest monitoring periodically by ninhydrin test until completion.
17. The poly-Arg(Pbf) stretch in the C-terminal region of this peptide can reduce cleavage efficiency, which is why cleavage time is extended.
18. When purifying HA2-TAT by RP-HPLC, we suggest maintaining the temperature at or below 50 °C to avoid split peaks arising, likely due to possible conformational changes in the peptide.
19. We recommend performing this reaction on a large (~10 mg or greater) scale, as we have found smaller-scale reactions tend to have lower yields of the desired conjugation product.
20. In order to verify that splicing will proceed successfully on chromatin, we suggest carrying out a preliminary *in nucleo* splicing experiments by isolating nuclei from transfected cells by hypotonic lysis, adding the intein cargo (without CPP), and monitoring splicing by western blot, taking time points up to 1 h post-delivery (Fig. 3). This is an important exercise to determine if any splicing difficulties might occur with the designed construct, and uncouple them from any cellular delivery issues.
21. If cells are to be used for live imaging microscopy, carry out the transfection in the appropriate glass-bottom plate rather than transfecting in standard cell culture dishes and reseeding Int^N-fusion expressing cells after transfection or delivery.

22. We utilize HEK293T cells in this example, but the method has also been applied successfully in HeLa, U2OS, and LF2 cell lines [13].
23. In the case of histones, the cells must be given at least a full cell cycle to express the histone-intein fusion and subsequently incorporate it into chromatin. However, for most other cellular proteins, expression times prior to delivery need not last so long.
24. Before proceeding with delivery, control experiments should be carried out to verify that the truncated intein-fused protein of interest is not deleterious to cellular homeostasis. We suggest, at a minimum, cell viability assays. In the case of histone H2B, because histones regulate all DNA-templated processes, we also suggest functional assays such as qPCR, RNA-seq or ChIP-based experiments. For other proteins, signaling or biochemical experiments may be necessary.
25. This can be performed at any scale, since the DMSO stock can be frozen and stored, but we generally dissolve approximately 5 mg of the conjugated CPP-Int^C cargo peptide at a time.
26. When diluting the peptide DMSO stock into cell media, do so gradually by adding media into the stock incrementally, or else the peptide may crash out of solution.
27. The optimal peptide concentrations for delivery may need to be determined empirically, as it can vary depending on the split intein pair in use and extein dependence.
28. Several analytical methods can be used to verify successful splicing depending on the cargo delivered, including western blot, immunofluorescence, live-cell imaging, or (co) immunoprecipitation.
29. The end-goal or final experimental protocol is likely to vary depending on the protein being studied. In the case of our histone delivery experiments, ChIP-based methods like ChIP-seq or ChIP-qPCR, or transcriptomics experiments (i.e., RNA-seq) are likely subsequent experiments.
30. Subsequent wash steps immediately after delivery should include iodoacetamide to rapidly alkylate the catalytic cysteine residues of any remaining unreacted intein.
31. At this point, in-gel or in-membrane fluorescence of TMR could be detected by imaging without proceeding further through the western blotting procedure.

Acknowledgments

Work in the David lab is supported by the Josie Robertson Foundation, the Pershing Square Sohn Cancer Alliance, Mr. William H. Goodwin and Mrs. Alice Goodwin and the Commonwealth Foundation for Cancer Research and the Center for Experimental Therapeutics at Memorial Sloan Kettering Cancer Center, the CCSG core grant P30 CA008748, and the NIH CEBRA award # DA044767. N.A.P. is supported by the NIH T32 GM115327-Tan chemistry-biology interface training grant and the National Science Foundation Graduate Research Fellowship Grant Number 2017239554.

References

1. Muir TW, Sondhi D, Cole PA (1998) Expressed protein ligation: a general method for protein engineering. Proc Natl Acad Sci U S A 95:6705–6710
2. Vila-Perelló M, Muir TW (2010) Biological applications of protein splicing. Cell 143:191–200
3. Perler FB (2002) InBase: the intein database. Nucleic Acids Res 30:383–384
4. Shah NH, Dann GP, Vila-Perelló M et al (2012) Ultrafast protein splicing is common among cyanobacterial split inteins: implications for protein engineering. J Am Chem Soc 134:11338–11341
5. Shah NH, Muir TW (2011) Split inteins: nature's protein ligases. Isr J Chem 51:854–861
6. Mootz HD (2009) Split inteins as versatile tools for protein semisynthesis. Chembiochem 10:2579–2589
7. Vila-Perelló M, Liu Z, Shah NH et al (2013) Streamlined expressed protein ligation using split inteins. J Am Chem Soc 135:286–292
8. Shah NH, Muir TW (2014) Inteins: nature's gift to protein chemists. Chem Sci 5:446–461
9. Juillerat A, Gronemeyer T, Keppler A et al (2003) Directed evolution of O6-alkylguanine-DNA alkyltransferase for efficient labeling of fusion proteins with small molecules in vivo. Chem Biol 10:313–317
10. Los GV, Encell LP, McDougall MG et al (2008) HaloTag: a novel protein labeling technology for cell imaging and protein analysis. ACS Chem Biol 3:373–382
11. Giriat I, Muir TW (2003) Protein semisynthesis in living cells. J Am Chem Soc 125:7180–7181
12. Mootz HD, Blum ES, Tyszkiewicz AB et al (2003) Conditional protein splicing: a new tool to control protein structure and function in vitro and in vivo. J Am Chem Soc 125:10561–10569
13. David Y, Vila-Perelló M, Verma S et al (2015) Chemical tagging and customizing of cellular chromatin states using ultrafast trans-splicing inteins. Nat Chem 7:394–402
14. Wadia JS, Stan RV, Dowdy SF (2004) Transducible TAT-HA fusogenic peptide enhances escape of TAT-fusion proteins after lipid raft macropinocytosis. Nat Med 10:310–315
15. Hoye AT, Davoren JE, Wipf P et al (2008) Targeting mitochondria. Acc Chem Res 41:87–97
16. Yu H-C, Lu M-C, Li C et al (2013) Targeted delivery of an antigenic peptide to the endoplasmic reticulum: application for development of a peptide therapy for ankylosing spondylitis. PLoS One 8:e77451
17. Stevens AJ, Sekar G, Shah NH et al (2017) A promiscuous split intein with expanded protein engineering applications. Proc Natl Acad Sci 114:8538–8543
18. Suhorutsenko J, Oskolkov N, Arukuusk P et al (2011) Cell-penetrating peptides, PepFects, show no evidence of toxicity and immunogenicity in vitro and in vivo. Bioconjug Chem 22:2255–2262
19. Guidotti G, Brambilla L, Rossi D (2017) Cell-penetrating peptides: from basic research to clinics. Trends Pharmacol Sci 38:406–424
20. García-Martín F, Quintanar-Audelo M, García-Ramos Y et al (2006) ChemMatrix, a Poly(ethylene glycol)-based support for the solid-phase synthesis of complex peptides. J Comb Chem 8:213–220
21. Stevens AJ, Brown ZZ, Shah NH et al (2016) Design of a split intein with exceptional protein splicing activity. J Am Chem Soc 138:2162–2165

Chapter 11

Site-Directed Modification of Yeast-Produced Proteins Using Expressed Protein Ligation

Benjamin J. Umlauf and Eric V. Shusta

Abstract

Expressed protein ligation (EPL), using non–self-cleaving inteins, allows for the site-specific addition of customized chemical moieties to the termini of proteins. In this way, protein activity can be preserved while functionalizing the target protein with a wide range of chemical handles. Here, we describe methods for EPL-based modification of proteins produced by yeast, employing an engineered, non–self-cleaving intein known as 202-08. Methods for EPL modification of both yeast surface displayed and secreted proteins with bioorthogonal chemical groups are described. These methods allow for the site-specific modification of intein-fused proteins produced in yeast.

Key words scFv, Yeast surface display, Yeast secretion system, Expressed protein ligation (EPL), 202-08 intein, Bioorthogonal click chemistry

1 Introduction

Many research and therapeutic applications of proteins require modification with bioorthogonal chemical handles, fluorescent molecules or therapeutic small molecules. However, nonspecific protein labeling strategies such as those employing NHS ester, maleimide, or imidoester chemistries often result in diminished protein activity and/or reduced binding affinity [1, 2]. Alternatively, expressed protein ligation (EPL) allows for appendage of a bioorthogonal handle specifically to the protein terminus to facilitate subsequent functionalization [3–5]. For yeast produced proteins, we previously described an engineered derivative of the non–self-cleaving Mxe GyrA intein, termed 202-08, which enhances protein production levels of intein–protein fusions from yeast [3, 4]. The 202-08 intein is genetically fused to the C-terminus of a target protein, and the fusion protein is produced either as a yeast surface fusion (yeast display) or as a secreted protein (Fig. 1). Performing EPL reactions using yeast displayed protein fusions is a rapid method for modification of small amounts of protein, while

Miquel Vila-Perelló (ed.), *Expressed Protein Ligation: Methods and Protocols*, Methods in Molecular Biology, vol. 2133, https://doi.org/10.1007/978-1-0716-0434-2_11,

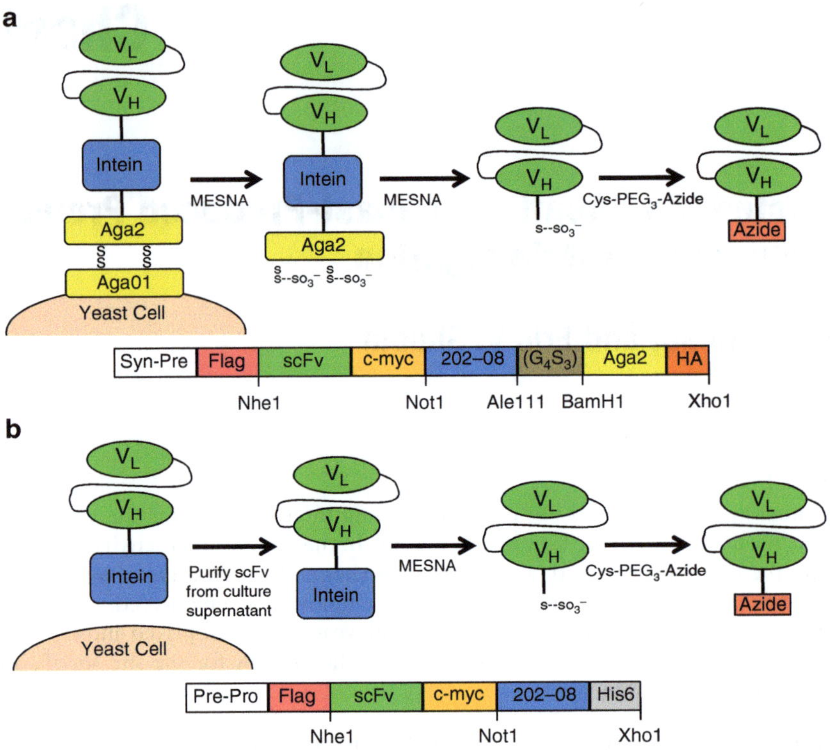

Fig. 1 Diagram of yeast EPL protocol. (**a**) Surface display protocol: scFv is fused to both 202-08 intein as well as Aga2p to tether scFv to the yeast surface. Next, MESNA is added to release the scFv fusion from the yeast surface as well as trigger intein cleavage yielding a C-terminal thioester that is reactive with a N-terminal cysteine-containing reagent. In this way, a cysteine-azide is appended. A block diagram depicting major elements of the pCT yeast surface display vector is included below the graphic protocol. As noted in Subheading 2.1, we recommend glycine or alanine in the −1 position upstream of the amino-terminal Cys of the 202-08 intein. (**b**) Secretion protocol: scFv is fused to 202-08 intein and secreted into the yeast culture supernatant and purified. Next, MESNA is added to supernatant to trigger intein cleavage. Finally, an azide handle is appended. A block diagram depicting major elements of the pRS316 yeast secretion vector is included below the graphic protocol. Full 202-08 intein nucleic acid and amino acid sequences are available in Ref. [16]

performing EPL reactions on secreted protein fusions allows for preparative scale protein modification. Oftentimes, the target protein is a single-chain antibody (scFv). The intein moiety undergoes an N- to S-acyl shift, and nucleophilic attack by a reducing agent such as MESNA yields a C-terminal thioester [6, 7]. Next, the desired functionality containing an N-terminal cysteine reacts with the thioester resulting in covalent attachment to the C-terminus of a protein (Fig. 1) [4, 8, 9]. Importantly, restricting modification to the C-terminus preserves scFv binding activity as well as ensures that a single functional moiety is appended to each scFv [5]. Using these approaches, it is possible to functionalize scFvs that are produced as yeast displayed constructs, as well as those secreted and purified from culture supernatants. Detailed EPL methodologies for both protein sources are presented below.

2 Materials

Yeast media can be stored at room temperature, while plates are stored at 4 °C. We recommend storing all suspended peptide and bioorthogonal modified reagents at −20 to −80 °C. After modification, we recommend storing scFv-azide at 4 °C.

2.1 Yeast Culture

1. EBY100 yeast (MATa AGA1::GAL1-AGA1::URA3 ura3–52 trp1 leu2Δ1 his3Δ200 pep4::HIS3 prb1Δ1.6R can1 GAL) [10] (*see* **Note 1**).
2. pCT plasmid backbone encoding for scFv fused to 202-08 Intein [3]. This plasmid contains ampicillin resistance for bacterial selection and TRP1 auxotrophic marker for yeast selection. The scFv-202-08 is fused to Aga2 protein for yeast surface display. We recommend glycine or alanine as the −1 amino acid preceding the 202-08 intein to prevent autocleavage [15].
3. SD-CAA media: 20.0 g/L dextrose, 6.7 g/L yeast nitrogen base, 5.0 g/L casamino acids, 10.19 g/L $Na_2HPO_4{\cdot}7\ H_2O$, 8.56 g/L $NaH_2HPO_4{\cdot}H_2O$, in water.
4. SG-CAA induction media: 20.0 g/L galactose, 6.7 g/L yeast nitrogen base, 5.0 g/L casamino acids, 10.19 g/L $Na_2HPO_4{\cdot}7\ H_2O$, 8.56 g/L $NaH_2HPO_4{\cdot}H_2O$, in water.
5. SD-CAA Agar Plates: SD-CAA media+15g/L Bacto agar.
6. YVH10 yeast (MATα PDI1::GAPDH-PDI1::LEU2 ura3–52 trp1 leu2Δ1 his3Δ200 pep4::HIS3 prb1Δ1.6R can1 GAL) [11] (*see* **Note 2**).
7. pRS316 plasmid backbone encoding for scFv fused to 202-08 Intein [3]. This plasmid contains ampicillin resistance for bacterial selection and URA3 auxotrophic marker for yeast selection. We recommend glycine or alanine as the −1 amino acid preceding the 202-08 intein to prevent autocleavage [15].
8. SD-2xSCAA+Trp media: 20 g/L dextrose, 6.7 g/L yeast nitrogenous base, 10.19 g/L $Na_2HPO_4{\cdot}7H_2O$, 8.56 g/L $NaH_2HPO_4{\cdot}H_2O$, 190 mg/L Arg, 108 mg/L Met, 52 mg/L Tyr, 290 mg/L Ile, 440 mg/L Lys, 200 mg/L Phe, 1260 mg/L Glu, 400 mg/L Asp, 480 mg/L Val, 220 mg/L Thr, 130 mg/L Gly, and 40 mg/L Trp. Please note the media lacks leucine and uracil to allow for auxotrophic selection.
9. SG-2xSCAA+Trp induction media: 20 g/L galactose, 6.7 g/L yeast nitrogenous base, 10.19 g/L $Na_2HPO_4{\cdot}7H_2O$, 8.56 g/

L $NaH_2HPO_4{\cdot}H_2O$, 190 mg/L Arg, 108 mg/L Met, 52 mg/L Tyr, 290 mg/L Ile, 440 mg/L Lys, 200 mg/L Phe, 1260 mg/L Glu, 400 mg/L Asp, 480 mg/L Val, 220 mg/L Thr, 130 mg/L Gly, and 40 mg/L Trp. Please note the media lacks leucine and uracil to allow for auxotrophic selection.

10. SD-SCAA+Trp plates: SD-SCAA+15g/L Bacto agar.
11. Bovine serum albumin (BSA).
12. Nhe1 and Not1 restriction enzymes.

2.2 scFv Purification

1. 0.22 μm polyethersulfone (PES) filter membranes.
2. 4000–6000 kDa MWCO dialysis tubing.
3. Tris-buffered saline (TBS): 25 mM Tris, 300 mM NaCl, 2 mM KCl, pH 7.9.
4. TBS + 250 mM imidazole, pH = 7.9. For 100 mL of solution, dissolve 1.702 g of imidazole into TBS.
5. Cobalt resin.
6. 0.5 mL 10 kDa PES MWCO spin filters (*see* **Note 3**).
7. 50 mM HEPES pH = 7.2.
8. Packable spin columns.

2.3 Synthesis of Cys-PEG$_3$-Azide

1. Fmoc-Cys (SASRIN™ resin)-OH.
2. Azido-PEG$_3$-amine.
3. Dimethylformamide (DMF).
4. *N*-methyl-morpholine (NMM).
5. 2-(6-Chloro-1H-benzotriazol-1-yl)-*N*,*N*,*N′*,*N′*-tetramethylaminium hexafluorophosphate (HCTU).
6. 20%(v/v) piperidine in DMF.
7. Dichloromethane (DCM).
8. Trifluoroacetic acid (TFA).
9. Triisopropylsilane (TIPs).
10. Diethyl Ether.
11. HPLC grade water + 0.01% TFA.
12. HPLC grade acetonitrile (ACN) + 0.01% TFA.

2.4 EPL Modification

1. 50 mM HEPES pH = 7.2.
2. 200 mM MESNA in 50 mM HEPES pH = 7.2.
3. 5 mM Cys-PEG$_3$-Azide in 50 mM HEPES pH = 7.2 (*see* **Note 4**).
4. 0.5 mL 3 kDa MWCO spin filters.

2.5 Analysis of Intein Cleavage and Bioorthogonal Handle

1. DBCO-PEG$_4$-Biotin.
2. DBCO-Cy5.
3. 4–12% Bis-Tris acrylamide gels.
4. 20× MES running Buffer.
5. 4× LDS Sample Buffer: 106 mM Tris–HCl, 141 mM Tris base, 2% LDS, 10% Glycerol, 0.51 mM EDTA, 0.22 mM SERVA Blue G250, 0.175 mM Phenol Red, pH 8.5.
6. Nitrocellulose.
7. Transfer Buffer.
8. TBS + 0.1% Tween 20, v/v (TBST).
9. Anti-FLAG M1 antibody.
10. Anti-biotin BTN.4 antibody.
11. Donkey anti-mouse HRP antibody.
12. 0.5 mL 3 kDa MWCO spin filters.
13. BCA protein quantification kit.

3 Methods

3.1 Yeast Culture

3.1.1 Yeast Surface Display

This method describes the basic steps for generating yeast expressing scFv-intein as a yeast surface-displayed fusion.

1. Clone desired scFv flanked by Nhe1 and Not1 restriction sites into pCT plasmid backbone using traditional cloning methods, such that the C-terminus of the scFv is fused to the 202-08 Intein. We recommend a glycine or alanine in the −1 position preceding the intein to prevent autocleavage [15].
2. Transform EBY100 yeast with pCT plasmid using a LiAc transformation protocol [12].
3. Plate transformed yeast on SD-CAA Agar plates and incubate at 30 °C for 24–48 h.
4. Inoculate a single colony into a 15 mL round bottom culture tube containing 3 mL of SD-CAA.
5. Incubate at 30 °C O.N. with 260 rpm shaking.
6. The following morning use the O.N. culture to inoculate a fresh SD-CAA culture to $OD_{600} = 0.3$. Ensure 5 times the headspace in the size of the culture flask compared to volume of media. We typically use 3–50 mL cultures depending on downstream experimental requirements.
7. Incubate at 30 °C for ~4–6 h with 260 rpm shaking till $OD_{600} = 0.8$.
8. Pellet yeast by centrifugation at 1100 × g for 5 min.

9. Remove media and resuspend yeast pellet with the same volume of SG-CAA induction media that was removed.
10. Incubate at 20 °C for 24–72 h with 260 rpm shaking to induce surface expression of scFv-Intein fusions.
11. Harvest yeast pellet by centrifuging cultures at 1100 × *g* for 5–10 min.
12. Yeast pellet is retained for immediate EPL modification described below in Subheading 3.4.1

3.1.2 Yeast Secretion

This method describes the basic steps for generating yeast producing scFv that is secreted into the growth medium.

1. Clone desired scFv flanked by Nhe1 and Not1 restriction sites into pRS316 plasmid backbone using traditional cloning methods such that the C-terminus of the scFv is fused to the 202-08 Intein. We recommend a glycine or alanine in the −1 position in front of the intein to prevent autocleavage [15].
2. Transform YVH10 yeast with pRS316 plasmid using the LiAc protocol [12].
3. Plate transformed yeast onto SD-SCAA+Trp agar plates and incubate at 30 °C for 48–72 h.
4. Inoculate a single YVH10 yeast colony transformed with pRS316-scFV-Intein into 3 mL of SD-2XSCAA + Trp media in a 15 mL round bottom tube.
5. Culture overnight at 30 °C with 260 rpm shaking.
6. Reset the following day to an $OD_{600\ nm} = 0.1$, by diluting to the desired culture volume in SD-2xSCAA+Trp. Ensure a minimum of 5 times headspace in the culture flask. We typically use 50–400 mL cultures depending on downstream experimental requirements.
7. Culture for 72 h at 30 °C with 260 rpm shaking.
8. Pellet yeast by centrifugation at 1100 × *g* for 5 min and remove supernatant.
9. Resuspend yeast pellet in induction media by adding an equivalent volume of SG-2xSCAA + Trp (20 g/L galactose replacing dextrose) containing 0.1% w/v bovine serum albumin (BSA) that was removed in **step 8**.
10. Culture for 72 h at 20 °C with 260 rpm shaking (*see* **Note 5**).
11. Centrifuge culture at 4000 × *g* for 10 min to pellet yeast and recover the culture supernatant containing the scFv-intein fusion protein.
12. Retain yeast culture supernatant for immediate scFv-intein fusion purification described in Subheading 3.2 below.

3.2 scFv Purification from Yeast Supernatants

After producing scFv-intein fusions using the secretion protocol above, scFv-intein fusion is purified from the culture supernatant using a metal chelation purification strategy.

1. Filter clarified supernatants from Subheading 3.1.2 Yeast Secretion protocol **step 12** (above) through 0.22 μm PES membrane (*see* **Note 3**).
2. Dialyze filtered supernatant against TBS using 4000–6000 kDa dialysis tubing using a minimum of 800 times the volume of supernatant being dialyzed or 6 h at room temperature or O.N. at 4 °C.
3. Add TBS+250 mM imidazole to dialyzed supernatant to a final concentration of 5 mM imidazole.
4. Dilute TBS+250 mM imidazole to TBS+5 mM imidazole with TBS and wash HisPur beads with a minimum of 10 times bead volume.
5. Add 500–1000 μL of cobalt HisPur resin to 400 mL of dialyzed supernatant.
6. Gently mix overnight at 4 °C on a rocker, ensuring beads remain in suspension.
7. Collect beads by centrifuging at 1000 × *g* for 10 min.
8. Aspirate 80–90% of the supernatant taking care not to disturb pelleted beads.
9. Transfer remaining supernatant and the beads to a packable spin column with frit and collect beads using filtration by applying vacuum.
10. Dilute TBS+250 mM imidazole to TBS+10 mM imidazole with TBS and wash beads with 25–50 mL of TBS+10 mM imidazole.
11. Centrifuge spin column containing washed beads at 1000 × *g* for 2 min to remove residual wash buffer.
12. Elute scFv-Intein by incubating beads with 300–500 μL of TBS+250 mM imidazole for 5 min.
13. Collect elution by centrifuging beads at 1000 × *g* for 2 min and recovering the protein-containing eluate.
14. Repeat **steps 11** and **12** two to three additional times to elute all protein from the cobalt beads. Elution fractions can be pooled at this point.
15. Buffer exchange pooled elutant into 50 mM HEPES pH = 7.2 using a 0.5 mL 10 kDa MWCO spin filter via manufacturer's protocol (*see* **Note 3**).
16. Retain purified protein for EPL modification described below in Subheading 3.4.2. ScFv-intein fusions are generally stable for up to 1 month when stored at 4 °C.

3.3 Synthesis of Cys-PEG3-Azide

The protocol below describes a peptide-based synthesis strategy [13, 14] for generating a Cys-bioorthogonal reagent for the EPL reaction. Custom peptide manufacturers could generate similar reagents for laboratories that do not have the chemical equipment or expertise required for this protocol.

1. For 200 μmol reaction, swell 444.4 mg of Fmoc-Cys (SASRIN™ resin)-OH (substitution level 0.45 mmol/g) in 2.5 mL DMF 2 times for 30 min each wash (*see* **Note 6**).
2. Dissolve 174.6 mg of Azido-PEG$_3$-Amine (800 μmols, 4 equivalents relative to the resin) in 1 mL of DMF.
3. Dissolve 82.74 mg of HCTU (200 μmols, 1 equivalent relative to the resin) and 44 μL of NMM into 1 mL of DMF for activation solution.
4. Add Azido-PEG$_3$-amine solution to swollen Fmoc-Cys (SASRIN™ resin)-OH followed by 1 mL of HCTU and NMM activation solution.
5. Gently agitate the reaction by rocking (or mix via bubbling with inert gas) for 3 h at room temperature.
6. Wash resin 9 times with 2.5 mL of DMF, 30 s each wash. Remove DMF by filtration or decanting.
7. Wash two times with 2.5 mL of 20% piperidine in DMF for 15 min each wash. Remove piperidine solution by filtration or decanting.
8. Wash six times with 2.5 mL of DMF, 30 s each wash. Remove DMF by filtration or decanting.
9. Wash nine times with 2.5 mL of DCM, 30 s each wash. Remove DCM by filtration or decanting.
10. Blow N_2 through the resin for 10 min to dry beads.
11. Transfer to a desiccator overnight.
12. Cleave peptide from the resin by adding 5–10 mL of 94% TFA with 2% TIPs and gently rocking resin for 3 h.
13. Separate peptide from resin via filtration.
14. Evaporate TFA solution under a stream of nitrogen to less than 1 mL.
15. Precipitate peptide with cold diethyl ether. Incubate at −80°C for ~30 min to ensure complete precipitation.
16. Pellet precipitated peptide by centrifugation at 4000 × *g* for 5 min.
17. Pour off cold ether.
18. Allow pellet to dry for 10–15 min in a fume hood.
19. Transfer crude peptide to desiccator to complete drying.
20. Dissolve peptide in 90:10 H_2O:ACN +0.01% TFA.

21. Purify peptide to >90% purity using reverse phase HPLC, C18 column, with a 90–60% gradient of H_2O:ACN + 0.01% TFA.
22. Lyophilize peptide to remove water and acetonitrile.
23. Confirm Cys-PEG_3-Azide peptide sequence via Mass Spectrometry.

3.4 EPL Modification

3.4.1 Surface Display

This protocol allows for modification of scFv directly released from the yeast surface with a bioorthogonal handle without requiring scFv purification.

1. Resuspend yeast pellet from Subheading 3.1.1 in 800 μL of 50 mM HEPES pH = 7.2 per 50 mL of culture (create a slurry).
2. Add 100 μL of 2M MESNA in 50 mM HEPES (10×, final concentration is 200 mM of MESNA).
3. Add 100 μL of 50 mM Cys-PEG_3-Azide in 50 mM HEPES for final concentration of 5 mM.
4. Shake reaction at 260 rpm for 45 min at room temperature.
5. Centrifuge reaction at ≥12,000 × *g* for 5 min to pellet yeast.
6. Harvest supernatant and transfer to clean tube.
7. Incubate harvested supernatant for an additional 2–18 h (*see* **Note 7**).
8. Remove excess Cys-PEG_3-Azide by buffer exchanging the reaction 2 times with 50 mM MESNA using 0.5 mL 3 kDa MWCO spin filter according to manufacturer's protocol.
9. Remove MESNA by washing 2–4 time with PBS using same 0.5 mL 3 kDa MWCO spin filter
10. Generally, protein is concentrated to <100 μL then the filter is washed with additional PBS to bring final volume to 150 μL. These volumes depend on reaction size and protein yield, and we typically use this elution volume for 3–50 mL of induced culture.

3.4.2 Yeast Secretion System

This protocol describes EPL modification of scFv-Intein fusions that were previously purified from yeast culture supernatants.

1. Add 2M MESNA in 50 mM HEPES to Co^{++} purified scFv-202-08 Intein fusion from Subheading 3.2 to final concentration of 200 mM.
2. Add 50 mM Cys-PEG3-Azide in 50 mM HEPES to reaction for final concentration of 5 mM Cys-PEG3-Azide.
3. Shake reaction at 260 rpm for 3–18 h at room temperature (*see* **Note 7**).
4. Remove excess Cys-PEG_3-Azide by buffer exchanging the reaction 2 times with 50 mM MESNA in 50 mM HEPES

using 0.5 mL 3 kDa MWCO spin filter according to manufacturer's protocol.

5. Remove MESNA by buffer exchanging 2–4 times with PBS using same 0.5 mL 3 kDa MWCO spin filter.
6. Generally, protein is concentrated to <100 μL then the filter is washed with additional PBS to bring final volume to 150 μL. These volumes depend on reaction size and protein yield; we typically use this elution volume for 50–200 mL of culture supernatant.

3.5 Analysis of Intein Cleavage and Bioorthogonal Handle Ligation

Here, two methods to analyze the amount and percentage of scFv modified with the bioorthogonal handle are described. The first uses western blotting methods to detect the percentage of cleaved scFv-intein by size and presence of bioorthogonal handle modified with biotin. The second method uses fluorescent probes to determine presence of bioorthogonal handle and BCA assay to determine overall protein yield.

3.5.1 Western Blot (Fig. 2)

1. Add 50 mM of DBCO-PEG$_4$-Biotin suspended in 50 mM HEPES to purified scFv-Azide to final concentration of 5 mM and incubate at 30 °C for 90 min with 260 rpm shaking to add a biotin to the C-terminus of the scFv-Azide via click chemistry. Ensure DBCO-biotin is in >10 to 100-fold excess of scFv-Azide to drive reaction to completion.

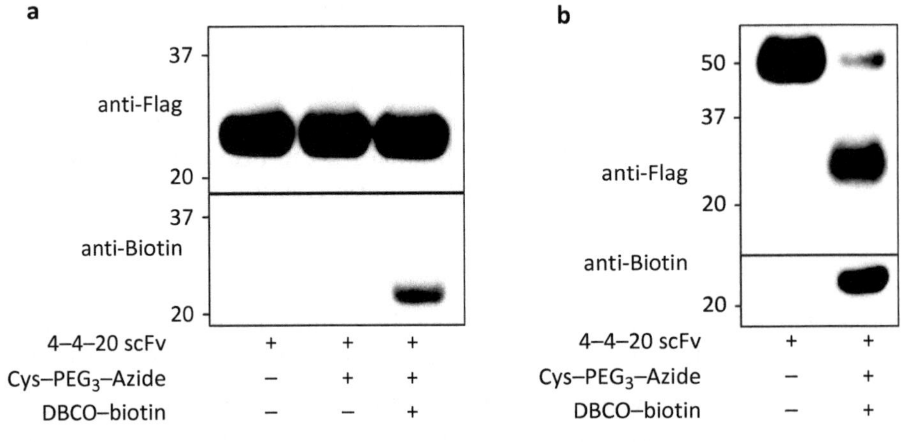

Fig. 2 Verification of scFv-Azide generation using the presented EPL protocol. (**a**) Western blots of 4-4-20-Intein scFv, produced using surface display protocol after MESNA cleavage, with the addition of Cys-PEG3-Azide and with subsequent reaction with DBCO-biotin. Conditions for each lane are indicated below the blots. The top blot is probed for a flag tag present of the N-terminus of the scFv. The lower blot is probed for scFv modification with biotin. (**b**) Western blots of 4-4-20 scFv, produced using the secretion protocol. The top blot is probed for a flag tag present at the N-terminus of the scFv. The lower blot is probed for biotin modification. Lane one is purified scFv-Intein prior to initiating intein cleavage. Lane two is scFv that has undergone the EPL protocol and subsequent reaction with DBCO-biotin. Conditions for each lane are indicated below the blots

2. For western blotting, add 0.1–1 μL of scFv-biotin reaction to PBS to reach final volume of 15 μL.
3. Add 5 μL of 4× LDS Sample Buffer with 5% BME to samples from **step 2** (20 μL final volume).
4. Denature samples by boiling for 10 min at 95 °C.
5. Load 20 μL of sample and run samples on 4–12% Bis-Tris gel using 1× MES running buffer.
6. Transfer proteins from the gel to nitrocellulose using manufacturer's protocol.
7. Block membrane in 5% BSA in TBST for a minimum of 15 min.
8. Probe blot with anti-FLAG M1 antibody diluted in TBST (1:1000–1:5000) to identify percentage of cleaved scFv-Azide product (will recognize both the full-length fusion and cleaved scFv products) (*see* **Note 8**).
9. Probe a replicate blot with BTN.4 antibody (1:1000) to quantify presence of Azide handle using biotin as a readout. Typically ~80% of the scFv cleaved fraction is modified with bioorthogonal handle.

3.5.2 Fluorescent Analysis

1. Add 50 mM of DBCO-Cy5 to purified scFv-Azide to final concentration of 5 mM to functionalize the scFv with Cy5. Ensure the amount of DBCO-Cy5 is in >10 to 100-fold excess of scFv-Azide to drive reaction.
2. Incubate at 30 °C for 90 min with 260 rpm shaking.
3. Remove excess DBCO-Cy5 via buffer exchange against PBS using a 0.5 mL 3 kDa MWCO spin filter according to manufacturer's protocol.
4. Determine scFv-Cy5 protein concentration using BCA assay.
5. Quantify Cy5 modification by determining the fluorescence.

4 Notes

1. EBY100 yeast is used for scFv surface display protocols.
2. YVH10 yeast is used for scFv secretion protocols.
3. Regenerated cellulose membranes contain high amounts of thiols that induce intein cleavage during isolation. We recommend using a PES membrane for buffer exchanges with secreted scFv-intein fusions prior to EPL.
4. Other bioorthogonal click handles can be appended to the scFv including alkyne and styrene moieties, however, many of the strain-based bioorthogonal reagents (e.g., DBCO) polymerize in the presence of high concentrations of free thiols used to facilitate intein cleavage and are not recommended to be

attached to the scFv when using this EPL method [5]. However, as demonstrated in Fig. 2, after MESNA and excess Cys is removed, scFv can undergo further modification with strain-based reagents such as DBCO in a stepwise fashion (e.g., Step 1 scFv-azide, Step 2 DBCO-PEG_4-biotin).

5. Refeeding cultures with fresh yeast nitrogen base and amino acids every 24 h during the induction phase may increase expression of protein-intein fusions. Details can be found in Ref. [15].
6. A disposable peptide reaction vessel is resistant to solvent and contain filters for both washing beads and eluting peptide. Using an automated synthesizer makes performing this protocol easier but is not essential. We do not recommend using a microwave based synthesizer due to the presence of PEG in the reaction.
7. Sample reaction at time points after 3 h to determine optimal yield of cleaved scFv-Azide.
8. For scFv produced and modified using the yeast secretion protocol, we typically observe 60–80% of the scFv is cleaved and modified with the bioorthogonal handle.

Acknowledgments

This work was funded by National Science Foundation grant CBET1403350 and a Falk Medical Research Trust Catalyst Award.

References

1. McCombs JR, Owen SC (2015) Antibody drug conjugates: design and selection of linker, payload and conjugation chemistry. AAPS J 17:339–351. https://doi.org/10.1208/s12248-014-9710-8
2. Flygare JA, Pillow TH, Aristoff P (2013) Antibody-drug conjugates for the treatment of cancer. Chem Biol Drug Des 81:113–121. https://doi.org/10.1111/cbdd.12085
3. Marshall CJ, Grosskopf VA, Moehling TJ, Tillotson BJ, Wiepz GJ, Abbott NL, Raines RT, Shusta EV (2015) An evolved Mxe GyrA intein for enhanced production of fusion proteins. ACS Chem Biol 10:527–538. https://doi.org/10.1021/cb500689g
4. Marshall CJ, Agarwal N, Kalia J, Grosskopf VA, McGrath NA, Abbott NL, Raines RT, Shusta EV (2013) Facile chemical functionalization of proteins through intein-linked yeast display. Bioconjugate Chem 24:1634–1644. https://doi.org/10.1021/bc4002618
5. Umlauf B, Mix KA, Grosskopf VA, Raines RT, Shusta EV (2018) Site-specific antibody functionalization using tetrazine-styrene cycloaddition. Bioconjugate Chem 29(5):1605–1613. https://doi.org/10.1021/acs.bioconjchem.8b00114
6. Sydor JR, Mariano M, Sideris S, Nock S (2002) Establishment of intein-mediated protein ligation under denaturing conditions: C-terminal labeling of a single-chain antibody for biochip screening. Bioconjugate Chem 13(4):707–712
7. Muir TW, Sondhi D, Cole PA (1998) Expressed protein ligation: a general method for protein engineering. Proc Natl Acad Sci U S A 95(12):6705–6710
8. Albertsen L, Shaw AC, Norrild JC, Strømgaard K (2013) Recombinant production of peptide C-terminal α-amides using an engineered intein. Bioconjugate Chem 24:1883–1894. https://doi.org/10.1021/bc4002689

9. Elias DR, Cheng Z, Tsourkas A (2010) An intein-mediated site-specific click conjugation strategy for improved tumor targeting of nanoparticle systems. Small 6:2460–2468. https://doi.org/10.1002/smll.201001095
10. Boder ET, Wittrup KD (1997) Yeast surface display for screening combinatorial polypeptide libraries. Nat Biotechnol 15:553–557. https://doi.org/10.1038/nbt0697-553
11. Shusta EV, Raines RT, Plückthun A, Wittrup KD (1998) Increasing the secretory capacity of Saccharomyces cerevisiae for production of single-chain antibody fragments. Nat Biotechnol 16:773–777. https://doi.org/10.1038/nbt0898-773
12. Gietz RD, Schiestl RH (2007) High-efficiency yeast transformation using the LiAc/SS carrier DNA/PEG method. Nat Protoc 2:31–34. https://doi.org/10.1038/nprot.2007.13
13. Brown KC (2010) Peptidic tumor targeting agents: the road from phage display peptide selections to clinical applications. Curr Pharm Des 16:1040–1054
14. Gray BP, Brown KC (2014) Combinatorial peptide libraries: mining for cell-binding peptides. Chem Rev 114:1020–1081. https://doi.org/10.1021/cr400166n
15. Goulatis LI, Ramanathan R, Shusta EV (2019) Impacts of the −1 Amino Acid on Yeast Production of Protein-Intein Fusions. Biotechnol Prog 35:e2736. https://doi.org/10.1002/btpr.2736
16. Shusta EV, Marshal CJ (2017) Engineered intein for improved production of protein-intein fusions. US Patent 9,771,412, 26 Sept 2017

Chapter 12

Site-Specific Phosphorylation of PDZ Domains

Sana Ma, Kristian Strømgaard, and Louise S. Clemmensen

Abstract

Classical approaches for probing protein phosphorylation events rely on phosphomimicking amino acids or enzymatic phosphorylation of proteins. In many cases, phosphomimicking amino acids inadequately imitate actual protein phosphorylation, whereas the latter method suffers from an inability to control site specificity and stoichiometry. To circumvent these shortcomings, chemical biological approaches have been developed to enable introduction of phosphorylated amino acids into proteins in a reliable and controlled way. Here, we describe methods to make semisynthetic, phosphorylated PDZ domains, covering expressed protein ligation (EPL) strategies involving modifications within the N-terminal or C-terminal regions. We also enclose protocols for the biophysical characterization of the semisynthetic phosphorylated PDZ domains to establish whether the introduced phosphorylation affects protein structure, stability, and function.

Key words PDZ domains, Posttranslational modification, Phosphorylation, Protein-protein interactions, Semisynthesis

1 Introduction

Protein phosphorylation is one of the most common posttranslational modifications (PTMs), and a key mechanism for regulating numerous cell functions, including signal transduction, cell cycle, cell differentiation, cell mobility, and metabolism. Thus, proper regulation of phosphorylation is critical to human health, and several human diseases are caused by aberrant protein phosphorylation. Protein phosphorylation occurs through protein kinases that catalyze the covalent attachment of a phosphate group to primarily Ser, Thr or Tyr residues (*see* Fig. 1). This mechanism is reversed by phosphatases. The transfer of a phosphate group to an amino acid side chain increases the size of the residue, adds negative charge and installs strong hydrogen bond acceptors. Such chemical changes can perturb the protein structure locally and induce conformational changes that ultimately lead to functional alterations [1–3]. In many cases, protein phosphorylation obstructs or promotes intra- or extracellular protein-protein interactions (PPIs),

Miquel Vila-Perelló (ed.), *Expressed Protein Ligation: Methods and Protocols*, Methods in Molecular Biology, vol. 2133, https://doi.org/10.1007/978-1-0716-0434-2_12, © Springer Science+Business Media, LLC, part of Springer Nature 2020

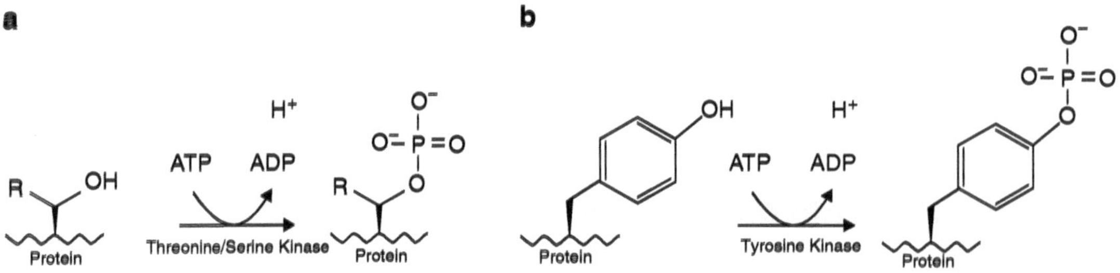

Fig. 1 Diagrams of posttranslational phosphorylation of Ser, Thr, and Tyr residues as catalyzed by kinases. (**a**) Phosphorylation of Ser or Thr. (**b**) Phosphorylation of Tyr

constituting a reversible switch that orchestrates the spatiotemporal dynamics of cellular networks [4–6]. Consequently, phosphopeptidomimetics that are peptide-like molecules specifically designed to mimic the phosphopeptide and act by blocking the normal phosphorylation process have been shown to prevent activation of oncogenic signaling pathways [7]. Moreover, several small molecules and antibodies that target protein kinases and the receptors that activates them are progressing through clinical development for various diseases, and indeed there are currently more than 20 kinase inhibitor drugs on the market [8–10].

Phosphoproteomics studies have identified hundreds of thousands of protein phosphorylation sites, but our understanding of the functions of the vast majority of these PTMs is lacking. Unraveling cellular phosphorylation actions is challenging. First, the reversible phosphorylation is often substoichiometric meaning that the protein is only partly phosphorylated at any given time. Second, the protein often contains several phosphorylation sites and may therefore exist in differentially phosphorylated isoforms simultaneously in this substoichiometric population. Finally, for many proteins the kinase responsible for phosphorylating a particular site remains unknown and the kinase may have several substrate sites within the same protein. Phosphomimetic mutations, such as a Glu or Asp, are often introduced in place of the phosphorylated residue to resemble the negative charge of the phosphate group. However, the chemical properties of phosphates and carboxylate groups differ substantially. Moreover, these phosphomimetic residues lack an aromatic moiety and have proven inefficient as substitutes for phosphotyrosine (pTyr) [6].

To evade the problems of enzymatic phosphorylation and phosphomimetic mutagenesis, innovative methodologies to site-specifically introduce phosphorylated amino acids have been developed. An elegant approach relies on unnatural amino acid mutagenesis where the phosphorylated amino acid is incorporated biosynthetically during ribosome-mediated translation. This technique was first employed successfully to incorporate phosphoserine (pSer) into proteins during recombinant expression [11], and recently expanded to incorporate phosphotyrosine (pTyr) and phosphothreonine (pThr) [12, 13]. Though a promising method, limitations exist in that not all sites are amenable to incorporation of the phosphorylated amino acid, and expression often results in product truncations and in low yields, not to mention a limited chemical space of modifications that can be introduced.

Expressed protein ligation (EPL) is a robust and widely used technique for incorporation of phosphorylated amino acids and their nonhydrolyzable phosphonate analogs. EPL allows the modification to be introduced to a protein via a synthetic peptide, which is ligated to a recombinant fragment of the protein. This allows for introduction of site-specific protein phosphorylation in a controlled and full stoichiometric manner. A major advantage of EPL is the nearly infinite chemical space of the introduced modification. Three relatively recent examples of EPL for studying phosphorylation of proteins are the Akt1 protein kinase, huntingtin exon1 (Httex1) and the PDZ (postsynaptic density protein, disks large, zonula occludens) domains of postsynaptic density protein (PSD-95). Using EPL, C-terminally phosphorylated forms of Akt1 were generated and pSer473 and pSer477/pThr479 were shown to activate Akt1 by relieving intramolecular autoinhibition via two different mechanisms [14]. For Httex1, pTyr3 was introduced N-terminally to the protein with EPL, where it was found to play a regulatory role in slowing oligomerization of polyQ repeats within the protein [15]. Finally, phosphorylated PDZ domains of PSD-95 were used to screen how phosphorylation tunes PPIs of this critical neuronal scaffolding protein [6].

As with other methods, there are also limitations to using EPL to study protein phosphorylation. EPL is best suited to modifications close to the N- and C-termini of the protein to avoid the need for multiple ligations. In addition, the requirement of a Cys residue at the ligation junction can impede the design of a semisynthetic strategy. For some proteins, a Cys residue may need to be introduced by mutagenesis and careful considerations are needed to ensure that the introduced mutation is not perturbing the structure and function of the protein. This limitation has to some extent been circumvented by development of ligation chemistries that avoid Cys or refined desulfurization strategies to convert Cys to Ala [16–18]. Finally, in some cases, EPL must be performed under denaturing conditions, imposing a requirement of the protein being able to refold. Despite these limitations, EPL offers a highly reliable and site-specific method to apply phosphorylation to the protein as evidenced by the many successful applications.

Here, we present protocols for site-specific introduction of phosphorylated amino acids into the N- and C-terminal regions of PDZ domains (*see* Figs. 2 and 3). In the former scenario, a synthetic phosphopeptide thioester, obtained from a peptide

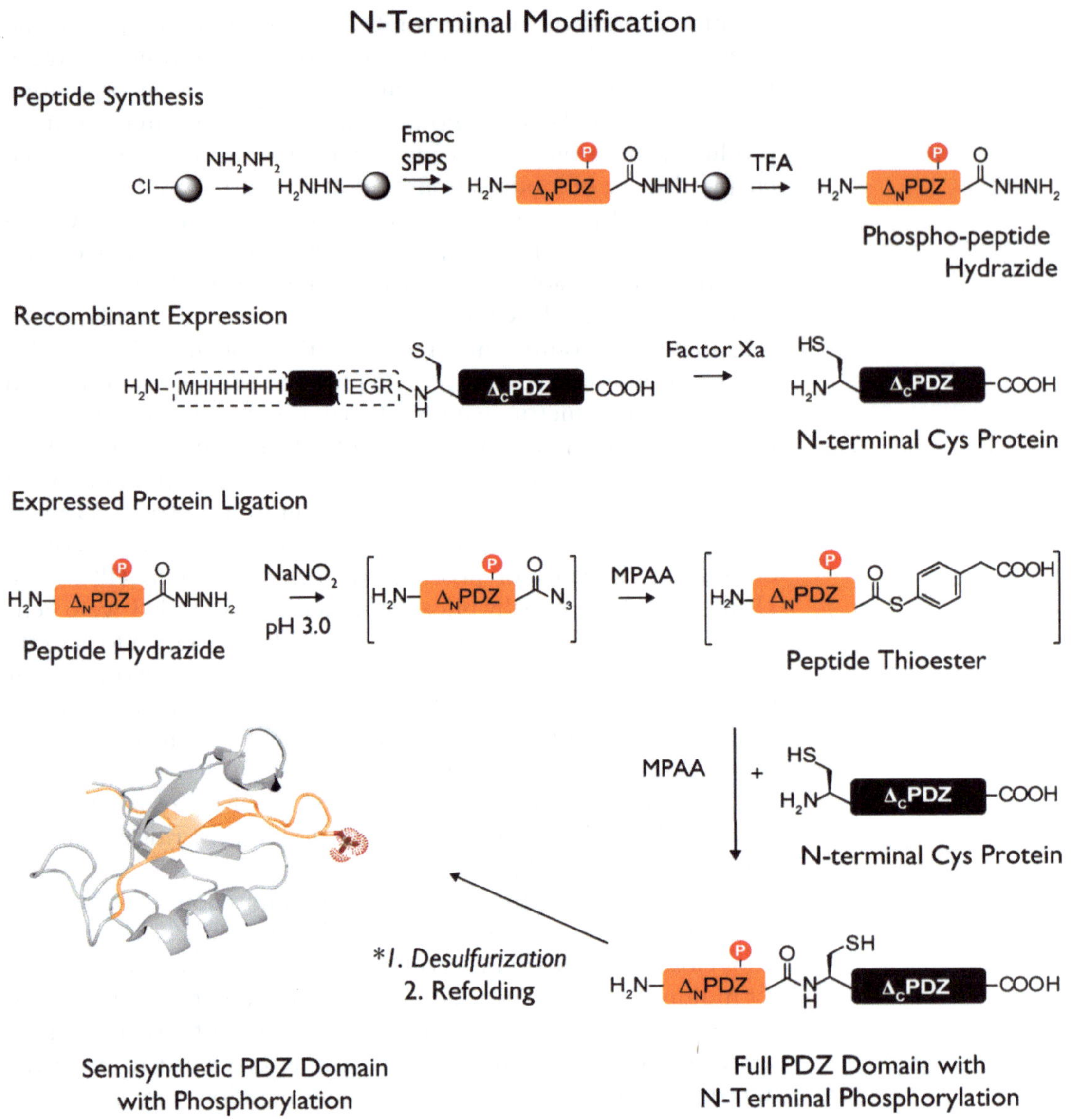

Fig. 2 Semisynthetic strategy for the introduction of N-terminal phosphorylations in PDZ domains. A synthetic peptide thioester, sΔ_NPDZ, with the site-specific phosphorylation is ligated to a recombinant fragment with an N-terminal Cys, rΔ_CPDZ. The recombinant fragment is produced by recombinant expression of a fusion protein that includes an intervening factor Xa protease cleavage site between an N-terminal polyhistidine tag and the target protein. Subsequently, the fusion protein is purified and exposed to enzymatic cleavage. The sΔ_NPDZ fragment is generated as a peptide hydrazide by SPPS using Fmoc-hydrazide resin. The peptide hydrazide is then oxidized with $NaNO_2$ and converted to a thioester with MPAA that is immediately reacted with the rΔ_CPDZ fragment in an EPL reaction. After ligation, the Cys at the ligation site is converted into Ala by desulfurization (starred italics) and the semisynthetic protein is refolded

hydrazide, is ligated to a recombinantly expressed protein containing an N-terminal Cys. For incorporation of C-terminal phosphorylations, a recombinantly expressed protein thioester is ligated with a synthetic N-terminal Cys peptide. Protocols for desulfurization of

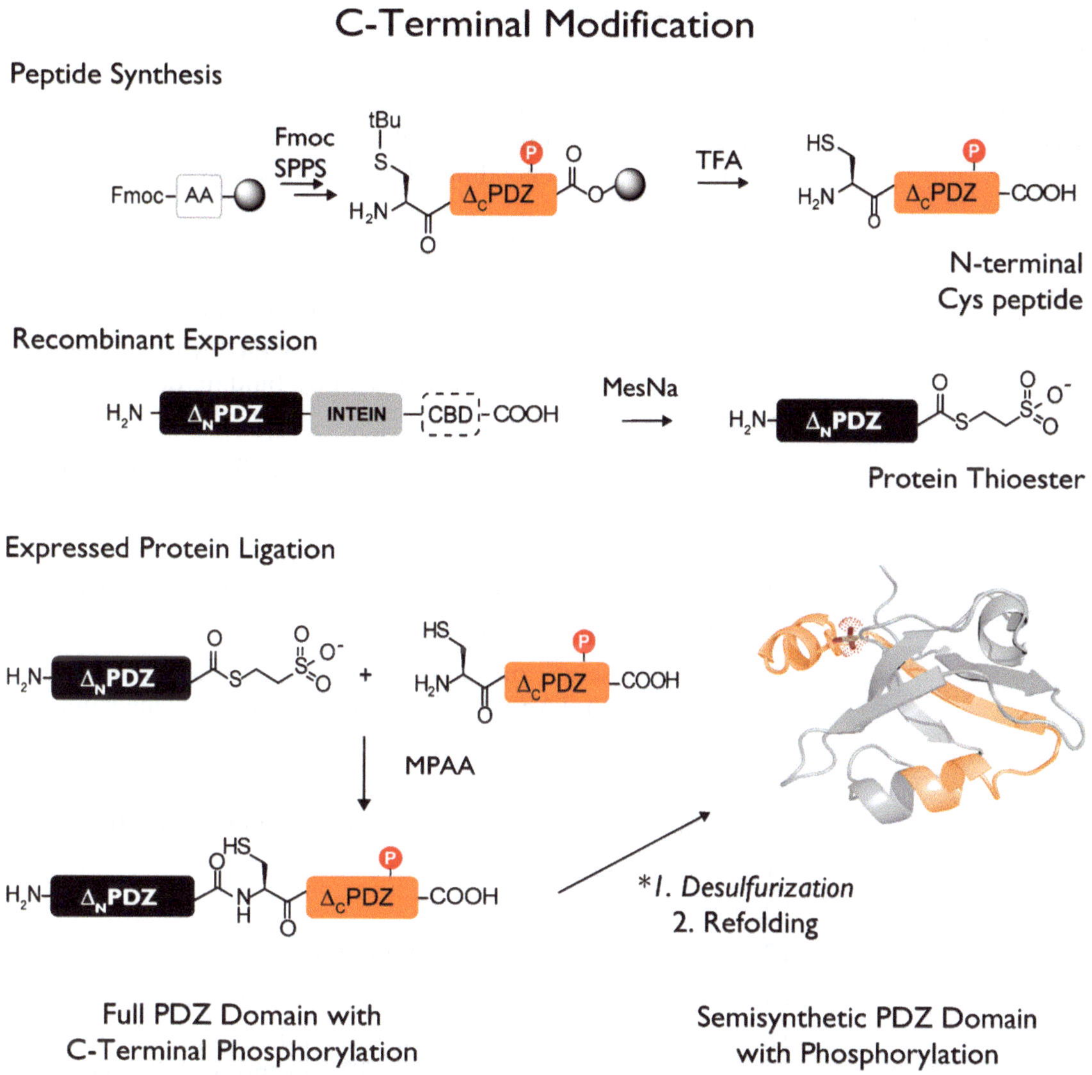

Fig. 3 Semisynthetic strategy for the introduction of C-terminal phosphorylations in PDZ domains. A recombinant protein thioester, rΔ_NPDZ, is ligated with a synthetic phosphopeptide, sΔ_CPDZ, containing an N-terminal Cys. The thioester moiety is introduced by thiolysis of an intein (*Mxe* GyrA) fusion protein with MESNa, and the phosphopeptide is synthesized using Fmoc-SPPS. The EPL reaction is carried out in the presence of MPAA. A subsequent desulfurization step (starred italics) can be included for traceless ligation prior to protein refolding

the Cys at the ligation junction are also provided. Finally, we outline some methodologies on how to evaluate the structural and functional consequences of phosphorylation of PDZ domains using circular dichroism and fluorescence polarization. The enclosed protocols are based on our experiences with PDZ domains but can easily be adapted to other proteins.

2 Materials

2.1 Molecular Biology

1. Bacterial IPTG-inducible expression vector with ampicillin resistance encoding the full-length PDZ domain. Here, a pRSET vector encoding the PDZ domain, pRSET(PDZ), is used.
2. Bacterial IPTG-inducible expression vector with ampicillin resistance encoding an intein fused to a purification tag. Here, we use a pTWIN1 vector, pTWIN1-His$_7$, which encodes the Mxe GyrA intein and has been modified with a C-terminal hepta-histidine tag in place of a chitin binding domain.
3. Nuclease-free water.
4. Oligonucleotide primer stock solutions: 10 μM in nuclease-free water (*see* Table 1).
5. Site-Directed Mutagenesis kit.
6. High-fidelity DNA polymerase and appropriate buffers.
7. NdeI and SapI restriction enzymes.
8. 1.5% agarose gel, prepared by melting solid ultrapure agarose suspended in TAE (Tris-base, acetic acid, ethylenediaminetetraacetic acid) buffer using a microwave oven.
9. PCR gel extraction kit.
10. Chemically competent *E. coli* cloning strain (TOP10).
11. S.O.C medium.
12. Plasmid Miniprep kit.
13. T4 DNA ligase and appropriate buffers.
14. Ampicillin stock solution: 25 mg/mL in ultrapure water.
15. Sterile Luria Broth (LB) agar plates: mix LB agar into ultrapure water. Autoclave at 121 °C for 30 min and cool down to 50 °C before adding ampicillin (100 μg/mL) and distributing on 92 mm × 16 mm sterile clear petri dishes. Store plates at 4 °C.

Table 1
Primers used for the construction of plasmids encoding recombinant EPL fragments

Primers for introduction of factor Xa site into pRSET(PDZ) placed adjacent to the N-terminal Cys[a, b]
Forward: 5′-(NNN)$_5$ **ATC GAG GGA AGG TGC** (NNN)$_5$-3′
Reverse: 5′-(NNN)$_5$ **GCA CCT TCC CTC GAT** (NNN)$_5$-3′
*Primers for PDZ cloning into pTWIN1-**His$_7$***
NdeI forward primer[c]: 5′-GGA ATT CAT ATG (NNN)$_{10}$-3′
SapI reverse primer[d]: 5′-GGT GGT TGC TCT TCC (NNN)$_6$-3′

[a]Bold letters highlight the gene encoding factor Xa site and a Cys residue
[b](NNN)*x* corresponds to flanking codons from the target gene, with *x* being the number of codons (*see* **Note 5**)
[c]NdeI site is underlined in the sequence
[d]SapI site is underlined in the sequence

2.2 Protein Expression and Purification

1. Chemically competent *E. coli* expression strain (BL21 (DE3) pLys).
2. Sterile LB agar plates (*see* Subheading 2.1).
3. Sterilized Luria-Bertani Broth (pH 7.2): dissolve 25 g of pre-mixed LB broth in 1 L of ultrapure water and autoclave at 121 °C for 30 min.
4. Sterilized 300 mL and 5 L culture flasks, autoclaved at 121 °C for 30 min.
5. Ampicillin stock solution: 25 mg/mL ampicillin in ultrapure water.
6. Isopropyl β-D-1-thiogalactopyranoside (IPTG) stock solution: 1 M in ultrapure water.
7. Cell lysis buffer (*see* Subheading 3.2): 50 mM Tris–HCl, 1 mM phenylmethanesulfonylfluoride, 25 μg/mL DNase, 40 mg/mL magnesium sulfate ($MgSO_4$) (pH 7.4). Cooled to 4 °C before use.
8. 5 mL HisTrap FF columns.
9. Nickel column equilibration buffer (*see* Subheading 3.2): 50 mM sodium phosphate, 150 mM NaCl, 20 mM imidazole (pH 7.4). Filtered (0.2 μm pore size). Cooled to 4 °C before use.
10. Nickel column elution buffer (*see* Subheading 3.2): 50 mM sodium phosphate, 150 mM NaCl, 250 mM imidazole (pH 7.4). Filtered (0.2 μm pore size). Cooled to 4 °C before use.
11. Dialysis membrane: dialysis tubing or device, 7 kDa MWCO.
12. Centrifugal Filters with MWCO of 10 kDa.
13. Factor Xa cleavage buffer (*see* Subheading 3.2): 50 mM Tris, 100 mM NaCl, 6 mM calcium chloride (pH 8). Cooled to 4 °C before use.
14. Factor Xa protease.
15. Thiolysis buffer (*see* Subheading 3.2): 1–2 M urea, 50 mM sodium phosphate, 150 mM NaCl (pH 6.8). Cooled to 4 °C before use.
16. Sodium 2-mercaptoethanesulfonate (MESNA).

2.3 Peptide Synthesis

1. 10 mL polypropylene disposable reaction tubes.
2. Dichloromethane (DCM).
3. Dimethylformamide (DMF).
4. Methanol (MeOH).
5. Preloaded Wang resin for Fmoc-based chemical synthesis of peptide thioesters.
6. 2-clorotitryl polystyrene resin for Fmoc-based chemical synthesis of peptide hydrazides.

7. Fmoc-hydrazide hydrochloride.
8. Fmoc-protected amino acids.
9. Fmoc-protected phosphorylated serine and tyrosine, O-benzyl protected.
10. 2-(1-benzotriazole-1-yl)-1,1,3,3-tetramethyluronium hexafluorophosphate (HBTU) stock solution: 0.5 M in DMF, stored in a light-protected bottle.
11. 2-(1H-7-azabenzotriazole-1-yl)-1,1,3,3-tetramethyluronium hexafluorophosphate (HATU) stock solution: 0.5 M in DMF, stored in a light-protected bottle.
12. *N*,*N*-diisopropylethylamine (DIPEA).
13. 20% (v/v) piperidine in DMF.
14. Kaiser test kit.
15. Capping solution: 8 mL of DMF, 1 mL of acetic anhydride, 1 mL of DIPEA. Prepare fresh before every use.
16. Cleavage cocktail: 80% v/v trifluoroacetic acid, 10% v/v water, 10% v/v triisopropylsilane.
17. Ice cold ether: store at −20 °C until usage.
18. 50 mM sodium phosphate buffer (pH 7.5).

2.4 Ligation

1. Sodium nitrite stock solution: 0.2 M stock in ultrapure water. Prepare fresh before every use.
2. Oxidation buffer: 6 M guanidine hydrochloride, 0.2 M sodium phosphate buffer (pH 3.0).
3. Ligation buffer: 6 M guanidine hydrochloride, 200 mM mercaptophenyl acetic acid (MPAA), 200 mM sodium phosphate, 1 mM TCEP (pH 7.0). Prepare fresh before every use.
4. Dithiothreitol (DTT) stock solution: 1 M stock in ultrapure water.

2.5 Desulfurization

1. Desulfurization buffer: 6 M guanidine hydrochloride, 200 mM sodium phosphate, 80 mM TCEP (pH 7). Prepared fresh before every use.
2. *Tert*-butylthiol (tBuSH) stock solution: 0.5 M in ultrapure water. Prepared fresh before every use.
3. VA-044 radical initiator stock solution: 0.5 M in ultrapure water. Prepared fresh before every use.

2.6 Protein Refolding

1. Dialysis membrane (*see* Subheading 2.2, **item 11**).
2. Solubilizing buffer: 6 M guanidine hydrochloride, 100 mM sodium phosphate (pH 7.5). Cooled to 4 °C before use.
3. Refolding buffer: 50 mM sodium phosphate (pH 7.5). Cooled to 4 °C before use.

2.7 Circular Dichroism

1. Circular dichroism (CD) buffer: 50 mM sodium phosphate (pH 7.5).
2. CD denaturing buffer stock: 8 M urea, 50 mM sodium phosphate (pH 7.5).
3. 1 mm quartz cuvettes, CD grade.

2.8 Fluorescence Polarization

1. Fluorescence polarization (FP) buffer: 50 mM sodium phosphate (pH 7.5).
2. Tetramethylrhodamine (TAMRA) conjugated peptide probe stock: 5 μM in FP buffer (*see* **Note 33**).
3. 384-well plate, black flat-bottom.
4. Analytical pipette channels, with nonstick pipette tips.
5. 10× Bovine Serum Albumin (BSA) stock solution: 0.1 g/mL in FP buffer.

2.9 Equipment

1. Mass Spectrometer with electron spray ionization (ESI) coupled to a HPLC system (ESI-LC/MS), equipped with C18 reverse phase column for the analysis of peptides and a C8 column for the characterization of proteins, using a linear gradient of a binary buffer system containing H_2O/MeCN/TFA (A: 95/5/0.1; B: 5/95/0.1) at a flow rate of 1 mL min^{-1} for proteins and H_2O/MeCN/formic acid (A: 95/5/0.1; B: 5/95/0.086) at a flow rate of 1 mL/min for peptides.
2. Analytical reverse phase ultra-high performance liquid chromatography (RP-UPLC) system with a C18 reverse phase column for the analysis of peptides and a C8 column for the analysis of proteins, using a binary buffer system consisting of H_2O/MeCN/TFA (A: 95/5/0.1; B: 5/95/0.1) at 0.45 mL/min.
3. Preparative reverse phase high-performance liquid chromatography (RP-HPLC) system with a C18 reverse phase column for the purification of peptides and a reverse phase C4 column for the purification of proteins, using a linear gradient of a binary solvent system of H_2O/MeCN/TFA (A: 95/5/0.1; B: 5/95/0.1) at a flow rate of 20 mL/min.
4. Fast protein liquid chromatography (FPLC) system, equipped with a gel filtration column for size-exclusion chromatography (SEC), operating at a flow rate of 1 mL/min.
5. Circular Dichroism Spectrometer.
6. Plate reader.
7. Cell disruptor system.

3 Methods

We provide protocols for the preparation of the required protein fragments to perform EPL, and their subsequent use in typical ligation reactions enabling N- or C-terminal modifications. Additionally, we provide protocols for the biophysical characterization of the semisynthetic phosphorylated proteins.

For introduction of phosphorylation in the N-terminal region of PDZ domains, the EPL reaction requires a synthetic phosphorylated PDZ peptide thioester (sΔ_NPDZ) and a recombinant protein containing an N-terminal Cys (rΔ_CPDZ) (*see* Fig. 2). Conversely, for C-terminal phosphorylations, the EPL reaction involves a recombinant thioester (rΔ_NPDZ) and a synthetic phosphorylated peptide containing an N-terminal Cys (sΔ_CPDZ) (*see* Fig. 3). As with any semisynthetic protocol that is limited to one single ligation step, the ligation junction should be placed near the native N or C terminus to reduce the length of the synthetic peptide fragment (*see* **Note 1**). Many PDZ domains do not contain a native Cys residue to enable EPL. Instead, we either introduce a Cys at an appropriate site by mutagenesis (*see* **Note 2**) or perform the ligation in a traceless manner converting the Cys into Ala. In both cases, it is important to choose a ligation site in nonstructured regions of the protein, while avoiding beta-branched residues (Ile, Thr, Val) in the position immediately upstream of the Cys to enable efficient reaction [19].

3.1 Vector Construction

The rΔ_CPDZ domain fragment is expressed with a factor Xa cleavage site (Ile-Glu-Gly-Arg) upstream to the N-terminal Cys residue so that upon enzymatic cleavage, the N-terminal Cys can be revealed (*see* Fig. 2). An easy way to introduce this cleavage site is to introduce the four additional amino acids simultaneously with the Cys mutation at the ligation junction into a bacterial expression vector encoding the full-length target protein. In the below protocol, a pRSET expression vector that already harbors a gene encoding the full-length PDZ domain, pRSET(PDZ), is used to generate pRSET(Δ_CPDZ) for expression of PDZ C-fragments Subheading 3.2, **item 1** (*see* **Note 3**). Typical primers for introducing the factor Xa site together with a Cys mutation are depicted in Table 1.

The rΔ_NPDZ thioester is produced from the pTWIN1-His$_7$(Δ_NPDZ) plasmid, which enables generation of the protein of interest fused to a C-terminal *Mxe GyrA* intein (*see* **Note 4**). The pTWIN1-His$_7$(Δ_NPDZ) plasmid is generated by subcloning the N-terminal PDZ fragment from the pRSET(PDZ) vector into the pTWIN1 vector. An important feature when cloning fragments for EPL reactions is to eliminate any intervening vector-derived amino acids at the ligation junction. When using pTWIN1 for protein thioester generation, it is critical to use a *SapI* restriction site at the 3′ end. Primers that were used to create pTWIN1-His$_7$(Δ_NPDZ) are listed in Table 1. The following protocol uses the modified pTWIN1-His$_7$ vector, which encodes a C-terminal hepta-histidine tag in place of the chitin binding domain, but the cloning protocol can also be used to clone into the original pTWIN1 vector for generation of protein thioesters using the IMPACT™ purification protocol.

3.1.1 Insertion of a Factor Xa Site Flanking a Cys for Generation of N-terminal Cys Fragments

1. Perform insertional mutagenesis directly on pRSET(PDZ) using the primer pair for factor Xa and Cys insertion (Table 1) and a site-directed mutagenesis kit according to the manufacturer's instructions (*see* **Note 5**). This generates the plasmid, pRSET(Δ_cPDZ).
2. Add 1 μL of DpnI and incubate overnight at 37 °C to digest DNA template, pRSET(PDZ)
3. Transform 1 μL of the DpnI digest directly into 10 μL of chemically competent *E. coli* TOP10 cells and grow in SOC medium for 30 min at 37 °C. Plate on LB agar plates containing ampicillin (100 μg/mL) and let colonies grow overnight at 37 °C.
4. Extract plasmid DNA, pRSET(Δ_cPDZ), from ampicillin-resistant colonies with a miniprep kit of a 5 mL overnight culture in LB medium containing ampicillin (100 μg/mL) and confirm mutation using DNA sequencing (*see* **Note 6**).

3.1.2 Gene Insertion into pTWIN1-His$_7$ for Generation of C-terminal Thioesters

1. Amplify the gene that encodes the rΔ_NPDZ domain fragment, using the primer pair for PDZ cloning (Table 1) and pRSET (PDZ) as DNA template, by following standard PCR conditions for a high-fidelity DNA polymerase (*see* **Note 6**).
2. Digest the pTWIN1-His$_7$ plasmid and PCR product with 1 μL each of NdeI and SapI restriction enzymes. Incubate digest at 37 °C for 2 h.
3. Use DNA gel electrophoresis to separate and isolate the bands of the digested pTWIN1 plasmid and PCR product. Extract the DNA products from agarose gel with a DNA gel extraction kit.
4. Ligate the purified PCR fragment to the digested pTWIN1 vector with T4 DNA ligase according to manufacturer's instruction to generate pTWIN1-His$_7$(Δ_NPDZ).
5. Follow Subheading 3.1.1, **steps 3** and **4** to transform ligation mixture into *E. coli* TOP10 cells and isolate pTWIN1-His$_7$(Δ_NPDZ) using a miniprep kit.

3.2 Expression and Purification of N-terminal Cys Proteins

1. Transform 1 μL of pRSET(Δ_CPDZ) that was obtained in Subheading 3.1.1 into 10 μL competent BL21 (DE3) pLys *E. coli* cells. After growing in SOC medium for 30 min at 37 °C, plate on LB agar plates containing ampicillin (100 μg/mL). Allow colonies to form overnight at 37 °C.
2. From the plate containing fresh colonies of the BL21 DE3 pLys pRESET(Δ_CPDZ) cells, inoculate 100 mL LB ampicillin (100 μg/mL) medium with a single ampicillin resistant colony in a sterile 300 mL flask and grow the starting culture overnight at 30 °C with 200 rpm shaking.
3. Prewarm 1 L sterile LB medium, prepared in a 5 L culture flask, to 37 °C in a shaking incubator with 200 rpm and add 100 μg/mL ampicillin.
4. Use the starting culture to inoculate the 1 L prewarmed LB-ampicillin medium to an optical density at 600 nm (OD_{600}) of 0.1.
5. Incubate flask at 37 °C with 200 rpm shaking until $OD_{600} = 0.6$. Then collect 1 mL of uninduced cell culture, harvest cells by centrifugation, discard supernatant and store cell pellet at –20 °C for SDS-PAGE analysis. Induce protein expression by adding IPTG at a final concentration of 1 mM.
6. Continue incubation for another 4 h at 37 °C. Then collect 1 mL of induced cell culture, harvest cells by centrifugation, discard supernatant and store cell pellet at –20 °C for SDS-PAGE analysis.
7. Harvest the remaining cells by centrifuging at 4 °C and 10,000 × *g* for 15 min and discard supernatant.
8. Resuspend cells in 50 mL cold lysis buffer and stir over ice until the cell pellet is completely resuspended.
9. Lyse cells using a cell-disrupter at a pressure of 26 kPsi.
10. Mix cell lysate gently at room temperature or approximately 20 °C for 15 min to allow the DNase in the cell lysis buffer to work, thereby reducing the viscosity.
11. Pellet the cellular debris by centrifugation at 25,000 × *g* for 35 min at 4 °C and carefully collect the clarified supernatant. Filter the clarified cell lysates using a 0.45 μm syringe filter.
12. With a peristaltic pump or manually with a syringe, equilibrate a 5 mL HisTrap FF/HP with 5 CV of HisTrap equilibration buffer. Load the filtered cell lysate onto the column at a flow rate of 1–2 mL/min. Wash column with 20 CV equilibration buffer to remove impurities. Elute protein into sterile conical tubes (on ice) in fractions of 5 mL with at least 10 CV of nickel column elution buffer.

13. Confirm the presence of the His-tagged target protein in the eluted fractions with SDS-PAGE or LC-MS with the reverse phase C18 column for protein analysis. Pool the cleanest fractions (*see* **Note 7**).
14. Exchange the purified protein into factor Xa cleavage buffer (*see* **Note 8**), either by dialysis or using centrifugal filters. If using centrifugal filters, concentrate protein sample by using a MWCO of 10 kDa and centrifugation at 3500 × *g*. Discard the flow-through and pour factor Xa buffer into the protein sample to replenish the displaced volume. Perform this centrifugation-buffer dilution cycle 6–10 times until buffer is replaced.
15. Add factor Xa protease at a 1:100 enzyme to substrate ratio and incubate at 6 °C in a cold room on a slow shaker for 12 h, closely monitoring the reaction with LC-MS (*see* **Note 9**).
16. Remove the cleaved hexahistidine tag fragment by 'reverse' immobilized metal affinity chromatography (IMAC) using the conditions described in **step 12**, but collecting the flow-through, which contains the Δ_CPDZ fragment.
17. With factor Xa cleavage buffer as elution buffer, use size-exclusion chromatography to further obtain high purity product to target ~80–90% purity, as measured on RP-UPLC using the C8 column for proteins at a flow rate of 0.45 mL/min.
18. Dialyze the purified Δ_CPDZ fragment into ultrapure water to remove salt and buffer constituents. If sample volume is too large for the dialysis device, reduce volume with centrifugal filters, targeting a protein concentration of ~10–20 μM for efficient dialysis. Freeze dry protein sample and lyophilize into white powder. Analyze with LC-MS (*see* Fig. 4).

3.3 Generation of Protein Thioesters

The transformation of pTWIN1-His$_7$(Δ_NPDZ) (*see* Subheading 3.1.2) into *E. coli* cells and subsequent heterologous protein expression (*see* **Notes 10** and **11**), preparation of crude cell extract, and first IMAC purification step to yield the intein fusion protein are the same as described in Subheading 3.2, **steps 1–13**. For certain PDZ domain fragments, the protein could be expressed as inclusion bodies when fused to an intein. This can be isolated and solubilized (*see* **Note 12**) before proceeding with the IMAC purification steps.

1. Using centrifugal filters or size-exclusion chromatography, exchange purified intein fusion proteins into thiolysis buffer to a target concentration of ~50 μM (*see* **Note 13**).
2. Dissolve 100 mM MesNa powder directly into the protein solution, and carefully adjust pH to 6.8 with NaOH/HCl (*see* **Note 14**).

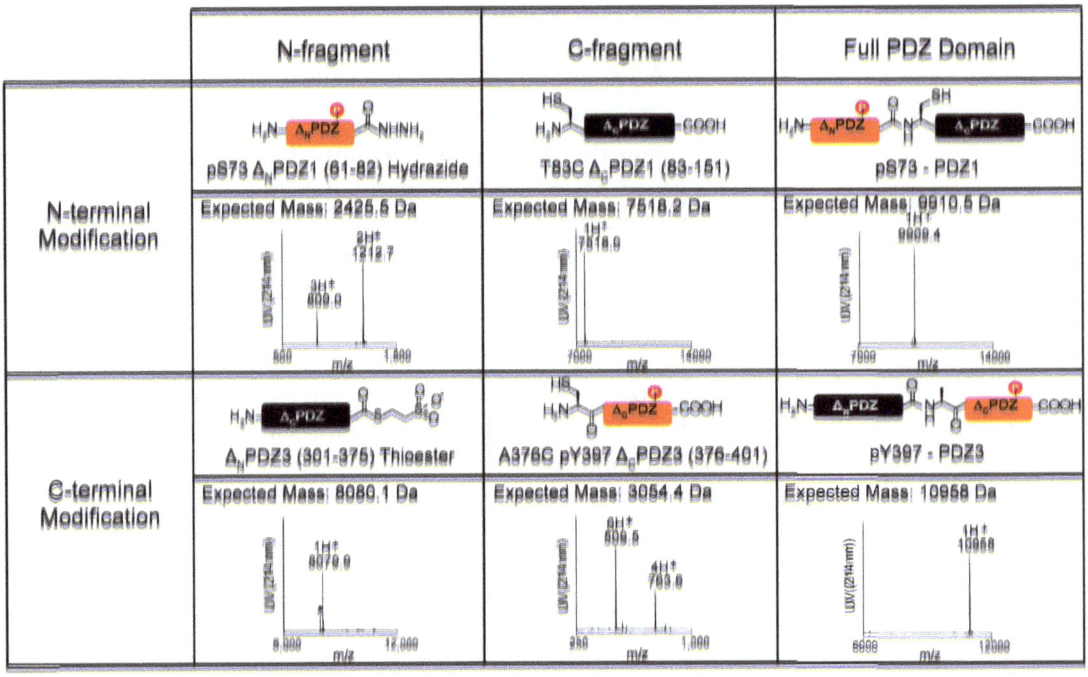

Fig. 4 Mass spectra of the peptide and protein fragments used for semisynthesis of N-terminally phosphorylated (pSer73) PSD-95 PDZ1 and C-terminally phosphorylated (pTyr397) PSD-95 PDZ3 as well as of the purified semisynthetic and phosphorylated PDZ domains. Deconvoluted spectra are shown for recombinant fragments and semisynthetic proteins

3. Incubate at 4 °C on a test tube rocker for 1–3 days until intein cleavage is complete (*see* **Note 15**).
4. With thiolysis buffer as elution buffer, purify final product by loading onto a 5 mL HisTrap FF/HP and collecting flow-through, following with SEC to achieve a pure product targeting ~80–90% purity, determined by RP-UPLC using the protein C8 column at 0.45 mL/min. Analyze fractions using LCMS and lyophilize protein.

3.4 Generation of Phosphopeptide Fragments

Fmoc-based SPPS is generally the preferred method for the synthesis of modified peptides because many PTMs, such as glycosylation and phosphorylation, are not stable in HF cleavage conditions required for Boc-based SPPS [20]. Manual Fmoc-SPPS is described here, but the peptide fragments can also be generated on an automated peptide synthesizer (*see* **Note 16**).

For the generation of thioester peptides, we recommend the peptide hydrazide method, but there are other Fmoc-based strategies such as "safety-catch" methods [21]. Preparation of Fmoc-hydrazine 2-chlorotrityl resin is presented below, which yields stable resin for long-term storage. Preloaded hydrazone pyruvyl resin is also commercially available (Iris Biotech GmbH) and can be used for the synthesis of peptide hydrazides.

3.4.1 Preparation of Fmoc-hydrazine Linked 2-Chlorotitryl Resin

1. In a 100 mL round-bottom flask equipped with a stir bar, suspend 2 g of 2-chlorotitryl resin in 16 mL of DCM. Gently mix at 0 °C for 30 min over an ice bath (*see* **Note 17**).
2. In a separate flask, mix 2 equivalents (relative to resin) of Fmoc-hydrazide, 5 equivalents of DIPEA, 25 mL DMF, and 5 mL of DCM until completely dissolved.
3. Add Fmoc-hydrazide solution drop-wise into resin suspension with gentle mixing while maintaining at 0 °C with an ice bath. Continue stirring reaction mixture overnight, allowing the ice bath to melt and increasing the reaction temperature from 0 °C to room temperature or approximately 20 °C.
4. Quench reaction by adding 320 μL of methanol.
5. Transfer resin to a filter vessel and drain all solvents. Wash resin successively with 20 mL of DMF, then 20 mL of water, 20 mL of methanol, and finally 20 mL of diethyl ether. Allow the resin to dry under vacuum.
6. To determine resin loading, with an analytical scale weigh dried resin in an Eppendorf tube to target between 1 and 2 mg and record exact dry weight. Deprotect resin with 1 mL 20% piperidine in DMF solution for 10 min, and remove solid resin with 0.2 μm syringe filter. Measure the optical density at 301 nm (OD_{301}) of filtrate, and calculate resin loading using the Fmoc extinction coefficient of 7800 M^{-1} cm^{-1} (*see* **Note 18**).

3.4.2 Fmoc Solid Phase Peptide Synthesis of Phosphopeptides

For procedures outlined in this section, all reagent equivalents are relative to resin.

1. For SPPS of peptide hydrazides, use Fmoc-hydrazine linked resin (*see* Subheading 3.4.1). For all other peptides, use pre-loaded Wang resin (*see* **Note 19**).
2. In a 10 mL disposable reaction column with filter, transfer 0.1 mmol of resin (*see* **Note 20**) and wash with 5 mL of DCM followed by 10 mL of DMF. Cap the tube and add 2 mL of DMF. Allow the resin to swell for 30 min.
3. Drain solvent and resuspend resin with 2 mL of Fmoc deprotection solution. Incubate on a shaker (500 rpm) at room temperature for 5 min. Drain solvent and wash resin with 10 washes of ~2 mL DMF.
4. In a glass scintillation vial, dissolve 4 equivalents of Fmoc amino acid derivative, 4 equivalent HBTU solution. Seal vial and vortex until completely dissolved.
5. Add 4 equivalent DIPEA to the amino acid solution and vortex. Incubate 2–3 min to preactivate the building block for efficient coupling.

6. Transfer amino acid solution to the reaction tube and cover with a Teflon stopper. Incubate at room temperature on a shaker at 500 rpm for 1 h.
7. Check that reaction is complete with Kaiser test, and proceed only when coupling is complete (*see* **Notes 21**). To perform the Kaiser test, wash a small aliquot (~1–2 mg) of resin beads with DCM in a glass filter tube. Place resin into a small glass test tube and apply test reagents according to manufacturer's instructions. Heat 2 min at 100 °C and observe the color of the beads. A dark purple color indicates incomplete coupling, while colorless beads indicate complete coupling. If test indicates that coupling is complete, proceed to the next steps.
8. Wash resin extensively with 20 mL of DMF. Repeat **steps 2–4** for the next amino acid until all building blocks have been coupled (*see* **Note 22**).
9. For the coupling of phosphorylated building blocks, use the same procedures as the coupling of standard amino acids in **steps 2–4**, but with 1.5 equivalent amino acid, 4 equivalent HATU instead of HBTU, and 4 equivalent DIPEA.
10. After extensive washing with 20 mL of DMF, add 5 mL cleavage cocktail to the reaction tube and incubate at room temperature on a shaker at 500 rpm for 3 h (*see* **Note 23**).
11. Filter resin by using air to push cleavage mixture through the disposable reaction tubes and collecting flow-through, which contains peptide and cleavage cocktail, in a sterile 50 mL conical tube.
12. Precipitate peptide by adding 50 mL cold diethyl ether, and centrifuge for 10 min at 3500 × *g*, in 4 °C. Discard supernatant.
13. Dissolve pellet with sodium phosphate buffer (*see* **Note 24**), and purify using preparative RP-HPLC at a flow rate of 20 mL/min.
14. Confirm final peptide mass with LC-MS using reverse-phase C18 column for peptide analysis. Analyze peptide purity with RP-UPLC using the C18 reverse phase column for peptide analysis. Lyophilize the final product.

3.5 Expressed Protein Ligation Reactions

3.5.1 Ligation of a Recombinant Thioester and a Synthetic Phosphopeptide

1. In a 1.5 mL Eppendorf tube, dissolve rΔ_NPDZ thioester into ligation buffer (*see* **Note 25**) to a target concentration of 1 mM (*see* **Note 26**).
2. In a separate 1.5 mL tube, add 1.2–2 equivalent of synthetic phosphopeptide powder. Pipette the thioester mixture directly into the tube with the phosphopeptide and adjust pH to 7.2 with NaOH/HCl (*see* **Note 27**).

3. Incubate on an Eppendorf mixer (500 rpm) at room temperature, or approximately ~20 °C, for 12 h or until reaction is complete, removing aliquots (10 μL) every 2–4 h to monitor reaction progress by LC-MS (*see* **Note 28**).
4. Quench reaction by adding 100 μL 1 M DTT followed by 5 mL of 20% aqueous acetonitrile with 1% TFA.
5. Purify ligated protein with preparative RP-HPLC using the C4 reverse phase column. Lyophilize the pure product and analyze by RP-UPLC and LC-MS with appropriate column and conditions for protein analysis (*see* Fig. 4).

3.5.2 Ligation of a Synthetic Phosphopeptide Thioester and a Recombinant N-Cys Protein

1. Dissolve lyophilized peptide hydrazide into oxidation buffer to a target concentration of 1 μM in a sterile 1.5 mL Eppendorf tube, at −15 °C maintained over an ice/NaCl bath.
2. Add 10 equivalent (relative to peptide hydrazide) sodium nitrite, and adjust pH to 3 with HCl. Incubate for 20 min at −15 °C on the ice-NaCl bath.
3. Adjust pH to 7 with NaOH, and add ligation buffer to the mixture to a target of 100 mM MPAA to generate the peptide thioester. Remove from the ice bath and bring reaction to room temperature, approximately 20 °C. Carefully adjust pH to 6.8–7.0 (*see* **Note 27**).
4. Weigh 1.2–2 equivalent lyophilized Δ_CPDZ to peptide ratio in a separate sterile 1.5 mL Eppendorf tube then add the peptide thioester mixture to the tube.
5. Follow Subheading 3.5.1, **steps 3–5** for subsequent ligation and purification steps.

3.6 Desulfurization

The following steps are used for traceless ligation where it is desired to recover an Ala residue at the ligation junction.

1. In a 1.5 mL Eppendorf tube, dissolve pure semisynthetic protein into desulfurization buffer to target a concentration of 1 μM.
2. Add a target of 300 mM *tert*-butylthiol and 9 mM VA-044 radical initiator from the corresponding stock solutions to the protein mixture. Adjust pH to 7.2 with NaOH/HCl.
3. Incubate at room temperature, approximately 20 °C, for 24 h on an Eppendorf mixer at 500 rpm, or until reaction is complete, removing aliquots (10 μL) every 1–2 h to monitor reaction by LC-MS (*see* **Note 29**).
4. Purify final product using preparative RP-HPLC. Lyophilize pure product.

3.7 Protein Refolding

We use the following steps to refold most semisynthetic PDZ domains, but refolding conditions are protein dependent and will require optimization.

1. Dissolve the lyophilized semisynthetic protein in solubilizing buffer to a target concentration of 10 μM. Incubate at room temperature, approximately 20 °C, for 30 min.
2. In a beaker equipped with a stir bar, prepare refolding buffer. Use a buffer volume at 1000× the volume of the protein sample to optimize dialysis efficiency.
3. Transfer the protein sample into dialysis tubing.
4. Dialyze solubilized protein against refolding buffer. Change buffer twice, once after 2 h, another after 4 h. After the last buffer change, continue to dialyze protein overnight at 4 °C (*see* **Note 30**).

3.8 Biophysical Characterization

Semisynthetic phosphorylated proteins can be characterized in various functional and structural assays. Here, we provide protocols for fluorescence polarization (FP) and circular dichroism (CD) experiments, which we have used to characterize semisynthetic PDZ domains.

3.8.1 Verification of Folding by Circular Dichroism

Below is a standard protocol for collecting CD spectral data of PDZ domains to determine that the phosphorylated PDZ domains are correctly folded (*see* Fig. 5), but this is also applicable for other proteins:

1. If not already in the appropriate CD analysis buffer, dialyze solubilized protein in dialysis membrane against CD analysis buffer overnight at 4 °C (*see* **Note 31**).
2. Dilute with CD analysis buffer to a concentration of 20 μM (*see* **Note 32**). Protein concentration can be determined with its extinction coefficient and A_{280}.
3. Transfer sample to a 1 mm path length quartz cuvette, and record spectral data from 190 to 250 nm wavelength at 20 °C in millidegrees of ellipticity, which can be converted to mean residue ellipticity (MRE). Repeat recordings and report the average of three consecutive readings per sample.
4. Prepare a urea-denatured protein sample to serve as control by adding the same protein to CD denaturing buffer to a target concentration of 20 μM, and transfer to a cuvette.
5. In a similar manner as the previous protein sample, record spectral data for the urea-denatured protein, but only for 200–250 nm wavelength, as urea interferes with spectral data below 200 nm wavelength.

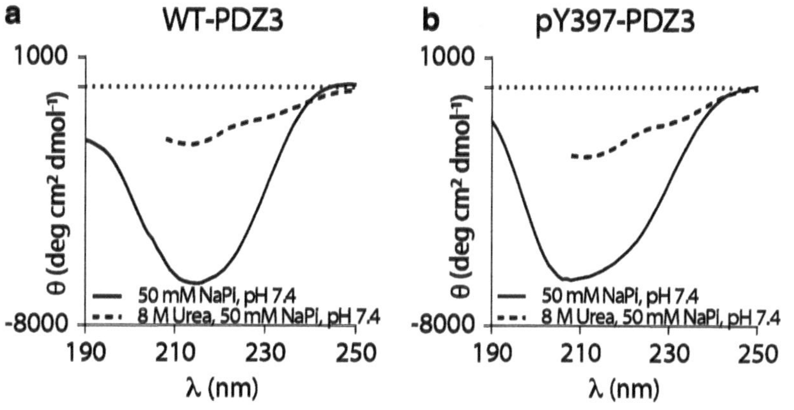

Fig. 5 Circular dichroism spectra of nondenatured (0 M urea) and denatured protein (>7.5 M urea, dashed lines). (**a**) CD spectra of recombinant wild-type PSD-95 PDZ3, showing an ellipticity minimum between 208 and 212 nm, which is characteristic for PDZ domains. (**b**) CD spectra of semisynthetic PSD-95 PDZ3 pTyr397, resembling the spectrum of the folded wild-type protein. In both cases, the secondary structure disappears in the presence of urea (dashed line). Each spectrum is an average of three recordings. Adapted with permission from (Pedersen S, Albertsen L, Moran G, Levesque B, Pedersen S, Bartels L et al. (2017) Site-Specific Phosphorylation of PSD-95 PDZ Domains Reveals Fine-Tuned Regulation of Protein–Protein Interactions. ACS Chemical Biology 12:2313–2323) [6]. Copyright (2017) American Chemical Society

3.8.2 Estimating the Thermodynamic Stability by Circular Dichroism

CD can also be used to estimate the thermodynamic stability of PDZ domains and assess whether phosphorylation introduces a destabilizing effect on protein folding. Varying degrees of protein unfolding is investigated by denaturing the protein in different concentrations of denaturant and measuring the CD signal at 222 nm as a function of the concentration of denaturant (*see* Fig. 6). Subsequently, the concentration of denaturant at which the protein is 50% unfolded ($[\text{Urea}]_{50\%}$), can be determined and applied as an indicator of protein stability. The following protocol is established for PDZ domains, but can be modified for the study of other proteins.

1. If not already in an appropriate buffer, dialyze semisynthetic and expressed wild-type proteins into CD analysis buffer overnight at 4 °C.
2. Generate a dilution series of varying urea concentration in the range of 0–8 M using CD denaturing buffer stock and CD analysis buffer. Add protein to each dilution at a target concentration of 20 μM.
3. For each sample, transfer protein into a 1 mm quartz cuvette and record spectral data at the wavelength of 222 nm in millidegrees of ellipticity. Convert data to MRE.

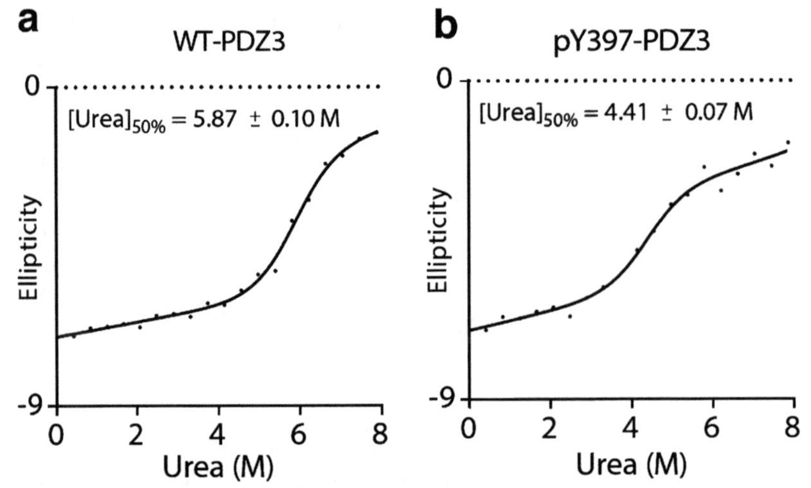

Fig. 6 Urea denaturing curves for recombinant wild-type and semisynthetic phosphorylated protein. (**a**) Denaturing curve of recombinant wild-type PSD-95 PDZ3. (**b**) Denaturing curve of semisynthetic PSD-95 PDZ3 pTyr397. The determined $[\text{Urea}]_{50\%}$ values are listed in each plot. The $[\text{Urea}]_{50\%}$ of PSD-95 PDZ3 pTyr397 is lower than that of the wild type, indicating that the phosphorylation destabilizes the protein. Adapted with permission from (Pedersen S, Albertsen L, Moran G, Levesque B, Pedersen S, Bartels L et al (2017) Site-Specific Phosphorylation of PSD-95 PDZ Domains Reveals Fine-Tuned Regulation of Protein--Protein Interactions. ACS Chemical Biology 12:2313-2323) [6]. Copyright (2017) American Chemical Society

4. Plot ellipticity of each protein as a function of urea concentration to obtain a denaturing curve (*see* Fig. 6) and determine the $[\text{Urea}]_{50\%}$ from the plot.

3.8.3 Fluorescence Polarization Assay

The semisynthetic PDZ domains can be used directly in any in vitro assay for detection of protein-protein interactions. Below, are protocols to perform an FP assay to assess the binding affinity of the semisynthetic protein with a TAMRA-labeled peptide ligand (*see* **Note 33**). An example of binding saturation curve is provided in Fig. 7, where a fixed concentration of the labeled peptide is combined with increasing concentrations of protein to obtain a saturation curve. The dissociation constant, K_D, that indicates the strength of the binding interaction between protein and ligand, is interpolated from the saturation curve.

1. Prepare a series of 12 twofold dilutions of semisynthetic PDZ domain, diluting from a starting concentration of 200 μM (or the protein concentration at which binding saturation is reached) with FP buffer. For triplicate measurements, transfer three times 15 μL of each dilution onto a 384-well black flat-bottom plate.

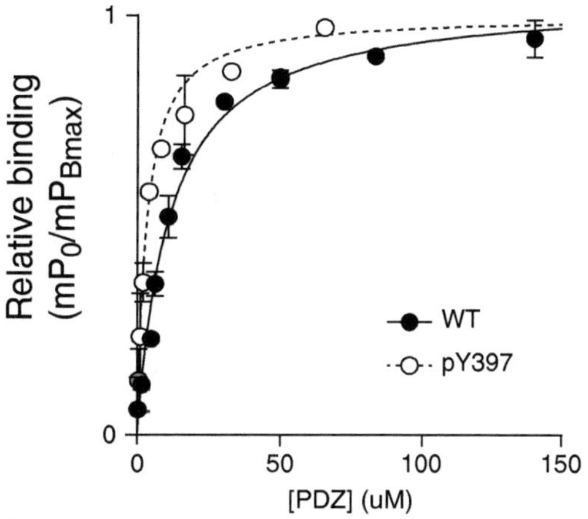

Fig. 7 Fluorescence polarization binding curves of wild-type PSD-95 PDZ3 and semisynthetic PSD-95 PDZ3 pY397 proteins to a TAMRA-labeled peptide ligand belonging to the C-terminal region of stargazin. A one-sided binding model (dashed and solid lines) was used to fit the data (circles), which was utilized to calculate the dissociation constants, K_D. The phosphorylated PDZ3 (dashed line) showed 4.5-fold increase in binding affinity compared to wild type (solid line). Each value is an average of three measurements. Adapted with permission from (Pedersen S, Albertsen L, Moran G, Levesque B, Pedersen S, Bartels L et al. (2017) Site-Specific Phosphorylation of PSD-95 PDZ Domains Reveals Fine-Tuned Regulation of Protein–Protein Interactions. ACS Chemical Biology 12:2313–2323) [6]. Copyright (2017) American Chemical Society

2. In an Eppendorf tube, prepare BSA blank by diluting 10× BSA stock solution into FP buffer. For triplicate measurements of BSA blank and FP buffer controls, transfer three times 30 μL BSA blank and 15 μL FP buffer, respectively, into empty wells of the 384-well plate.
3. In an Eppendorf tube, prepare 2× probe stock by diluting 10× BSA stock and a TAMRA conjugated peptide ligand to a final assay concentration of 200 nM with FP buffer (*see* **Note 34**). Add 15 μL of the 2× probe stock to each of the protein and buffer control wells of the 384-well plate. Incubate plate at room temperature for 15 min.
4. Analyze samples with a plate reader and adjust instrumental z-factor at maximum signal. Calibrate g-factor with BSA blanks and buffer controls to target an initial millipolarization (mP) of 20 mP.
5. Record measurements at excitation/emission wavelengths of λ545 nm/λ575 nm at 25 °C. After subtracting background of 20 mP, fit saturation data using one-site specific binding model with GraphPad Prism mathematical software. Determine K_D by identifying protein concentration at half-saturation.

4 Notes

1. For applications that require the selective introduction of probes into the middle of a protein, or the incorporation of probes within two or more remote parts of the sequence, it will be necessary to assemble the protein from a minimum of three peptide segments.
2. A nonnative Cys must be introduced with care and the mutation should be designed to be as conservative as possible (for example, Ala or Ser to Cys mutations). In addition, mutating highly conserved residues should be avoided. It is advised to evaluate the effect of one or more Cys mutation(s) on protein function prior to beginning semisynthesis.
3. Our PDZ domains are cloned into a pRSET vector, but other vectors should be applicable to the described protocol. Also, insertion of the factor Xa cleavage site and the Cys residue into the protein sequence could have some impact on protein expression even though the wild-type protein expresses stably in the same vector.
4. We use a modified pTWIN1 in order to fuse our PDZ domain fragments with the *Mxe GyrA* intein, which resulted in stable expression of the resulting fusion proteins. However, other inteins exist and could be examined if unsatisfied expression is observed.
5. For example, for preparation of the vector encoding the C-terminal fragment of the PDZ1 domain of PSD-95, the factor Xa cleavage site was introduced upon mutating PDZ1 Thr83 to Cys in pRSET(PDZ1) using the forward primer (5-catt gcg ggt ggc ATC GAG GGA AGG TGC gat aat ccg cat attg-3) and the reverse primer (5-caat atg cgg att atc GCA CCT TCC CTC GAT gcc acc cgc aatg-3), with the lowercase letters indicating the flanking PDZ1 nucleotides. This yielded pRSET(Δ_cPDZ1). For the subcloning of PDZ3 from pRSET (PDZ3) into the pTWIN1-His_7 vector to generate pTWIN1-His_7(Δ_NPDZ3), the forward primer (5-GGA ATT CAT ATG atc gag gga agg tgt tta ggg gaa gaa gat-3) was used to introduce an NdeI cleavage site to the 5′ end while the reverse primer (5-GGT GGT TGC TCT TCC GCA tgc ctg ttc atg gct-3) was used to introduce SapI to the 3′ end. The flanking nucleotide sequences of rΔ_NPDZ3 are represented in lowercase letters in both primers.
6. Verification of the mutation in pRSET constructs and nucleotide sequences of the inserts cloned into pTWIN1-His_7 can be accomplished with T7 promoter forward primer (5-TAATACGACTCACTATAGGG-3) and the T7 terminator reverse primer (5-GCTAGTTATTGCTCAGCGG-3).

7. Typically, the protein is pure enough to move forward (~80–90% purity), but there could still be some impurities in the protein sample due to nonspecific binding to the nickel column. In this case, it is possible to further purify the protein by SEC, using a FPLC system with a gel filtration column.
8. Addition of reducing agents could be necessary in order to stabilize the N-terminal Cys residue of the EPL fragment. If that is the case, only add the minimal amount, as it will compromise factor Xa activity. Otherwise, an alternative protease can be considered (e.g., SUMO protease).
9. Nonspecific cleavage may occur and can be ameliorated by reducing factor Xa concentration and temperature, and purifying the mixture immediately after cleavage is complete. Conversely, incomplete cleavage may also be a problem, so increasing enzyme concentration may be necessary to improve conversion rate. Factor Xa activity is affected by freeze–thaw cycles, so we recommend storing the enzyme in single-use aliquots.
10. We have used these expression conditions with success for many intein fusion proteins. If unsuccessful, lowering expression temperature to 18 °C with extended incubation time (ca. 12 h) could be attempted and will usually minimize formation of insoluble proteins.
11. The *Mxe GyrA* intein is used commonly to generate recombinant protein thioesters, but in some cases these intein fusion proteins have low expression level or are expressed as inclusion bodies. Furthermore, we often notice premature cleavage during expression, which may complicate the subsequent purification. Optimizing expression conditions, such as lowering the temperature, can fix these issues, as well as screening other inteins.
12. To break inclusion bodies, isolate the insoluble protein by collecting the protein pellet rather than the supernatant after cell lysis in Subheading 3.2, **step 11**. Dissolve insoluble protein pellet in lysis buffer, but containing high concentrations of denaturant (e.g., 6 M guanidine hydrochloride). Incubate for 1 h until all particles are visually solubilized. Clarify lysate by centrifugation, at 50,000 × *g* for 60 min and collect supernatant. The lysate can then be purified using the same IMAC steps as outlined in Subheadings 3.2 and 3.3, but with elution buffers containing the same concentration of denaturant in lysis buffer.
13. Optimal urea concentration in the thiolysis buffer depends on the reactivity of the intein and the thioester's susceptibility to hydrolysis. A higher concentration of up to 2 M of urea can be used with the benefit of slowing down the reaction rate and

thereby reducing the amount of hydrolyzed side product. If problems persist, a lower pH can also be attempted to reduce the amount of hydrolysis side product.

14. Thioesters can hydrolyze at extreme pH ranges, so it is critical that pH is adjusted very carefully at this step.
15. Immediately end the reaction and purify the protein thioester fragment upon completion to avoid thioester hydrolysis. We typically monitor the reaction with LC-MS every hour or two, depending on reaction speed.
16. When using microwave-based synthesizers, microwave-assisted Fmoc deprotection cannot be used for phosphopeptides and consequently longer deprotection times are necessary.
17. Most resins can be easily damaged by vigorous mixing with magnetic stirring, so it is critical that only a very gentle mixing speed is applied during this step. Bubbling nitrogen gas into the resin suspension is also possible, if that is available.
18. Dilution of resin filtrate with MeOH is typically required for accurate measurements of OD_{301}, and therefore two samples are prepared with 5× and 10× dilution. Measurements are typically recorded in triplicate per sample for reliable calculation of resin loading, with MeOH as the blank.
19. Preloaded resins are available and are often more convenient, facilitating high reproducibility of your syntheses, and in certain cases circumvent serious racemization issues.
20. The solvent amounts are based on our typical scale of 0.1 mmol. Larger reaction sizes will require a corresponding scale-up of solvent amounts as well.
21. If incomplete coupling is detected, extend coupling time up to 12 h. If coupling is still incomplete, redoing the coupling again is usually sufficient. If repeated reactions still result in insufficient coupling, unreacted amino functionalities can be capped. To do this, wash resin extensively with DMF before adding 2 mL of capping solution to the resin and incubate on the shaker at room temperature for 10 min. Let the capping solution drain, and repeat the capping reaction. When complete, wash the resin extensively before continuing with subsequent Fmoc deprotection.
22. Coupling efficiency usually becomes much lower at increasing peptide length, and it is recommended to perform test cleavages periodically (about every 5–10 amino acid couplings) to evaluate the quality of the peptide. To perform the test cleavage, add a small amount, roughly 1–2 mg, of resin into a 1.5 mL Eppendorf tube after the deprotection step, and wash with DCM. Add 0.5 mL of cleavage solution, and put the reaction tube on a shaker for 2 h at room temperature.

Evaporate liquid with a stream of nitrogen gas until about 100 μL of liquid remain. Add 1 mL of cold ether and centrifuge for 4 min at 11,000 × *g*. Discard ether layer. Redissolve pellet with 10% or less aqueous acetonitrile, and filter with a 0.2 μm syringe filter. Analyze using LC-MS and UPLC.

23. Fmoc deprotection times should be kept within 2–5 min, since there is a risk of the phosphoryl groups hydrolyzing in those conditions.
24. Some proteins may require some vortexing or incubation at room temperature for 15 min or more before it can be completely solubilized.
25. GnHCl is preferred over urea as the denaturant in ligation reactions to avoid carbamylation. Conversely, GnHCl cannot be used in SDS-PAGE analysis, hence LC-MS is preferred to monitor reaction progression.
26. Before weighing lyophilized powders, let sample warm to room temperature and use ionizing cartridges and other anti-static solutions to improve weighing accuracy. Avoid the use of latex gloves to prevent freeze-dried powder from clinging to gloved hands.
27. Carefully adjust and maintain pH below 7.2 because a higher pH would increase the risk of thioester hydrolysis.
28. Expressed protein ligation can also be performed at 4 °C for proteins that are unstable at room temperature. The ligation reaction, however, would require a longer period of time.
29. Desulfurization reaction can be performed at 4 °C for proteins that are unstable, but the reaction time has to be increased.
30. In general, for PDZ domains, it is sufficient to dialyze them overnight at 4 °C. For proteins that are difficult to refold, a stepwise dialysis protocol can be examined, where the protein is brought to equilibrium at decreasing denaturant. Step size and dialysis times must be optimized for specific proteins. It is also possible to adjust the components and pH in the refolding buffer. Adding salt and reducing agents could increase the stability of certain proteins.
31. It is critical to select the buffer for CD measurements with care as many buffer components absorb UV light (particularly in the far UV region (<250 nm) and interfere with the CD signal. Many buffer components that are commonly used for protein storage are known to interfere with CD measurement and should be avoided, such as certain electrolytes and buffers [22]. However, most PDZ domains are stable in 50 mM sodium phosphate, which is a suitable buffer to use for recording CD spectra.

32. Protein concentration may have to be optimized in order to obtain a good signal. Optimal concentration for single PDZ domains is typically around 20 μM, larger proteins may need to be scanned at lower concentrations.
33. The fluorophore peptide can either be synthesized by Fmoc-SPPS or procured commercially. In the provided example, the stargazin peptide, with the sequence "NTANRRTTPV" is coupled to TAMRA. For PDZ domains, we also coupled "NNG" amino acid sequence in between the peptide ligand and TAMRA for optimal assay results, but other sequences, such as "AAG," have been used in assays for other proteins.
34. The selected probe concentration of 200 nM has to be sufficiently below K_D in order for the one-sided binding model to be applicable. For binding studies of PDZ domains and most interacting partners, probe concentrations in the range of 50–200 nM have been employed successfully.

Acknowledgments

This work is supported by the Danish Council for Independent Research, Medical Sciences to S.W.P. (4092-00402B), EU Horizon 2020 RIA under the Marie Skłodowska-Curie grant agreement No. 675341, and Agnes and Poul Friis Fund.

References

1. Yaffe MB (2002) Phosphotyrosine-binding domains in signal transduction. Nat Rev Mol Cell Biol 3:177–186
2. Cohen P (2000) The regulation of protein function by multisite phosphorylation. Trends Biochem Sci 25:596–601
3. Yaffe MB, Elia AE (2001) Phosphoserine/threonine-binding domains. Curr Opin Cell Biol 13:131–138
4. Lu W, Gong D, Bar-Sagi D, Cole PA (2001) Site-specific incorporation of a phosphotyrosine mimetic reveals a role for tyrosine phosphorylation of SHP-2 in cell signalling. Mol Cell 8:759–769
5. Ottesen JJ, Huse M, Sekedat MD, Muir TW (2004) Semisynthesis of phosphovariants of Smad2 reveals a substrate preference of the activated TGF beta RI kinase. Biochemistry 43:5698–5706
6. Pedersen S, Albertsen L, Moran G, Levesque B, Pedersen S, Bartels L et al (2017) Site-specific phosphorylation of PSD-95 PDZ domains reveals fine-tuned regulation of protein–protein interactions. ACS Chem Biol 12:2313–2323
7. Mandal P, Gao F, Lu Z, Ren Z, Ramesh R, Birtwistle J et al (2011) Potent and selective phosphopeptide mimetic prodrugs targeted to the Src homology 2 (SH2) domain of signal transducer and activator of transcription 3. J Med Chem 54:3549–3563
8. Cohen P, Alessi D (2012) Kinase drug discovery – what's next in the field? ACS Chem Biol 8:96–104
9. Ferguson F, Gray N (2018) Kinase inhibitors: the road ahead. Nat Rev Drug Discovery 17:353–377
10. Wu P, Nielsen T, Clausen M (2016) Small-molecule kinase inhibitors: an analysis of FDA-approved drugs. Drug Discovery Today 21:5–10
11. Rogerson D, Sachdeva A, Wang K, Haq T, Kazlauskaite A, Hancock S et al (2015) Efficient genetic encoding of phosphoserine and its nonhydrolyzable analog. Nat Chem Biol 11:496–503

12. Luo X, Fu G, Wang R, Zhu X, Zambaldo C, Liu R et al (2017) Genetically encoding phosphotyrosine and its nonhydrolyzable analog in bacteria. Nat Chem Biol 13:845–849
13. Zhang M, Brunner S, Huguenin-Dezot N, Liang A, Schmied W, Rogerson D, Chin J (2017) Biosynthesis and genetic encoding of phosphothreonine through parallel selection and deep sequencing. Nat Methods 14:729–736
14. Chu N, Salguero A, Liu A, Chen Z, Dempsey D, Ficarro S et al (2018) Akt kinase activation mechanisms revealed using protein semisynthesis. Cell 174:897–907.e14
15. Ansaloni A, Wang Z, Jeong J, Ruggeri F, Dietler G, Lashuel H (2014) One-pot semisynthesis of Exon 1 of the Huntingtin protein: new tools for elucidating the role of posttranslational modifications in the pathogenesis of huntington's disease. Angewandte Chemie Int Edition 53:1928–1933
16. Canne L, Bark S, Kent S (1996) Extending the applicability of native chemical ligation. J Am Chem Soc 118:5891–5896
17. Nilsson B, Kiessling L, Raines R (2001) High-yielding Staudinger ligation of a phosphinothioester and azide to form a peptide. Organ Lett 3:9–12
18. Yan L, Dawson P (2001) Synthesis of peptides and proteins without cysteine residues by native chemical ligation combined with desulfurization. J Am Chem Soc 123:526–533
19. Pedersen S, Pedersen S, Anker L, Hultqvist G, Kristensen A, Jemth P, Strømgaard K (2014) Probing backbone hydrogen bonding in PDZ/ligand interactions by protein amide-to-ester mutations. Nat Commun 5:3215
20. Pedersen S, Armishaw C, Strømgaard K (2013) Synthesis of peptides using Tert-Butyloxycarbonyl (Boc) as the α-Amino protection group. Meth Mol Biol 1047:65–80
21. Mende F, Seitz O (2010) 9-Fluorenylmethoxycarbonyl-based solid-phase synthesis of peptide α-thioesters. Angewandte Chemie Int Edition 50:1232–1240
22. Kelly S, Price N (2000) The use of circular dichroism in the investigation of protein structure and function. Curr Protein Peptide Sci 1:349–384

Chapter 13

Semisynthesis and Reconstitution of Nucleosomes Carrying Asymmetric Histone Modifications

Nora Guidotti and Beat Fierz

Abstract

Nucleosomes, the basic unit of chromatin, contain a protein core of histone proteins, which are heavily posttranslationally modified. These modifications form a combinatorial language which defines the functional state of the underlying genome. As each histone type exists in two copies in a nucleosome, the modification patterns can differ between the individual histones, resulting in asymmetry and increasing combinatorial complexity. To systematically explore the regulation of chromatin regulatory enzymes (writers, erasers, or readers), chemically defined nucleosomes are required. We have developed strategies to chemically modify histones and control nucleosome assembly, thereby enabling the reconstitution of asymmetric histone modification patterns. Here, we report a detailed protocol for the modular assembly of such nucleosomes. Employing a three-segment ligation strategy for the semisynthesis of H3, coupled with the use of the protease cleavable "*lnc*-tag," we provide an efficient and traceless method for the controlled semisynthesis and reconstitution of asymmetrically modified nucleosomes.

Key words Protein semisynthesis, Native chemical ligation, Expressed protein ligation, Asymmetric nucleosomes, Chromatin, Epigenetics, Histone modifications, Bivalent domains

1 Introduction

In eukaryotic cells, the genome is organized in a nucleoprotein complex called chromatin. The nucleosome is the basic building block of chromatin, and it is composed of 147 base pairs (bp) of double-stranded DNA which wraps around a protein core formed from histone proteins. Each nucleosome contains two copies of each histone proteins H3, H4, H2A, and H2B [1]. Histones can be extensively modified, especially on their N-terminal tails which protrude from the nucleosome structure [2]. Such posttranslational modifications (PTMs) of histone proteins form a chemical language of high combinatorial complexity which controls the activity of the underlying genome [3]. Combinations of histone PTMs are read out by effector proteins which may contain multiple reader domains, specific for histone PTMs [4]. Moreover,

Miquel Vila-Perelló (ed.), *Expressed Protein Ligation: Methods and Protocols*, Methods in Molecular Biology, vol. 2133, https://doi.org/10.1007/978-1-0716-0434-2_13,

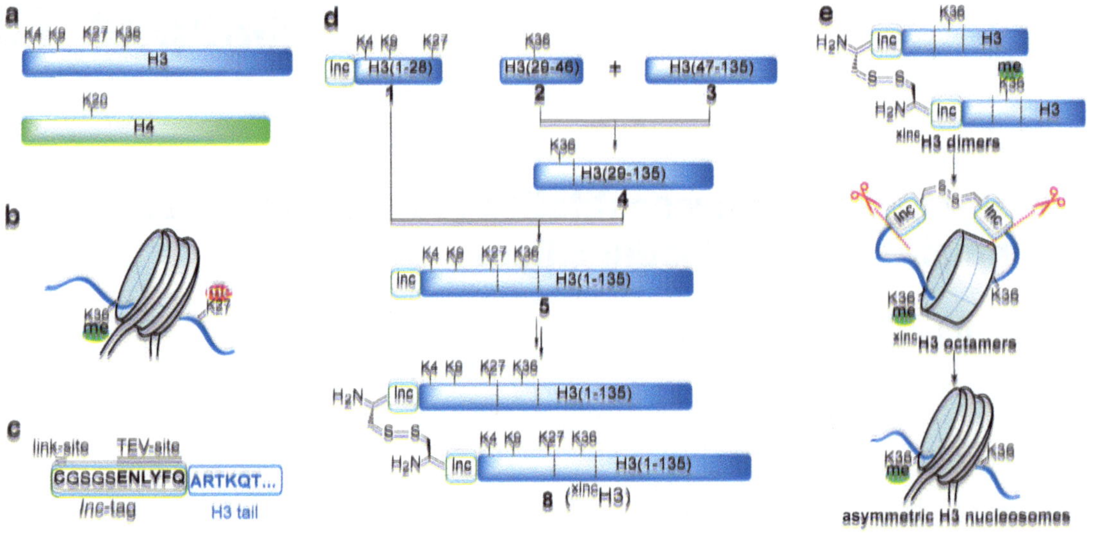

Fig. 1 (**a**) Scheme of H3 and H4 showing the position of the most relevant lysine residues. (**b**) Scheme of a nucleosome carrying asymmetric PTMs on its H3 tails. The H3K27me3 and the H3K36me3 marks are deposited on different H3 tails. (**c**) Sequence of the *Inc*-tagged N-terminal H3 tail. (**d**) Schematic representation of three-piece ligation strategy employed for the synthesis of asymmetric H3 dimers. (**e**) Nucleosomes asymmetrically modified at K36 are obtained by refolding of asymmetric histone octamers, followed by nucleosome reconstitution, disulfide bond reduction and *Inc*-tag cleavage by TEV protease

preexisting histone PTM patterns modulate the activity of chromatin enzymes which write or erase modifications [5]. Together, these mechanisms enable complex regulatory systems that establish and maintain chromatin states and control gene activity.

Histone H3 is the carrier of a number of a series of biologically important PTMs (Fig. 1a). H3 trimethylated at lysine 4 (H3K4me3) is a PTM associated with active genes and generally deposited by mixed-lineage leukemia (MLL) family of lysine methyltransferases (KMTs) [6]. H3K9me3 is a major marker for transcriptionally silent heterochromatin, and the binding site of heterochromatin protein 1 (HP1) family proteins [7]. H3K27me3 is a key PTM involved in gene repression by the polycomb system and is deposited by polycomb repressive complex 2 (PRC2) [6]. H3K36me3 is mostly associated with active transcription but has also further more complex roles [8]. Finally, a critical methylation site exists also on histone H4 at K20, with widespread roles in gene repression and chromatin compaction [9] and DNA repair [10], among others.

In particular, the transcription associated PTMs H3K4me3/H3K36me3 and H3K27me3 are mostly found in distinct genomic regions, due to their antagonistic role [11]. In embryonic stem cells and in particular cancer cells, however, these PTMs have been found to coexist on key developmental promoters, forming "bivalent domains" [12–16]. Recent studies have further revealed that the PTMs in bivalent chromatin exist in an asymmetrical configuration [13, 17]: the two copies for each histone within a nucleosome do not carry the same PTM pattern but are decorated with different modifications (Fig. 1b). Such asymmetry greatly expands the combinatorial space of histone marks, yet little is known about their establishment, maintenance, and function in embryonic stem cells.

Biochemical studies of the functional roles of asymmetric PTMs patterns, for example, of their regulatory effects on writer enzyme complexes, require access to defined and pure asymmetrically modified nucleosomes (Fig. 1a). Here, we describe a modular synthetic route based on peptide–protein ligation chemistry [18, 19] which was developed for the assembly of nucleosomes containing asymmetrically modified H3 [20, 21]. This method enables to access all key methylation sites in H3, namely, K4, K9, K27, and K36. We provide here a protocol for the synthesis of nucleosomes asymmetrically methylated at H3K36 [21] and also indicate how other modification patterns can be installed throughout the protocol. We further mention structural and practical differences that one must consider when applying the strategy to asymmetric H4 nucleosomes instead [22].

Our strategy is traceless and modular and provides access to the major methylation sites in H3 up to K36 via a three-piece ligation scheme (Fig. 1d). The desired modifications are introduced into peptide segments **1** and **2** during solid phase peptide synthesis (SPPS) as modified amino acids. Subsequent ligation enables facile introduction of combinatorial PTM patterns. Asymmetry is induced by the "link-and-cut" tag (*lnc*-tag), a short peptide sequence installed at the N-terminus of the histone tail in peptide **1**, that allows to transiently cross-link two histone proteins via a disulfide bond (Fig. 1c, e). This temporary linkage constrains two differentially modified histones to be included in the same octamer. Importantly, after nucleosome reconstitution, the tag can be removed by protease digestion, ensuring the traceless nature of the procedure (Fig. 1e).

The detailed chemical strategy (Fig. 2) to synthesize asymmetric H3K36me3, the example protein discussed in this protocol, begins with SPPS or recombinant expression of the different building blocks. The following protocols describe the production of each of the required building blocks:

1. **Synthesis of peptides 1 and 2:** The synthetic peptide **1** includes residues 1–28 of H3 and the *lnc*-tag, which provides the cysteine utilized for the transient cross-linking of the two H3 molecules. The N-terminal Cys is introduced protected with an acetamidomethyl (Acm) protecting group. Moreover, the *lnc*-tag contains a Tobacco Etch Virus (TEV) cleavable sequence that enables tag removal and a traceless procedure.

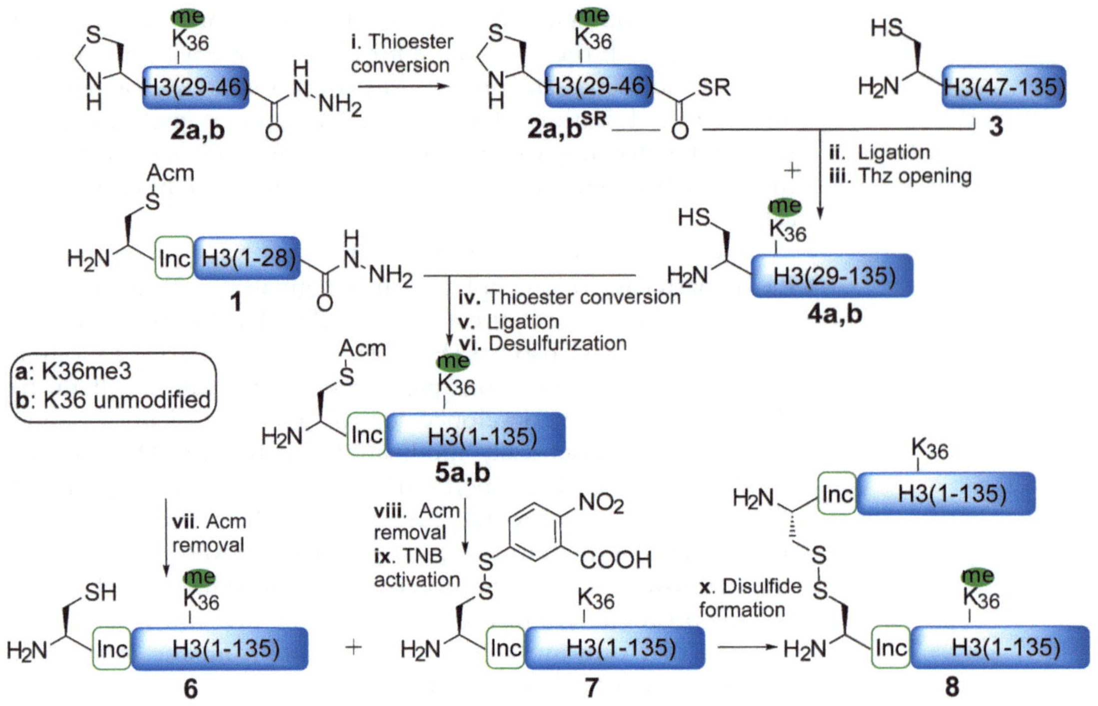

Fig. 2 Synthetic strategy for the formation of H3 heterodisulfide dimers asymmetrically modified at K36. (i) MPAA thioester conversion of peptide **2** hydrazide. (ii) Ligation of peptide thioester $\mathbf{2}^{SR'}$ to recombinant truncated H3Δ46 protein **3**. (iii) In situ thiazolidine to cysteine conversion. (iv) MTG thioester conversion of N-terminal *Inc*-tagged peptide hydrazide **1**. (v) Ligation of $\mathbf{1}^{SR}$ to truncated H3Δ28 protein **4**. (vi) In situ desulfurization of Cys47 and Cys29. (vii) Acm deprotection of the *Inc*-tag cysteine. (viii, ix) Acm deprotection and in situ 2-nitro-5-thiobenzoic acid (TNB) activation of the *Inc*-tag cysteine. x) Heterodisulfide formation

The peptide is synthesized by Fmoc-SPPS with a C-terminal hydrazide as a thioester precursor. If desired, K4, K9, or K27 can be included as methylated residues (other histone PTMs would also be possible).

Peptide **2** contains H3 residues 29–46 and is synthetically produced as a C-terminal hydrazide. Moreover, Ala29 at its N-terminus is mutated to cysteine and protected either as a thiazolidine (Thz) or alternatively by Acm. Finally, H3K36 can be incorporated in different methyl-states (e.g., H3K36me3) (resulting in peptides **2a** or **b**, depending on the presence or absence of the trimethyl-mark).

2. **Recombinant expression and purification of protein 3, H3Δ(1–46)A47C:** The remainder of H3 lacking the N-terminal 46 amino acids, protein **3**, is recombinantly expressed and subsequently purified from bacterial culture. It further carries an A47C mutation to provide the cysteine needed for ligation.

These three fragments enable facile synthesis of cross-linked H3 molecules for asymmetric nucleosome assembly. The synthesis and assembly steps are described in the following protocols:

1. **MPAA thioester conversion of crude peptide hydrazide 2:** The first reaction prepares peptide **2**, which carries the desired PTMs (making it either **2a** or **2b** depending on the modification state), for ligation to protein **3**. To this end, its C-terminal hydrazide has to be converted to a thioester.
2. **One-pot MPAA ligation and Thz deprotection—synthesis of 4a and 4b:** In a next step, peptides **2a** or **2b** are ligated to protein **3**, followed by deprotection of the N-terminal Thz in one-pot, liberating a cys residue (generating proteins **4a** or **4b**).
3. **One-pot MTG ligation and desulfurization—synthesis of 5a and 5b:** In a second ligation step, peptide **1** (containing further PTMs if required) is connected to **4a** or **4b** and in situ free radical desulfurization of the cysteines at the ligation junctions is performed to obtain the native H3 sequence.
4. **Cys(Acm) deprotection—synthesis of protein 6:** The second ligation is followed by the deprotection of the Cys residue within the *lnc*-tag using AgOAc, ending the histone assembly for the proteins carrying trimethylated Lys36.
5. **One-pot Cys(Acm) deprotection and TNB activation—synthesis of 7:** A second subset of histone proteins (carrying an unmodified lysine residue at position 36 in this protocol) is however activated by Ellman's reagent (5,5′-dithiobis-(2-nitrobenzoic acid, DTNB).
6. **Generation of heterodisulfide dimer proteins 8:** At this point, any activated histone can be combined with any other *lnc*-tagged histone with a free Cys and carrying different PTMs to form a heterodisulfide dimer that will force the two differentially modified H3 to be included in the same nucleosome. Here, we will describe the assembly of asymmetric octamers with one H3K36me3 modification.
7. **Refolding of asymmetric histone octamers**: The heterodisulfide dimers are then reconstructed into histone octamers with recombinantly expressed histone H4, H2A and H2B under denaturing conditions.
8. **Nucleosome reconstitution and TEV cleavage**: By combining the asymmetric octamers with a DNA segment containing the high-affinity 601 nucleosome positioning sequence [23] (or any other suitable sequence), nucleosomes are assembled by a gradual dialysis from high to low salt concentration. Finally, dithiothreitol (DTT) is used to reduce the disulfide bond and the *lnc*-tag is removed by TEV protease digestion.

9. **Synthesis of nucleosomes asymmetrically modified on H4:** In a final protocol we detail the changes that have to be implemented to modify H4 instead.

Using this methodology, we have assembled a library of asymmetric nucleosomes comprising a large fraction of asymmetric PTMs found in stem cells, which has enabled enzymatic studies of PRC2 and Set2 regulation by preexisting PTMs [20–22].

2 Materials

2.1 Instrumentation

1. Analytical and semipreparative HPLC: analytical HPLC system with a quaternary pump, an autoinjector module, and UV-Vis detection.
2. Preparative HPLC: preparative HPLC system equipped with an autosampler module and UV-Vis detection.
3. ESI-MS analyses: Mass spectra are acquired by electrospray ionization in positive ion mode in the mass range of 200–2000 m/z.
4. Peptide synthesis: Fmoc-SPPS performed on an automated peptide synthesizer.
5. Size exclusion chromatography: FPLC system equipped with an autosampler module.

2.2 General RP-HPLC Analyses and Purifications

1. HPLC Solvent A: 0.1% (v/v) trifluoroacetic acid (TFA) in H_2O.
2. HPLC Solvent B: 0.1% TFA, 90% (v/v) acetonitrile (ACN) in H_2O.
3. HPLC solubilization buffer: 6 M guanidine hydrochloride, pH 2 (*see* **Note 1**).
4. Analytical RP-HPLC column: C18 column (150 × 4.6 mm, 5 μm particle size). Flow rate 1 mL/min.
5. Semipreparative RP-HPLC column: C18 column (250 × 9.4 mm, 5 μm particle size). Flow rate 4 mL/min.
6. Preparative RP-HPLC column: C18 column (250 × 21.2 mm, 7 μm particle size). Flow rate 20 mL/min.

2.3 Synthesis of Peptides

1. 30 mL glass reaction vessel for manual SPPS.
2. 2-Chlorotrityl chloride (2-Cl Trt) resin.
3. Hydrazine solution: 2.45 M *N,N*-diisopropylethylamine (DIEA), 1.65 M hydrazine monohydrate in dimethylformamide (DMF). For a 1 mL solution dissolve 427 μL DIEA (2.45 mmol) and 80 μL hydrazine monohydrate (1.65 mmol) in DMF.

4. HCTU solution: 0.5 M O-(6-chlorobenzotriazol-1-yl)-*N,N,N′,N′*-tetramethyluronium hexafluorophosphate (HCTU) in DMF (*see* **Note 2**). Prepare fresh.
5. DIEA solution: 0.7 M DIEA in DMF.
6. Piperidine solution: 20% (v/v) piperidine in DMF.
7. Fmoc N-protected amino acids with standard acid labile protecting groups.
8. Special amino acids: N-Boc-Thiazolidine-2-carboxylic acid, Fmoc-Lys(Me3)-OH chloride, Fmoc-Cys(Acm)-OH.
9. Pseudoproline dipeptides: Fmoc-Ala-Thr(psiMe,Me pro)-OH, Fmoc-Lys(Boc)-Ser(psiMe,Me pro)-OH, Fmoc-Gln(Trt)-Thr (psiMe,Me pro)-OH.
10. TIS cleavage solution: 2.5% (v/v) triisopropylsilane (TIS), 2.5% (v/v) H_2O in TFA.

2.4 Protein Expression and Purification

1. Plasmids encoding for His_6-SUMO-H3Δ(1–46)A47C and His_6-Ulp1 in Pet30 vector, *see* Ref. [21].
2. E. coli BL21(DE3) cells transfected with vectors for His_6-SUMO-H3Δ(1–46)A47C and His_6-Ulp1, see ref. [20].
3. Kanamycin stock solution: 50 mg/mL kanamycin in H_2O. For a 30 mL solution, dissolve 1.80 g kanamycin sulfate in 30 mL H_2O. Filter through a 0.22 μm filter and aliquot. Store at −20 °C.
4. KAN LB medium: 50 μg/mL kanamycin in lysogeny broth (LB) medium. Add 1 mL kanamycin stock solution to 1 L autoclaved LB medium.
5. IPTG solution: 1 M isopropyl β-D-1-thiogalactopyranoside (IPTG) in H_2O. For a 20 mL solution dissolve 4.77 g IPTG in 20 mL H_2O. Filter through a 0.22 μm filter and aliquot. Store at −20 °C.
6. Lysis buffer: 20 mM Tris–HCl, 200 mM NaCl, 1 mM EDTA, 1 mM 2-mercaptoethanol, pH 7.5. Add 2-mercaptoethanol freshly before use.
7. Triton X-100 Lysis buffer: 1% (v/v) Triton X-100 in Lysis buffer.
8. Ni-NTA agarose affinity chromatography matrix (50% (v/v) beads in 20% ethanol).
9. Ni-NTA purification buffer A: 6 M guanidine hydrochloride, 50 mM Tris–HCl, 100 mM NaCl, 5 mM imidazole, pH 7.5.
10. Ni-NTA purification buffer B: 6 M urea, 150 mM Tris–HCl, 150 mM NaCl, 50 mM imidazole, pH 7.5. To avoid carbamoylation of proteins, prepare fresh.

11. Ni-NTA purification buffer C: 6 M urea, 150 mM Tris–HCl, 150 mM NaCl, 500 mM imidazole, pH 7.5. To avoid carbamoylation of proteins, prepare fresh.
12. SDS-PAGE loading buffer (2×): 60 mM Tris–HCl pH 6.8, 25% (v/v) glycerol, 2% (w/v) sodium dodecyl sulfate (SDS), 1% (w/v) bromophenol blue. Add 5% (v/v) 2-mercaptoethanol freshly before use.
13. Dialysis buffer: 1 M urea, 75 mM Tris–HCl, 150 mM NaCl, 5 mM 1,4-dithiothreitol (DTT), 25 mM L-arginine, 5 mM L-cysteine, pH 7.5. Add DTT freshly before use.
14. Acetic acid solution I: 1% (v/v) acetic acid in H_2O.

2.5 Peptide and Protein Ligations

2.5.1 Conversion to MPAA Thioester from Crude

1. Ligation buffer I: 6 M guanidine hydrochloride, 0.2 M NaH_2PO_4, pH 3.
2. $NaNO_2$ solution: 0.5 M $NaNO_2$ in H_2O. For a 50 μL solution dissolve 1.73 mg $NaNO_2$ in 50 μL H_2O. Prepare fresh.
3. MPAA solution I: 0.5 M 4-mercaptophenylacetic acid (MPAA) in Ligation buffer I. For a 50 μL solution dissolve 4.2 mg MPAA in 50 μL Ligation buffer I.

2.5.2 MPAA Ligation and One-Pot Thz Opening

1. Ligation buffer II: 6 M guanidine hydrochloride, 0.2 M NaH_2PO_4/Na_2HPO_4, pH 7.4.
2. TCEP solution: 0.5 M tris(2-carboxyethyl)phosphine hydrochloride (TCEP) in ligation buffer II. For a 50 μL solution dissolve 7.2 mg TCEP in 25 μL Ligation buffer II. Readjust the pH to 7.4 with 5 M NaOH. Bring the volume to 50 μL with Ligation buffer II. Prepare fresh.
3. MPAA solution II: 1 M MPAA in Ligation buffer II. For a 50 μL solution dissolve 8.41 mg MPAA in 50 μL Ligation buffer II.
4. $[Pd(allyl)Cl]_2$ solution: 100 mM $[Pd(allyl)Cl]_2$ in Ligation buffer II. For a 100 μL solution dissolve 3.66 mg $[Pd(allyl)Cl]_2$ in 100 μL Ligation buffer II.
5. DTT solution I: 0.5 M DTT in H_2O. For a 200 μL solution dissolve 15.4 mg DTT in 200 μL H_2O. Prepare fresh or store as frozen aliquot.

2.5.3 MTG-Ligation and One-Pot Desulfurization

1. Ligation buffer I: *see* Subheading 2.5.1, **item 1**.
2. $NaNO_2$ solution: *see* Subheading 2.5.1, **item 2**.
3. Desulfurization buffer: 0.5 M TCEP, 6 M guanidine hydrochloride, 0.2 M NaH_2PO_4/Na_2HPO_4, pH 7. Prepare fresh.
4. VA-044 radical initiator solution: 0.3 M 2,2′-azobis[2-(2-imidazolin-2-yl)propane]dihydrochloride (VA-044) radical initiator in H_2O. For a 50 μL solution dissolve 4.8 mg VA-044 in 50 μL H_2O.

5. GSH hydrogen donor solution: 0.4 M L-glutathione reduced (GSH) in Desulfurization buffer. For a 50 μL solution dissolve 6.15 mg GSH in 50 μL Desulfurization buffer.

2.6 Acm Deprotection and 2-Nitro-5-Thiobenzoic Acid (TNB) Activation

1. Acetic acid solution II: 50% (v/v) acetic acid in H_2O.
2. AgOAc solution: 30 mM AgOAc in acetic acid solution II. For a 200 μL solution dissolve 1.02 mg AgOAc in 200 μL acetic acid solution II.
3. DTT solution I: *see* Subheading 2.5.2, **item 5**.
4. Activation buffer: 6 M guanidine hydrochloride, 0.2 M NaH_2PO_4/Na_2HPO_4, pH 7.
5. DTNB solution: 60 mM DTNB in Activation buffer. For a 500 μL solution dissolve 11.9 mg DTNB in 500 μL Activation buffer.

2.7 Heterodisulfide Dimer Formation

1. Reaction buffer: 6 M guanidine hydrochloride, 0.2 M NaH_2PO_4/Na_2HPO_4, pH 6, degassed and flushed with argon.

2.8 Histone Octamer Refolding

1. Unfolding buffer: 6 M guanidine hydrochloride, 20 mM Tris–HCl, 1 mM EDTA, pH 7.5.
2. Recombinant human lyophilized histones. *See* Ref. [24].
3. Dialysis cassettes (7K molecular weight cutoff, [MWCO]).
4. Refolding buffer: 2 M NaCl, 20 mM Tris–HCl, 1 mM EDTA, pH 7.5.
5. Centrifugal concentrators (10K MWCO).
6. Superdex S200 10/300GL prepacked size exclusion column.
7. SDS-PAGE loading buffer (2×): *see* Subheading 2.4, **item 12**.

2.9 Nucleosome Reconstitution and TEV Cleavage

1. 153 bp 601 DNA at a concentration of about 10–50 μM in H_2O. For the sequence, *see* Ref. [23], and for preparation *see* Ref. [25].
2. Start buffer: 2 M KCl, 10 mM Tris–HCl, 1 mM EDTA, pH 7.5. Prepare 200 mL and adjust the pH to 7.5 at 4 °C.
3. End buffer: 10 mM KCl, 10 mM Tris–HCl, 1 mM EDTA, pH 7.5. Prepare 2 L and adjust the pH to 7.5 at 4 °C.
4. Dialysis buttons (7K MWCO, 500 μL).
5. DTT solution II: 1 M DTT in H_2O. For a 100 μL solution dissolve 15.4 mg DTT in 100 μL H_2O. Prepare fresh or store as frozen aliquot.
6. TEV protease.
7. Sucrose solution: 25% (v/v) sucrose in H_2O.

8. SDS-PAGE loading buffer (5×): 150 mM Tris–HCl pH 6.8, 62.5% (v/v) glycerol, 5% (w/v) SDS, 2.5% (w/v) bromophenol blue. Add 12.5% (v/v) 2-mercaptoethanol freshly before use.

3 Methods

3.1 Synthesis of Peptides 1 and 2

In this protocol we describe the synthesis of the two peptides **1** and **2**, consisting of the N-terminal tail portion of H3 as well as a part of the globular domain. Histone modifications can be directly incorporated at this synthesis step. Both peptides are synthesized as C-terminal hydrazides. Moreover, both peptides **1** and **2** contain an N-terminal cysteine which must be protected. In our protocol, we describe the synthesis of peptide **1** using Acm protection and peptide **2** using a Thz protection scheme to cover both methodologies.

3.1.1 Resin Preparation

In order to synthesize C-terminal hydrazide peptides **1** and **2**, the 2-Cl-Trt resin must be functionalized with hydrazine. However, since the hydrazine resin is unstable and prone to hydrolysis, the first amino acid coupling must be performed before storage.

1. Weigh 0.5 g of 2-Cl-Trt resin (substitution: 1.63 mmol/g, 0.82 mmol, 1 eq.) into a 10 mL glass bottle (*see* **Note 3**). Add 3 mL DMF and swell the resin for 15 min at room temperature (RT) using a magnetic stirrer (*see* **Note 4**).
2. Cool the resin to 0 °C in an ice bath and add 1 mL of Hydrazine solution. Allow the resin to reach RT and stir 1 h at RT.
3. Subsequently, add 100 μL methanol (MeOH) to quench the reaction. Stir 10 min at RT.
4. Transfer the resin to a reaction vessel for manual peptide synthesis and wash thoroughly with DMF.
5. Weigh the desired Fmoc-protected amino acid (2.05 mmol, 2.5 eq.) into a 10 mL glass bottle (*see* **Note 5**). Add 3.9 mL HCTU solution (1.95 mmol, 2.38 eq.) and incubate for 2 min at RT. Subsequently, add 714 μL DIEA (4.1 mmol, 5 eq.) and let the amino acid activation reaction proceed for 1 min at RT.
6. To initiate the coupling step, add the activated amino acid to the resin. Let react for 30 min at RT using nitrogen bubbling to homogenize the reaction mixture (*see* **Note 6**). Wash thoroughly with DMF, then rinse with DCM and MeOH and finally dry under vacuum. The amino acid-resin can be stored at −20 °C.

7. To determine the resin substitution: Weigh 3 × 1 mg of dry resin into three glass test tubes. Add 4 mL piperidine solution in DMF and incubate for 30 min at RT under gentle nutation. Take 200 μL of the solution and dilute 1:10 in piperidine solution (*see* **Note 7**) Measure the absorbance at 301 nm with a UV-vis spectrophotometer and calculate the resin loading as the average of the three measurements (*see* **Note 8**).

3.1.2 Peptide Synthesis

With the preloaded hydrazide resin in hand, the synthesis of peptides **1** and **2** can be performed using automated Fmoc-SPPS, followed by HPLC purification (peptide sequences are reported in **Note 9**). At this step, all desired histone modifications can be included into the peptides using the appropriate building blocks. Here, we provide the protocols for the synthesis of unmodified peptide **1**, as well as of peptide **2** carrying K36me3 (**2a**) or without modification (**2b**). In general, peptide synthesis is performed on an automated peptide synthesizer at a synthesis scale of 0.1 mmol (1 eq.).

1. Using the peptide synthesizer, swell the resin in 5 mL DMF under light agitation for 30 min at RT.
2. 1 mL HCTU solution (0.48 mmol, 4.8 eq.) is pumped to the desired Fmoc-N-protected amino acid (0.5 mmol, 5 eq.) followed by 1 mL of 0.7 M DIEA solution (1 mmol, 10 eq.) and the reaction proceeds for 1 min at RT.
3. The activated amino acid is delivered to the resin and the coupling is performed for 30 min at RT under light agitation (*see* **Note 9**).
4. The resin is washed with 3 × 5 mL DMF and drained.
5. Fmoc-deprotection reaction is achieved by delivering 5 mL piperidine solution to the resin. The reaction is allowed to proceed for 3 min at RT under light agitation.
6. **Step 5** is repeated to ensure an efficient Fmoc deprotection.
7. The resin is washed with 3 × 5 mL DMF and drained.
8. **Steps 2–7** are repeated until the last N-terminal amino acid has been coupled (*see* **Note 10**).
9. Following the peptide synthesis, the peptidyl-resin is removed from the synthesizer, washed with DMF, DCM, MeOH and dried under vacuum. The peptidyl-resin can be stored at −20 °C.
10. In a next step, the peptide is cleaved from the resin. To this end, weigh the desired amount of peptidyl-resin in a 50 mL tube (*see* **Note 11**). Add 1 mL TIS cleavage solution per 100 mg of peptidyl-resin, and let react at RT for 3 h under gentle agitation.

11. For collection of the cleaved peptide, add 3 volumes of ice-cold diethyl ether to the cleaved peptide. The solution should turn white and cloudy immediately, indicating the precipitation of the peptide. Incubate the mixture for 10 min on ice to allow precipitation. Centrifuge at 5000 × *g* for 10 min at 4 °C to pellet the peptide.
12. Decant the supernatant and resuspend the white peptide pellet (which still contains resin beads) in 3 volumes of ice-cold diethyl ether. Then repeat this step for a total of 3 times to wash away all hydrophobic impurities, including the cleaved protecting groups. Finally, dry the pellet with argon flow.

3.1.3 Peptide Purification

After synthesis, peptides **1** and **2** (and their modified variants) are purified by preparative HPLC. Note, however, that to increase the overall yield, for peptide **2** we generally skip initial purification and directly move to thioester conversion (Subheading 3.3, **steps 1–4**).

1. Dissolve the peptide in 5–10 mL 50% ACN in H_2O and filter the solution to remove the beads (filter pore size: 35 μm). Collect the solution in a tared 50 mL tube. Take a sample of 5 μL for analytical HPLC analysis. Flash-freeze the dissolved peptide with liquid N_2 and lyophilize. The crude, lyophilized peptide can be stored at −20 °C.
2. Analyze the crude peptide by analytical RP-HPLC. Dilute the 5 μL sample taken in **step 1** by 20-fold to 100 μL with HPLC solvent A. Centrifuge the sample for 5 min at 21,000 × *g*. Inject 80 μL onto the C-18 column and run a gradient from 0 to 70% HPLC solvent B over 30 min. Collect the peaks and analyze them by ESI-MS for identification.
3. After complete lyophilization of the peptide, weigh the tube containing the peptide powder for quantification. Subsequently, redissolve the peptide in 5–10 mL HPLC solvent A. Centrifuge at 21,000 × *g* for 10 min. Decant the supernatant.
4. Purify the peptide by preparative RP-HPLC using a C-18 reversed phase preparative HPLC column over a gradient of 10–40% HPLC solvent B over 40 min for the N-terminal tail peptide **1** (expected retention time of the product: 15min), and of 10–35% HPLC solvent B over 50 min for peptide **2** (expected retention time of the product: 11 min) (*see* **Note 12**).
5. Pool fractions containing pure peptide. Flash-freeze with liquid N_2 and lyophilize. The pure peptides can be stored at −20 °C (MW peptide **1**: 4268.8 Da. MW peptide **2a**: 1993.4 Da, MW peptide **2b**: 1951.3 Da).

3.2 Recombinant Expression and Purification of Protein 3, H3Δ (1–46)A47C

The truncated histone H3Δ(1–46)A47C (**3**) is expressed as N-terminal fusion to a His_6-SUMO tag in BL21(DE3) cells. The use of the His_6-SUMO tag increases yield and enables facile liberation of the N-terminal cysteine for ligation. Nevertheless, overexpressed histone proteins are insoluble and will be purified from inclusion bodies.

3.2.1 Recombinant Expression and Inclusion Bodies Preparation

1. Transfect BL21(DE3) *E. coli* cells with the plasmid containing His_6-SUMO-H3Δ(1–46)A47C.
2. Inoculate 6 ×1 L KAN LB medium with the transfected bacterial culture and grow at 37 °C, with agitation at 200 rounds per minute (rpm) until the OD_{600} reaches 0.6–0.8.
3. Induce protein expression with 6 × 150 μL 1 M IPTG solution to a final concentration of 0.15 mM IPTG, and continue expression at 37 °C for 4 h.
4. Transfer the culture into 1 L centrifugation tubes and harvest the cells by centrifugation at 4000 × *g* at 4 °C for 20 min.
5. Resuspend the cell pellets in a minimal amount of Lysis buffer and transfer the suspension to 50 mL tubes (for a 6 L culture, a total volume of 40 mL distributed equally into two tubes should suffice). The resuspended cell pellets can be stored frozen at −80 °C, or directly used in **step 6**.
6. If frozen, thaw the cell suspension. Lyse the cells by sonication using a pulse pattern of 15 s pulse on, 45 s pulse off. Keep the solution surrounded by ice during sonication to avoid heating. Pulse for 10 min to ensure good lysis, as judged by a slight change in color and viscosity of the lysate.
7. Centrifuge the cell lysate at 15,000 × *g* at 4 °C for 30 min. As the histone proteins are contained in inclusion bodies, they are collected in the cell pellet. Thus, the supernatant can be discarded.
8. Wash the pelleted inclusion bodies as follows: resuspend each pellet with 30 mL Lysis buffer by pipetting. Centrifuge the sample at 15,000 × *g* at 4 °C for 20 min. Repeat the wash. Subsequently, wash each pelleted inclusion body sample with 30 mL Triton X-100 Lysis buffer, followed by a single wash with 30 mL Lysis buffer. After a final centrifugation at 15,000 × *g* at 4 °C for 20 min, the collected inclusion bodies can be stored at −80 °C.

3.2.2 Protein Purification by Ni-NTA Chromatography

1. Solubilize each washed inclusion bodies pellet in 30 mL Ni-NTA purification buffer A. To fully solubilize the protein, vortex thoroughly and stir at 4 °C for 2 h. Centrifuge the samples at 15,000 × *g* at 4 °C for 30 min and pool the supernatants (60 mL).

2. Pipet 20 mL of Ni-NTA slurry (10 mL bed volume = 1 CV) to a 100 mL empty glass chromatography column. Wash twice with 2 CV of H_2O and equilibrate the beads with 3 CV Lysis buffer. Apply the solubilized protein to the equilibrated Ni-NTA resin and allow binding to occur overnight at 4 °C under gentle nutation.
3. Collect the flow-through and wash the beads with 5 CV Ni-NTA purification buffer A, then with 2 × 5 CV Ni-NTA purification buffer B. Finally, elute the protein with 4 × 1 CV Ni-NTA purification buffer C.
4. Analyze samples taken from flow-through, washing and elution fractions by SDS-PAGE: Mix 10 μL sample with 10 μL SDS-PAGE loading buffer (2×), boil the sample at 100 °C for 5 min and allow to cool down to RT. Load 10 μL per lane on a 15% polyacrylamide gel and analyze by electrophoresis (*see* **Note 13**). Pool elution fractions containing the pure His_6-SUMO protein **3** as judged by SDS-PAGE (Expected MW: 25 kDa).

3.2.3 Cleavage of His_6-SUMO Tag and Preparative RP-HPLC Purification

To cleave the N-terminal His-SUMO tag, we employ Ulp1 SUMO protease. The protease is expressed as N-terminal fusion to His_6-tag and purified similarly to the histones as inclusion body pellets following the steps described under Subheadings 3.2.1 and 3.2.2. Since the amount of purified Ulp1 SUMO protease obtained from 1 L bacterial culture is normally enough to completely cleave the His_6-SUMO-H3Δ(1–46)A47C obtained from 6 L bacterial culture, quantities given in Subheadings 3.2.1 and 3.2.2 can be scaled down accordingly.

1. Express and purify Ulp1 SUMO protease in 1 L bacterial culture as described under Subheadings 3.2.1 and 3.2.2.
2. Mix the Ni-NTA elution fractions of His_6-SUMO-H3Δ(1–46) A47C protein (from 6 L bacterial culture, ~250 mg estimated protein amount) and His_6-Ulp1 (from 1 L bacterial culture, ~60 mg estimated protein amount), resulting in a mass-ratio of about 4:1. Aim for a concentration of about 5–10 mg/mL of His_6-SUMO-H3Δ(1–46)A47C and transfer the cleavage mixture to a dialysis tube (MWCO 3500). Dialyze overnight against 2 L of Dialysis buffer at 4 °C.
3. Confirm that the protein has been completely cleaved by RP-HPLC and ESI-MS. Dilute a 50 μL sample twofold to 100 μL with HPLC solvent A and follow the HPLC instructions described under Subheading 3.1.3, **step 2** to analyze the protein (MW protein **3** H3Δ(1–46)A47C: 10,387 Da).
4. If completely cleaved, dialyze the protein mixture (~25–50 mL) against 3 × 500 mL acetic acid solution I over 16 h (i.e., dialysis against 500 mL 1% acetic acid, 3 changes of

4–6 h each). If not completely cleaved, extend reaction time or add more Ulpl SUMO protease (*see* **Note 14**). After dialysis, flash-freeze the protein solution with liquid N_2 and lyophilize. The crude protein (expected amount ~200 mg) can be stored at −20 °C.

5. Dissolve lyophilized powder (100 mg) in 5 mL HPLC solvent A. Centrifuge at 21,000 × *g* for 10 min to remove undissolved or precipitated protein. Inject the cleared solution onto the preparative C-18 column and run a gradient from 30 to 70% HPLC solvent B over 30 min (expected retention time of the product: 27 min). Pool fractions containing pure protein, flash-freeze with liquid N_2 and lyophilize. The pure protein **3** (expected yield ~15 mg) can be stored as a dry powder at −20 °C.

3.3 MPAA Thioester Conversion of Crude Peptide Hydrazide 2

With the truncated histone protein **3** in hand, the next step is the ligation of peptide **2** (here described for peptides with K36me3, **2a**, or without modification **2b**) to protein **3**. In order to allow this ligation, the N-terminal hydrazide of peptide **2** is converted to a MPAA thioester. Since the SPPS of peptide **2** is normally very efficient and the crude peptide recovered after ether precipitation is sufficiently pure (>75%), one can skip the preparative RP-HPLC purification step (Subheading 3.1.3, **steps 3–5**) and directly proceed with the thioester conversion reaction.

1. Weigh 6.0 mg of crude peptide **2a** or 5.85 mg of crude **2b** (3 μmol, 1 eq.) in a 2 mL tube. Add 300 μL Ligation buffer I to a final peptide concentration of 10 mM and vortex to dissolve completely. Cool the peptide solution to −18 °C in an ice-salt bath and stir the solution with a small magnetic bar.
2. Add 5.4 μL 0.5 M $NaNO_2$ solution (2.7 μmol, 0.9 eq.) and incubate at −18 °C for 20 min while stirring (*see* **Note 15**). Subsequently, add 60 μL MPAA solution I (30 μmol, 10 eq.) and adjust the pH to 7 with 1 M NaOH (*see* **Note 16**). Allow the reaction mixture to reach RT and stir for 10 min.
3. To confirm the thioester conversion, prepare a sample for analytical RP-HPLC and ESI-MS. Dilute a 1 μL sample 100-fold to 100 μL with HPLC solvent A and follow the procedure described under Subheading 3.1.3, **step 2**.
4. Purify by semipreparative RP-HPLC. Dilute the reaction mixture to 2 mL with HPLC solvent A. Centrifuge for 5 min at 21,000 × *g* to eliminate any insoluble material. Inject onto the semipreparative C-18 column and run a gradient from 20 to 40% HPLC solvent B over 40 min (expected retention time of the product: 12 min). Characterize the collected fraction by analytical RP-HPLC and ESI-MS and pool the desired fractions into a tared tube. Flash-freeze the purified peptide in

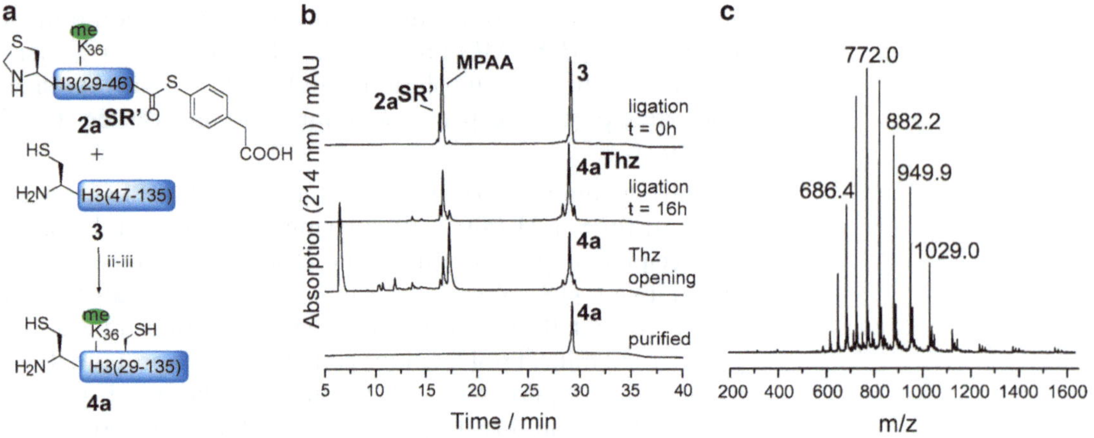

Fig. 3 Scheme and HPLC chromatograms of the reactions involved in the synthesis of protein **4a**. (a) Scheme of the reaction involving the following steps: ligation of peptide thioester $\mathbf{2a}^{\mathbf{SR'}}$ to protein **3** (ii) and thiazolidine opening (iii). (b) RP-HPLC chromatograms. From top to bottom: ligation of peptide thioester $\mathbf{2a}^{\mathbf{SR'}}$ to protein **3** at $t = 0$; ligation at $t = 16$ h; thiazolidine to cysteine conversion to afford **4a**; purified product **4a**. (c) ESI-MS spectrum of purified product **4a** (expected MW = 12,336 Da, observed MW = 12,336 Da)

liquid N_2 and lyophilize. The peptide thioester $\mathbf{2a}^{\mathbf{SR}}$, $\mathbf{b}^{\mathbf{SR}}$ can be stored as a dry powder at −20 °C (Typical yield 40%, MW of peptide $\mathbf{2a}^{\mathbf{SR}}$: 2128.5 Da, MW of peptide $\mathbf{2b}^{\mathbf{SR}}$: 2086.4 Da).

3.4 One-Pot MPAA Ligation and Thz Deprotection—Synthesis of 4a and 4b

The following protocol describes the ligation of MPAA peptide thioester $\mathbf{2a}^{\mathbf{SR}}$, $\mathbf{2b}^{\mathbf{SR}}$ to the recombinant truncated histone **3**, followed by N-terminal Thz opening (Fig. 3a). The opening of Thz provides the N-terminal free cysteine needed for the subsequent ligation step (Subheading 3.5). Alternatively, Acm can be used as the N-terminal protecting group. In this case, refer for Acm deprotection to Subheading 3.6, **steps 1–3**.

3.4.1 MPAA Ligation

1. Dissolve 2.5 mg of peptide thioester $\mathbf{2a}^{\mathbf{SR}}$ or 2.45 mg of $\mathbf{2b}^{\mathbf{SR}}$ (1.2 μmol, 3 eq.) in 80 μL Ligation buffer II to a final peptide concentration of 15 mM. Vortex until completely dissolved. Add 8 μL 0.5 M TCEP solution (4 μmol, 10 eq.) to a final TCEP concentration of 50 mM.
2. In a separate 1.5 mL tube, weigh 4.15 mg of the recombinant truncated H3 **3** (0.4 μmol, 1 eq.). Add the peptide thioester solution (*see* **Note 17**), check that the pH is 7.4 using pH paper (adjust if necessary by adding 5 M NaOH in 0.5 μL aliquots) and shake at 200 rpm at 25 °C for 16 h (*see* **Note 18**).
3. Monitor the progress of the ligation reaction by analytical RP-HPLC and ESI-MS (Fig. 3b). Dilute a 0.5 μL sample 200-fold to 100 μL with HPLC solvent A and follow the procedure described under Subheading 3.1.3, **step 2** (*see* **Note 19**) (MW of $\mathbf{4a}^{\mathbf{Thz}}$: 12,348 Da, MW of $\mathbf{4b}^{\mathbf{Thz}}$:

12,306 Da). When the reaction is complete (i.e., ideally all of protein **3** is consumed, or no further reaction progress can be detected over the course of a few hours), proceed with the in situ Thz to Cys conversion.

3.4.2 Thz Opening and Semipreparative RP-HPLC Purification

1. To the ligation mixture, add 4 μL TCEP solution (2 μmol, 5 eq) and 9 μL 1 M MPAA solution II (9 μmol, 22.5 eq.). Subsequently, add 90 μL 100 mM $[Pd(allyl)Cl]_2$ solution (9 μmol, 22.5 eq.) and stir the reaction mixture at 37 °C for 45 min using a small magnetic bar.
2. Monitor the progress of the Thz opening reaction by analytical RP-HPLC and ESI-MS (Fig. 3b). Dilute a 1 μL sample 100-fold to 100 μL with HPLC solvent A and follow the procedure described under Subheading 3.1.3, **step 2**. When the reaction is complete, proceed with semipreparative RP-HPLC purification.
3. To precipitate the Pd catalyst, add 190 μL DTT solution I to a final concentration of 250 mM and vortex for 1 min. Centrifuge for 10 min at 21,000 × *g* (*see* **Note 20**). Dilute the supernatant to 2 mL with HPLC solvent A. Centrifuge for 5 min at 21,000 × *g*. Inject onto the semipreparative C-18 column and run a gradient from 45 to 70% HPLC solvent B over 60 min (expected retention time of the product: 28 min). Characterize the collected fraction by analytical RP-HPLC and ESI-MS and pool the desired fractions into a tared tube. Flash-freeze the purified protein in liquid N_2 and lyophilize. The proteins **4a, b** can be stored as a dry powder at −20 °C (Typical yield 35%, MW of **4a**: 12,336 Da, MW of **4b**: 12,294 Da, Fig. 3c).

3.5 One-Pot MTG Ligation and Desulfurization—Synthesis of 5a and 5b

In order to initiate the ligation reaction between peptide **1** and protein **4a, 4b**, the N-terminal *lnc*-tagged peptide hydrazide **1** is first converted to thioester using methyl thioglycolate (MTG). MTG thiol additive is used instead of MPAA because the latter is not compatible with the following free radical desulfurization step. The desulfurization reaction allows for the concomitant conversion of Cys29 and Cys47 to alanine (Fig. 4). Here we describe the use of unmodified peptide **1**. However, at this point the combinatorial complexity of histone PTMs can be increased by using peptide **1** variants containing further desired PTM patterns.

3.5.1 MTG Thioester Conversion and Ligation

1. Dissolve 1.56 mg of peptide **1** (0.36 μmol, 3 eq.) in 36 μL Ligation buffer I to a final concentration of 10 mM and vortex to dissolve completely. Cool the peptide solution to −18 °C in an ice-salt bath and stir the solution with a small magnetic bar.

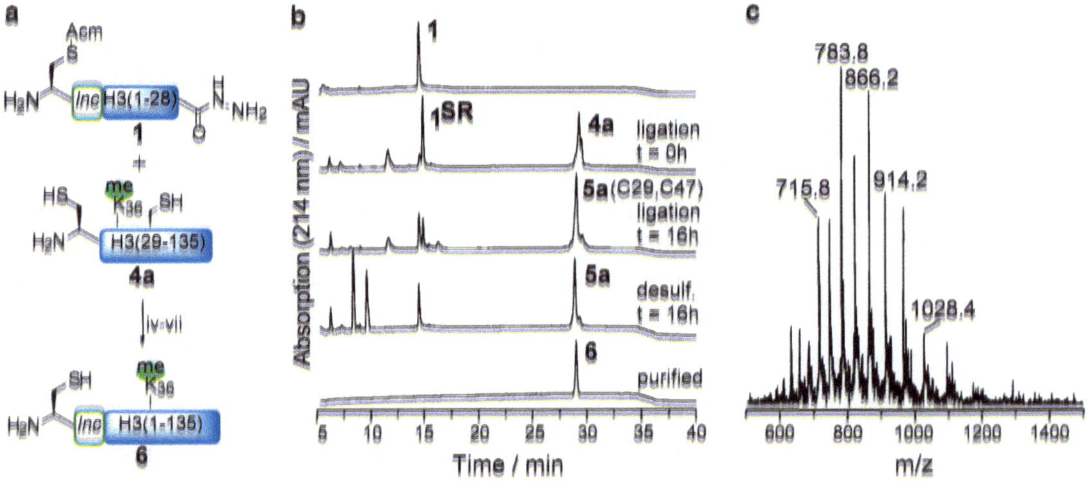

Fig. 4 Scheme and HPLC chromatograms of the reactions involved in the synthesis protein **6**. (**a**) Scheme of the reaction involving the following steps: MTG thioester conversion of peptide **1** hydrazide (iv), ligation of peptide thioester $\mathbf{1^{SR}}$ to protein **4a** (v), desulfurization of Cys29 and Cys47 (vi) and Acm deprotection (vii). (**b**) RP-HPLC chromatograms. From top to bottom: peptide **1**; ligation of peptide $\mathbf{1^{SR}}$ to protein **4a** at $t = 0$; ligation at $t = 16$ h; desulfurization at $t = 16$ h; Acm deprotection of **5a** to afford **6**. (**c**) ESI-MS spectrum of purified product **6** (expected MW = 16,438 Da, observed MW = 16,439 Da)

2. Add 1.08 μL 0.5 M $NaNO_2$ solution (0.54 μmol, 4.5 eq.) and incubate the reaction mixture at −18 °C for 20 min while stirring. Subsequently, add 0.24 μL MTG (2.7 μmol, 22.5 eq.) and adjust the pH to 7 with 1 M NaOH (*see* **Note 16**). Allow the reaction mixture to reach RT and stir for 10 min.
3. In a separate 1.5 mL tube, weigh 1.48 mg protein **4a** or 1.47 mg **4b** (0.12 μmol, 1 eq.) Subsequently, add the peptide thioester solution (*see* **Note 17**) and 1.8 μL TCEP solution (0.9 μmol, 7.5 eq.) and check that the pH is 7 using pH paper (adjust if necessary by adding 5 M NaOH in 0.5 μL aliquots). Shake at 200 rpm at 25 °C for 16 h (*see* **Note 18**).
4. Monitor the progress of the ligation reaction by analytical RP-HPLC and ESI-MS (Fig. 4b). Dilute a 1 μL sample 100-fold to 100 μL with HPLC solvent A and follow the procedure described under Subheading 3.1.3, **step 2** (*see* **Note 19**). When the reaction is complete, proceed with the in situ desulfurization reaction (MW of $\mathbf{5a}^{C29,C47}$: 16,542 Da, MW of $\mathbf{5b}^{C29,C47}$: 16,500 Da).

3.5.2 Desulfurization and Semipreparative RP-HPLC Purification

1. Add 8 μL of VA-044 radical initiator solution (2.4 μmol, 20 eq.), 8 μL of GSH hydrogen donor solution (3.2 μmol, 26.7 eq.) and complete the reaction volume to 100 μL with Desulfurization buffer. Shake at 200 rpm at 42 °C for 16 h (*see* **Note 18**).

2. Monitor the progress of ligation by analytical RP-HPLC and ESI-MS (Fig. 4b). Dilute a 2.5 μL sample 40-fold to 100 μL with HPLC solvent A and follow the procedure described under Subheading 3.1.3, **step 2**. When the reaction is complete, proceed with the semipreparative RP-HPLC purification.
3. Dilute the supernatant to 2 mL with HPLC solvent A. Centrifuge for 5 min at 21,000 × *g*. Inject onto the semipreparative C-18 column and run a gradient from 30 to 70% HPLC solvent B over 45 min (expected retention time of the product: 33 min). Characterize the collected fraction by analytical RP-HPLC and ESI-MS and pool the desired fractions into a tared tube. Flash-freeze the purified protein in liquid N_2 and lyophilize. Protein **5a, 5b** can be stored as a dry powder at −20 °C (Typical yield 55%, MW of **5a**: 16,509 Da, MW of **5b**: 16,467 Da).

3.6 Cys(Acm) Deprotection—Synthesis of Protein 6

At this point, to prepare asymmetric nucleosomes, a subset of semisynthetic *lnc*(Acm)-H3 histones is deprotected to generate variants with a free N-terminal Cys (**6**), whereas a separate subset is activated at their N-terminal cysteine using DTNB to generate variants **7**. This protocol illustrates the removal of the Acm protecting group from the N-terminal cysteine of protein **5a** to give histone **6** (Fig. 4a), whereas the DNTB activation of **5b** is described under Subheading 3.7, **steps 1–5**.

1. Dissolve 1.1 mg of the Cys(Acm) protected histone protein **5a** in 67 μL Acetic acid solution II (66.63 nmol, 1 eq.) to a final protein concentration of 1 mM. Vortex to dissolve completely.
2. To initiate the Acm deprotection reaction, add 67 μL of AgOAc solution (2.01 μmol, 30 eq.) to a final AgOAc concentration of 15 mM. Shake at 37 °C at 200 rpm for 6 h.
3. Monitor the progress of the deprotection by analytical RP-HPLC and ESI-MS (Fig. 4b). Take 2.5 μL of the reaction mixture and mix with 2.5 μL of DTT solution I in order to precipitate the Ag. Vortex for 30 s and centrifuge for 5 min at 21,000 × *g*. Dilute 4 μL of the supernatant by 25-fold to 100 μL with HPLC solvent A and follow the procedure described under Subheading 3.1.3, **step 2**. When the reaction is complete proceed with semipreparative RP-HPLC purification.
4. Add 134 μL of DTT solution I to a final DTT concentration of 250 mM to the reaction mixture in order to precipitate the Ag and vortex for 1 min. Centrifuge for 10 min at 21,000 × *g* (*see* **Note 20**). Transfer the supernatant to a new 2 mL tube. Dilute the supernatant to 2 mL with HPLC solvent A. Centrifuge for 5 min at 21,000 × *g*. Inject onto the semipreparative C-18 column and run a gradient from 30 to 70% HPLC solvent B

over 40 min (expected retention time of the product: 30 min). Characterize the collected fraction by analytical RP-HPLC and ESI-MS and pool the desired fractions into a tared tube. Flash-freeze the purified protein in liquid N_2 and lyophilize. The protein **6** can be stored as a dry powder at −20 °C (Typical yield 68%, MW of 6: 16,438 Da, Fig. 4c).

3.7 One-Pot Cys(Acm) Deprotection and TNB Activation—Synthesis of 7

This protocol describes the deprotection of cysteine from the Acm group followed by its activation by Ellman's reagent to obtain an activated heterodisulfide. Here, the reaction is performed on protein **5b**; however, protein **5a** may be activated instead (the same applies for any other modified variants).

1. Follow the instructions given under Subheading 3.6, **steps 1–3**, using Cys(Acm) protected histone protein **5b**. Instead of proceeding with RP-HPLC purification after Acm removal, the free cysteine is activated with Ellman's reagent.
2. Add to the reaction mixture 134 μL of DTNB solution (8.04 μmol, 120 eq.) to a final DTNB concentration of 30 mM and vortex for 1 min. Incubate for 5 min at RT to allow the disulfide–thiol exchange reaction to occur between DTNB and the free cysteine of protein **5b**. A rapid coloration of the solution to a bright yellow indicates reaction progress. Subsequently, centrifuge the solution for 5 min at 21,000 × *g* (*see* **Note 21**). A dark pellet corresponding to the precipitated Ag should be visible at the bottom of the tube. Transfer the supernatant to a new 2 mL tube.
3. Dilute the supernatant to 2 mL with HPLC solvent A. Centrifuge for 5 min at 21,000 × *g*. Inject onto the semipreparative C-18 column and run a gradient from 30 to 70% HPLC solvent B over 45 min (expected retention time of the product: 31 min). Characterize the collected fraction by analytical RP-HPLC and ESI-MS and pool the desired fractions into a tared tube. Flash-freeze the purified protein in liquid N_2 and lyophilize. The protein 7 can be stored as a dry powder at −20 °C (Typical yield 67%, MW of 7: 16,592 Da).

3.8 Generation of Heterodisulfide Dimer Proteins 8

The steps involved in the generation and purification of heterodisulfide dimers (Fig. 5a) have to be performed relatively quickly to avoid the exchange of disulfide partners and formation of homodimers.

1. Dissolve 0.75 mg Cys(TNB)-protein 7 in 22.6 μL Reaction buffer to a final protein concentration of 2 mM (45.2 nmol, 1.1 eq.) in a 1.5 mL tube.
2. Dissolve 0.67 mg Cys-protein **6** in 20.2 μL Reaction buffer to a final protein concentration of 2 mM (40.7 nmol, 1 eq.) in another 1.5 mL tube (*see* **Note 22**). Save a sample for analytical HPLC analysis.

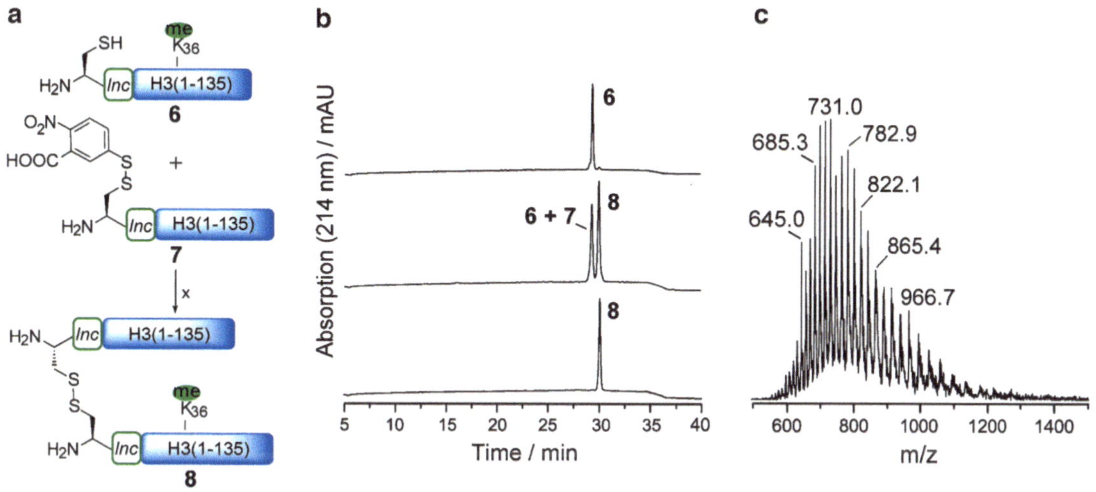

Fig. 5 Scheme and HPLC chromatograms of the reaction involved in the synthesis of H3K36me3 heterodisulfide dimers **8**. (**a**) Scheme of the heterodisulfide bond formation between protein **6** and protein **7** (×). (**b**) RP-HPLC chromatograms. From top to bottom: histone **6**; heterodisulfide dimer formation between **6** and **7**, purified product **8**. (**c**) ESI-MS spectrum of purified product **8** (expected MW = 32,832 Da, observed MW = 32,844 ± 12 Da)

3. Mix the two proteins, vortex shortly to homogenize the solution and incubate at RT for 30 s. Dilute immediately the reaction mixture to 2 mL with HPLC solvent A (*see* **Note 23**). Yellow coloration of the solution indicates reaction progress. Before purification, save a sample for analytical HPLC analysis. Centrifuge for 5 min at 21,000 × *g*. Inject onto the semipreparative C-18 column and run a gradient from 30 to 70% HPLC solvent B over 45 min (expected retention time of the product: 35.5 min). Characterize the reaction samples (from **steps 2** to **3**) as well as the collected fraction by analytical RP-HPLC and ESI-MS (Fig. 5b) and pool the desired fractions into a tared tube. Flash-freeze the purified dimers in liquid N_2 and lyophilize. The heterodisulfide dimers **8** can be stored as a dry powder at −20 °C (Typical yield 57%, MW of **8:** 32,832 Da, Fig. 5c).

3.9 Refolding of Asymmetric Histone Octamers

By incubating the H3 heterodisulfide dimers **8** (^{xlnc}H3) with the other core histone proteins under denaturing conditions and dialyzing them against native buffer, asymmetric octamers are refolded, followed by purification by size exclusion chromatography.

1. Weigh ~0.5–1 mg of H2A, H2B and H3 recombinant histones (*see* **Note 24**) and semisynthetic ^{xlnc}H3 histone dimers **8** each in four different 1.5 mL tube. Dissolve the histones in Unfolding buffer to a final concentration of ~2 mg/mL. Centrifuge for 5 min at 21,000 × *g* to remove any aggregates and transfer the solution to a new 1.5 mL tube.

Table 1
Extinction coefficients at 280 nm and molecular weight of the recombinant and semisynthetic histones used for the refolding of asymmetric H3K36me3 octamers

Histone	Extinction coefficient (280 nm)	Molecular weight
hH2A	4470 $M^{-1}cm^{-1}$	13,964 g/mol
hH2B	7450 $M^{-1}cm^{-1}$	13,759 g/mol
hH4	5970 $M^{-1}cm^{-1}$	11,236 g/mol
^{xlnc}H3K36me3/unmodified (**8**)	12,045 $M^{-1}cm^{-1}$	11,236 g/mol

2. Measure the exact protein concentration by UV-vis absorbance measurement at 280 nm using the following extinction coefficient values (Table 1):
3. Mix the H3 heterodisulfide dimers **8** (11.6 nmol, 1 eq.) with H4 (23.2 nmol, 2 eq.), H2A (25.5 nmol, 2.2 eq.) and H2B (25.5 nmol, 2.2 eq.) and dilute with Unfolding buffer to a final total protein concentration of 0.5 mg/mL (*see* **Note 25**). Transfer the solution in a dialysis cassette and dialyze against 3 × 500 mL Refolding buffer over a total of 16 h at 4 °C.
4. Concentrate the octamers to a final concentration of 40 μM (*see* **Notes 26** and **27**) and transfer to a new tube. Centrifuge for 5 min at 21,000 × *g* at 4 °C to remove any aggregates.
5. Purify by size exclusion chromatography (SEC) in Refolding buffer at 4 °C (Fig. 6a). The elution volume of the octamers from the Superdex S200 10/300GL column is approximately 12 mL. Protein eluting at the solvent front (~8 mL) is generally aggregated H3 or H4, a peak at ~13 mL are H3–H4 tetramers, whereas a peak at ~14.5 mL corresponds to H2A–H2B dimers.
6. Analyze the collected fractions by SDS-PAGE: Mix 10 μL of the fraction with 10 μL SDS-PAGE loading buffer (2×), boil the sample at 100 °C for 5 min and allow to cool down to RT. Load 10 μL per lane on a 17% polyacrylamide gel and analyze by electrophoresis. Pool fractions containing the pure asymmetric ^{xlnc}H3 octamers (*see* **Note 28**) (Fig. 6b).
7. Concentrate to a final concentration of 30–40 μM (minimum volume 100 μL, **Note 27**) and add glycerol to a final glycerol concentration of 50% (v/v). The octamers can be stored at −20 °C (Typical yield: 15–20%).

3.10 Nucleosome Reconstitution and TEV Cleavage

The last step of the procedure involves the reconstitution of asymmetric nucleosomes. The reconstitution is performed in the absence of any reducing agents to ensure the integrity of the disulfide bond and prevent exchange of histone proteins. After

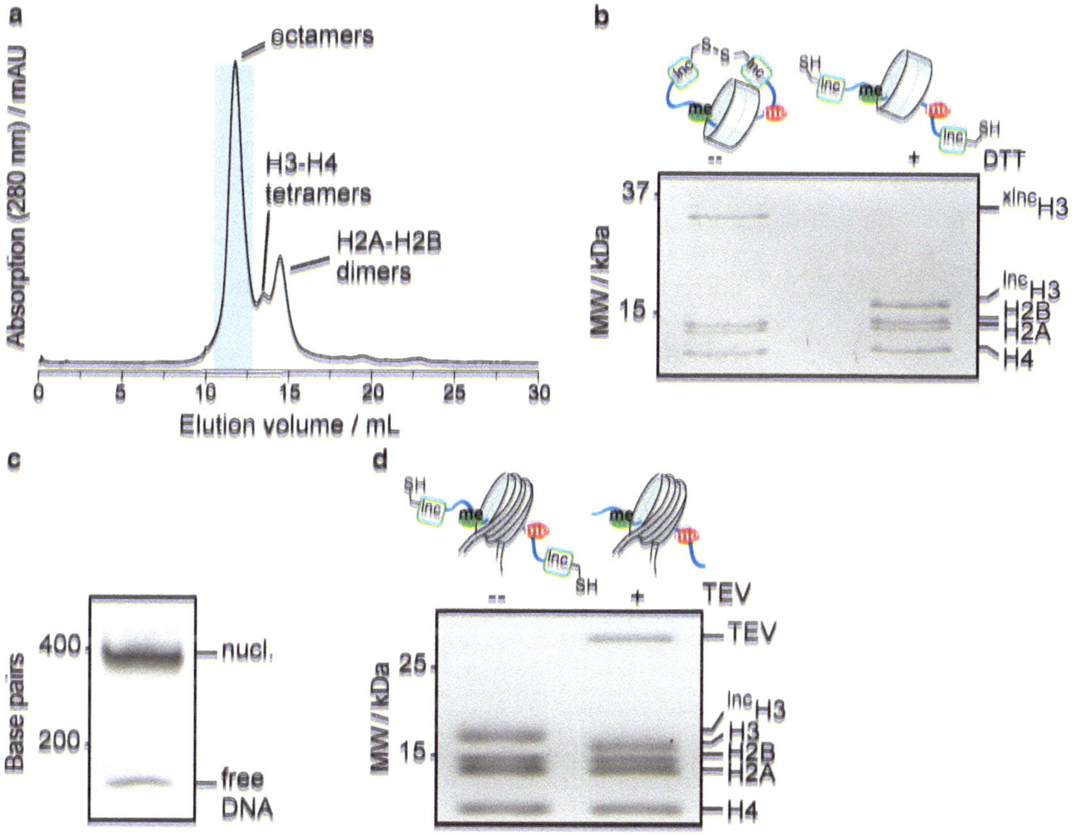

Fig. 6 Asymmetric octamer refolding, nucleosome reconstitution, and TEV cleavage. (**a**) SEC chromatogram of the purification of asH3K36me3 octamers. Octamers are eluted at 12 mL elution volume (blue area), H3–H4 tetramers at an elution volume of 13 mL and H2A–H2B dimers at 14.5 mL. (**b**) SDS-PAGE analysis of purified asymmetric histone octamers after SEC in the presence (+) or absence (−) of the reducing agent DTT. ^{xlnc}H3: lnc-tagged H3 dimers linked through a disulfide bond. ^{lnc}H3: lnc-tagged H3 monomer after reduction of the disulfide bond. (**c**) Native-PAGE analysis of reconstituted nucleosomes. (**d**) SDS-PAGE analysis of the TEV cleavage efficiency after disulfide bond reduction

the reconstitution, the disulfide bond is reduced and TEV protease is added to remove the *lnc*-tag and obtain traceless asymmetric nucleosomes.

1. Add the histone octamers (1–1.2 eq.) to an aqueous solution of 601 DNA (100 pmol, 1 eq., final concentration 2 μM) (*see* **Note 29**). Before adding the octamers, adjust the salt concentration to 2 M with 4 M NaCl solution. In fact, in the absence of DNA, disassembly of the histone octamer can occur at low salt concentration, due to the overall positive charge of the histone proteins.
2. Transfer the solution to the dialysis buttons (previously equilibrated in Start buffer) and gradually dialyze them from Start buffer to End buffer at 4 °C over 16 h (*see* **Note 30**, Ref. [26]). Dialyze against End buffer for an additional 2 h at 4 °C.

3. Transfer the reconstituted nucleosomes to 1.5 mL tubes and add 1M DTT to a final concentration of 100 mM to reduce the disulfide bond between the H3 pairs. Incubate at 4 °C for 30 min.
4. Analyze the nucleosomes by native-PAGE (Fig. 6c): Mix 4 μL of the nucleosome samples with 2 μL of 25% sucrose. Load the 6 μL on a 5% polyacrylamide gel and analyze by electrophoresis at 4 °C. A single strong band at around 400 bp indicates the correct reconstitution of the nucleosomes. A weaker lower band corresponding to free DNA is often observed and is considered acceptable when it does not exceed ~10% of the nucleosome band intensity (*see* **Note 31**).
5. Take a 10 μL sample for SDS-PAGE. To the rest of the reduced nucleosomes, add 1.6 μg TEV protease per 100 pmol nucleosomes and incubate at 4 °C for 16 h. Take another sample of 10 μL for SDS-PAGE.
6. Analyze the cleavage efficiency by SDS-PAGE (Fig. 6d): Mix 10 μL of the nucleosome samples with 2 μL of SDS-PAGE loading buffer (5×). Boil the sample at 100 °C for 5 min and allow to cool down to RT. Load 12 μL per lane on a 17% polyacrylamide gel and analyze by electrophoresis.
7. The asymmetric nucleosomes are ready to use and can be stored in the fridge up to one month.

3.11 Synthesis of Nucleosomes Asymmetrically Modified on H4

When asymmetric H4 nucleosomes have to be synthesized, a similar procedure can be adopted. However, one has to consider a few additional points [22]. First, in cells H4 is generally acetylated at the N-terminus. This prevents the attachment of the *lnc*-tag at that position. The *lnc*-tag can thus be incorporated via an isopeptide bond at the Lys5 position on the H4 N-terminal tail. For a detailed description, *see* Ref. [22]. Second, the H4 histone is much more prone to precipitation than H3. For this reason, the desulfurization reactions are performed in larger volumes and at lower concentrations (for optimal results, use a dilution factor ×4 compared to H3). Finally, the TEV protease cleaves isopeptide bonds less efficiently than normal peptide bonds. Longer incubation times (24 h), multiple additions of TEV protease (4 × 1 μg per 100 pmol nucleosome) and higher temperatures (RT) are needed to completely cleave the tag.

4 Notes

1. When preparing the samples for HPLC analysis or purification, it is possible to observe precipitation of the peptide/protein after addition of HPLC solvent A. HPLC solubilization buffer can be added accordingly to improve solubility.

2. For the preparation of 0.5 M HCTU solutions in DMF, weigh the calculated amount of activating reagent and add the entire volume of solvent, that is, for 100 mL of 0.5 M HCTU solution: weigh 20.68 mg HCTU, add 500 mL DMF (do not make up to volume). This ensures a slightly lower concentration than calculated, which will reduce the likelihood of guanidinylation of the growing peptide N-terminus, thus allowing for higher yields.
3. Resin substitution changes from batch to batch: calculate the amount of resin needed accordingly.
4. Always use fresh DMF as it decomposes in dimethylamine over time.
5. It is assumed that 50% of the 2Cl-Trt resin is loaded with the hydrazine (0.41 mmol, 0.5 eq.).
6. To ensure high yields, the amino acid coupling is repeated.
7. Make sure to properly homogenize the 20% piperidine solution by pipetting up and down, then wait for the beads to settle down before taking the 200 μL sample.
8. Extinction coefficient of dibenzofulven at 301 $\varepsilon_{301} = 7800\ M^{-1}cm^{-1}$. Formula for calculating the resin loading: $((A_{301} * \text{mL piperidine solution})/(\varepsilon_{301} * \text{g of resin})) * \text{dilution factor} = \text{resin loading in [mmol/g]}$.
9. To increase coupling efficiencies and purity, the underlined amino acids were double coupled (Repeat Subheading 3.1.2, **steps 2–3**). In addition, to avoid aggregation of the growing peptide, pseudoproline dipeptides were employed where necessary (bold), as they break up peptide structure on the resin:
   (a) N-terminal fragment **1**:
   - C(Acm)GSGSENLYFQARTK**QT**ARKSTGG-KAPRKQL**AT**KAARKS-NHNH$_2$

   (b) Middle fragments **2a,b**:
   - C(Thz)PATGGVK(me3)KPHRYRPGTV
   - C(Thz)PATGGVKKPHRYRPGTV
10. As the N-terminal peptide amino acid, one can use a Boc-N-protected amino acid (i.e., N-Boc-Thiazolidine-2-carboxylic acid is used in this protocol for the synthesis of peptides **2a** and **2b**). In this case the final Fmoc-deprotection (Subheading 3.1.2, **steps 5–7**) can be omitted.
11. You might consider performing a test cleavage on a small number of beads before cleaving the whole peptide from the resin. Follow Subheading 3.1.2, **step 2**, and scale down the amounts accordingly. After drying the peptide pellet with Ar, dissolve it in 50 μL 50% ACN in H_2O. Centrifuge to let the

beads settle and take 10 μL of the supernatant. Analyze by analytical RP-HPLC and ESI-MS as described under Subheading 3.1.3, **step 2**.

12. The incorporation of histone PTMs may alter the retention times on the RP-HPLC column. Consider adjusting the gradient and run time accordingly to achieve a good peak separation and high yield purification.
13. Perform trichloroacetic acid (TCA) precipitation of guanidine-containing samples prior to SDS-PAGE analysis: add 2 volumes of 20% (v/v) TCA in H_2O to the samples, vortex. Incubate for 10 min on ice. Centrifuge at 21,000 × *g* at 4 °C for 5 min. Discard supernatant. Wash the pellet with 400 μL ice-cold ether/ethanol (1:1). Centrifuge at 21,000 × *g* at 4 °C for 5 min. Discard supernatant. Dry the pellet.
14. Before adding additional Ulp1 to the cleavage mixture, the protease needs to be refolded separately. This is achieved by dialyzing overnight His_6-Ulp1 (from 1 L bacterial culture) Ni-NTA elution fractions against 2 L of Dialysis buffer at 4 °C.
15. To avoid formation of nitrosamine side products on the Thz moiety, which is observed when more than 1 eq. $NaNO_2$ is present in the solution, one can slightly reduce the amount of $NaNO_2$ added to the reaction mixture. Alternatively, nitrosamine formation can be avoided by using an Acm-protected cysteine in peptide **2**.
16. Be careful not to over titrate. Thioester hydrolysis occurs at basic pH.
17. Do not pipet up and down but vortex: the truncated histone might take some time to dissolve and might stick to the pipet tip.
18. To preserve TCEP reduction activity, flush the reaction mixture with Argon and seal the tube with Parafilm.
19. If a slow reaction rate is observed, consider rechecking the pH of the reaction mixture, as a too low pH will slow down the transthioesterification reaction. Alternatively, add some more peptide thioester to the reaction mixture to increase its concentration.
20. To recover the maximal amount of the peptide, we normally wash the pellet twice with 100 μL 250 mM DTT solution and pool the supernatants.
21. To recover the maximal amount of the peptide, we normally wash the pellet twice with 100 μL 60 mM DTNB solution and pool the supernatants.

22. The TNB-Cys activated histone protein is in slight excess with respect to the free-Cys protein to immediately consume all the free H3 thiol and thus avoid formation of homodisulfide dimers.
23. The reaction must be quickly quenched and diluted and the product isolated to avoid disulfide–thiolate interchange.
24. For the recombinant expression and purification of human histone proteins H3(C110A), H2A and H2B see ref. [24] (to avoid complications arising due to additional Cys residues, a H3 mutant devoid of Cys is generally used).
25. The histone concentration is kept low (0.5 mg/mL) to avoid the incorporation of the two H3 present in the cross-linked dimer into two different histone octamers.
26. Do not allow the volume to be less than 100 μL to prevent sticking of the octamers to the filter. Centrifuge at 8000 × *g* at 4 °C. After every centrifugation, mix the concentrated octamers with the rest of the solution to homogenize the concentration. Rinse carefully the filter each time without making bubbles. Always keep on ice.
27. The octamer concentration is determined by UV-vis absorbance measurement at 280 nm using as an extinction coefficient the sum of the ε_{280} of the single histones: ε_{280}(octamer) = [ε_{280}(hH2A) + ε_{280}(hH2B) + ε_{280}(hH4)] * 2 + ε_{280}(^{xlnc}H3K36me3/unmodified) = 47,825 $M^{-1}cm^{-1}$.
28. Only pool fractions showing four histone bands with the same intensity on the SDS-PAGE gel. Lanes with more intense upper and lower bands reveal the presence of ^{lnc}H3–H4 tetramers (elution volume 13 mL), while stronger middle bands indicate H2A–H2B dimer enrichment (elution volume 14 mL).
29. To determine the appropriate DNA–octamer molar ratio, perform a microscale nucleosome reconstitution as described in Ref. [26].
30. To set up the gradual dialysis, place the dialysis buttons with the nucleosome assembly mixture, in a beaker containing 200 mL Start buffer (beaker I). Use a magnetic stirrer to continuously stir the buffer in beaker I. Place a second beaker nearby containing a reservoir of End buffer (beaker II). Using a two-channel peristaltic pump, transfer End buffer from beaker II to beaker I at a flow rate of 1 mL/min. Simultaneously, use the second channel of the pump to remove solvent from beaker I at the same rate, such that the solvent level in beaker I remains constant. Thereby, the buffer in beaker I is slowly exchanged from Start to End buffer.

31. A too high octamer–DNA molar ratio results in a band at the bottom of the well. Since the histone proteins are positively charged, the octamer excess does not enter into the gel and accumulates at that position. Nucleosomes tend to precipitate in the presence of excess histone proteins, leading to a low yield reconstitution. On the other hand, a too low octamer–DNA molar ratio results in a strong lower band. Importantly, DNA is stained by intercalating DNA dye much more efficiently than the DNA engaged in the nucleosomes and the band intensity on the gel is thus not directly proportional to the amount of DNA. Keeping the free DNA band intensity lower than 10% of the nucleosome band ensures that only a negligible amount of free DNA is present.

Acknowledgments

We thank Dr. Carolin Lechner for having established some of the procedures described and Dr. Andreas Bachmann for the compounds used in the reactions. This work was supported by funding by the SystemsX program and EPFL.

References

1. van Holde K (1989) Chromatin. In: Springer series in molecular biology. Springer, New York
2. Luger K, Mader AW, Richmond RK, Sargent DF, Richmond TJ (1997) Crystal structure of the nucleosome core particle at 2.8 A resolution. Nature 389(6648):251–260
3. Strahl BD, Allis CD (2000) The language of covalent histone modifications. Nature 403 (6765):41–45
4. Ruthenburg AJ, Li H, Patel DJ, Allis CD (2007) Multivalent engagement of chromatin modifications by linked binding modules. Nat Rev Mol Cell Biol 8(12):983–994
5. Torres IO, Fujimori DG (2015) Functional coupling between writers, erasers and readers of histone and DNA methylation. Curr Opin Struct Biol 35:68–75. https://doi.org/10.1016/j.sbi.2015.09.007
6. Schuettengruber B, Chourrout D, Vervoort M, Leblanc B, Cavalli G (2007) Genome regulation by polycomb and trithorax proteins. Cell 128(4):735–745
7. Lachner M, O'Carroll D, Rea S, Mechtler K, Jenuwein T (2001) Methylation of histone H3 lysine 9 creates a binding site for HP1 proteins. Nature 410(6824):116–120. https://doi.org/10.1038/35065132
8. Wagner EJ, Carpenter PB (2012) Understanding the language of Lys36 methylation at histone H3. Nat Rev Mol Cell Biol 13 (2):115–126. https://doi.org/10.1038/nrm3274
9. Brejc K, Bian Q, Uzawa S, Wheeler BS, Anderson EC, King DS, Kranzusch PJ, Preston CG, Meyer BJ (2017) Dynamic control of X chromosome conformation and repression by a histone H4K20 demethylase. Cell 171 (1):85–102.e123. https://doi.org/10.1016/j.cell.2017.07.041
10. Jorgensen S, Schotta G, Sorensen CS (2013) Histone H4 lysine 20 methylation: key player in epigenetic regulation of genomic integrity. Nucleic Acids Res 41(5):2797–2806. https://doi.org/10.1093/nar/gkt012
11. Ringrose L, Paro R (2004) Epigenetic regulation of cellular memory by the Polycomb and Trithorax group proteins. Annu Rev Genet 38:413–443
12. Bernstein BE, Mikkelsen TS, Xie X, Kamal M, Huebert DJ, Cuff J, Fry B, Meissner A, Wernig M, Plath K, Jaenisch R, Wagschal A, Feil R, Schreiber SL, Lander ES (2006) A bivalent chromatin structure marks key developmental genes in embryonic stem cells. Cell 125(2):315–326

13. Voigt P, LeRoy G, Drury WJ 3rd, Zee BM, Son J, Beck DB, Young NL, Garcia BA, Reinberg D (2012) Asymmetrically modified nucleosomes. Cell 151(1):181–193. https://doi.org/10.1016/j.cell.2012.09.002
14. Voigt P, Tee WW, Reinberg D (2013) A double take on bivalent promoters. Genes Dev 27 (12):1318–1338. https://doi.org/10.1101/gad.219626.113
15. Rodriguez J, Munoz M, Vives L, Frangou CG, Groudine M, Peinado MA (2008) Bivalent domains enforce transcriptional memory of DNA methylated genes in cancer cells. Proc Natl Acad Sci U S A 105(50):19809–19814
16. Bapat SA, Jin V, Berry N, Balch C, Sharma N, Kurrey N, Zhang S, Fang F, Lan X, Li M, Kennedy B, Bigsby RM, Huang TH, Nephew KP (2010) Multivalent epigenetic marks confer microenvironment-responsive epigenetic plasticity to ovarian cancer cells. Epigenetics 5 (8):716–729
17. Shema E, Jones D, Shoresh N, Donohue L, Ram O, Bernstein BE (2016) Single-molecule decoding of combinatorially modified nucleosomes. Science 352(6286):717–721. https://doi.org/10.1126/science.aad7701
18. Dawson PE, Muir TW, Clark-Lewis I, Kent SBH (1994) Synthesis of proteins by native chemical ligation. Science 266:776–779
19. Muir TW, Sondhi D, Cole PA (1998) Expressed protein ligation: a general method for protein engineering. Proc Natl Acad Sci U S A 95(12):6705–6710
20. Lechner CC, Agashe ND, Fierz B (2016) Traceless synthesis of asymmetrically modified bivalent nucleosomes. Angew Chem Int Ed Engl 55(8):2903–2906. https://doi.org/10.1002/anie.201510996
21. Guidotti N, Lechner CC, Bachmann AL, Fierz B (2019) A modular ligation strategy for asymmetric bivalent nucleosomes trimethylated at K36 and K27. Chembiochem 20 (9):1124–1128. https://doi.org/10.1002/cbic.201800744
22. Guidotti N, Lechner CC, Fierz B (2017) Controlling the supramolecular assembly of nucleosomes asymmetrically modified on H4. Chem Commun 53(74):10267–10270. https://doi.org/10.1039/c7cc06180c
23. Lowary PT, Widom J (1998) New DNA sequence rules for high affinity binding to histone octamer and sequence-directed nucleosome positioning. J Mol Biol 276(1):19–42
24. Kilic S, Bachmann AL, Bryan LC, Fierz B (2015) Multivalency governs HP1alpha association dynamics with the silent chromatin state. Nat Commun 6:7313. https://doi.org/10.1038/ncomms8313
25. Fierz B, Chatterjee C, McGinty RK, Bar-Dagan M, Raleigh DP, Muir TW (2011) Histone H2B ubiquitylation disrupts local and higher-order chromatin compaction. Nat Chem Biol 7(2):113–119. https://doi.org/10.1038/nchembio.501
26. Dyer PN, Edayathumangalam RS, White CL, Bao Y, Chakravarthy S, Muthurajan UM, Luger K (2004) Reconstitution of nucleosome core particles from recombinant histones and DNA. Methods Enzymol 375:23–44

Chapter 14

Facile Semisynthesis of Ubiquitylated Peptides with the Ligation Auxiliary 2-Aminooxyethanethiol

Caroline E. Weller and Champak Chatterjee

Abstract

The posttranslational modification of cellular proteins by ubiquitin (Ub), called ubiquitylation, is indispensable for the normal growth and development of eukaryotic organisms. In order to conduct studies that elucidate the precise mechanistic roles for Ub, access to site-specifically and homogenously ubiquitylated proteins and peptides is critical. However, the low abundance, heterogeneity, and dynamic nature of protein ubiquitylation are significant limitations toward such studies. Here we provide a facile expressed protein ligation method that does not require specialized apparatus and permits the rapid semisynthesis of ubiquitylated peptides by using the atom-efficient ligation auxiliary 2-aminooxyethanethiol.

Key words Ubiquitin, Ligation, Auxiliary, Semisynthesis, Peptide, Ubiquitylation, Sumoylation

1 Introduction

Ubiquitin (Ub) is a 76 amino acid–long highly conserved protein that has been found in all eukaryotes studied to date [1]. Similar to chemical protein modifications such as phosphorylation, acetylation and methylation, Ub is best known as a modifier of protein side-chains. Cellular processes regulated by ubiquitylation (modification by Ub) include the proteasomal degradation of oxidatively damaged and terminally misfolded proteins, receptor trafficking [2], protein aggregation [3], DNA-damage repair and gene transcription [4]. In a vast majority of known ubiquitylated proteins, Ub is conjugated to specific lysine side-chain ε-amines by means of an isopeptide linkage through its C-terminal α-carboxylic acid. As Ub, and indeed most proteins contain multiple lysines, their site-specific ubiquitylation leads to a large diversity of modification states in eukaryotic cells [5]. Even within the same protein, ubiquitylation at different positions may be associated with differing functional outcomes [3]. In humans, a family of over 630 Ub-ligase enzymes (designated as E1-E3 proteins) and about 100 deubiquitylating enzymes (DUBs) are tasked with imparting substrate and

Miquel Vila-Perelló (ed.), *Expressed Protein Ligation: Methods and Protocols*, Methods in Molecular Biology, vol. 2133, https://doi.org/10.1007/978-1-0716-0434-2_14,

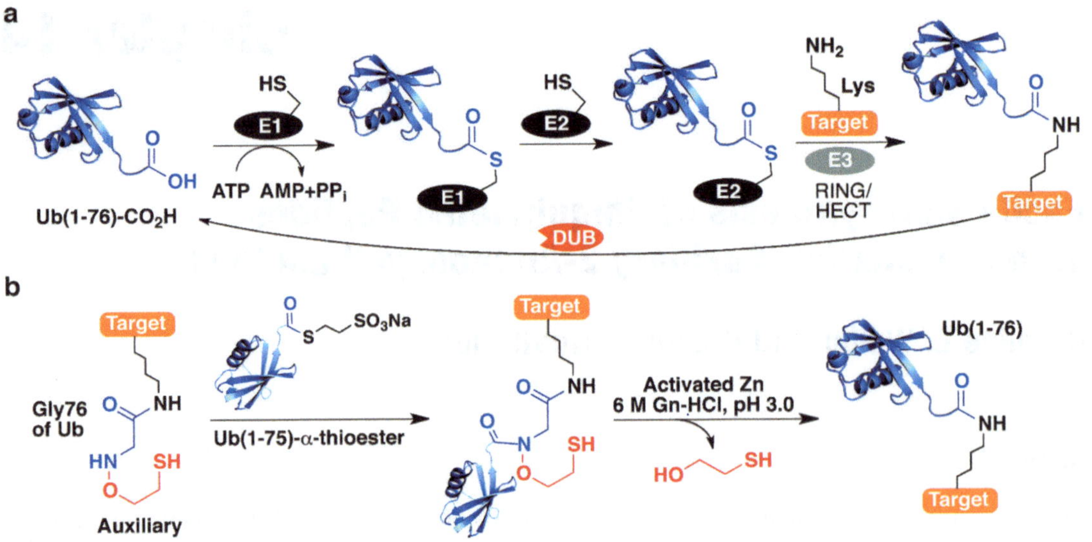

Fig. 1 Biological and chemical protein ubiquitylation. (**a**) Site-specific ubiquitylation of protein lysine side-chains by the E1-E3 family of ubiquitin ligases. The deubiquitinase (DUB) family of proteases remove Ub from mono- and polyubiquitylated target proteins. (**b**) Chemical strategy for site-specific ubiquitylation of peptides employing the temporary ligation auxiliary, 2-aminooxyethanethiol, at the desired site of ubiquitin attachment in substrates

lysine-site specificity to ubiquitylation, and in ensuring the synchronization of protein ubiquitylation states with varying cellular demands (Fig. 1a) [6].

Given its central role in eukaryotic biology, understanding the context-specific functions of protein ubiquitylation is a pressing goal for modern molecular biology. One significant challenge toward studies aimed at deciphering mechanistic roles for Ub is access to sufficient quantities of homogenously and site-specifically ubiquitylated proteins for in vitro biochemical and biophysical assays [7]. The highly dynamic nature of protein ubiquitylation, coupled with the heterogeneity of protein posttranslational modifications in vivo, has rendered access to uniformly ubiquitylated proteins limited to instances where the substrate-specific Ub ligases are known and found to be active in vitro. In some instances, sufficient quantities of the uniformly ubiquitylated protein may be isolated from cultured cells [8]. However, most studies seeking to ascribe mechanistic roles to site-specific ubiquitylation require significant quantities of Ub-modified proteins that are not readily accessible from cells, or from in vitro ubiquitylation reactions.

Here we present a chemical strategy that provides facile access to ubiquitylated peptides (Fig. 1b), which, in principle, are readily extendable to protein substrates that are amenable to semisynthesis [9]. Key aspects of our methodology include (a) the synthesis of a ligation auxiliary, 2-aminooxyethanethiol, and its installation at the Lys site of ubiquitylation in peptide substrates [10], (b) expressed protein ligation of the auxiliary-bearing peptide with a C-terminally truncated Ub α-thioester [11], and (c) removal of the auxiliary from the ligation product by reduction with metallic Zn. The need for the 2-aminooxyethanethiol ligation auxiliary arises from the fact that the Ub C-terminus is conjugated to lysine side-chain ε-amines by means of an isopeptide linkage. This side-chain linkage cannot be generated by standard Cys-directed ligation techniques that, despite their tremendous utility, are limited to the generation of native peptide bonds [12].

2 Materials

2.1 Synthesis of the S-Trityl-Protected Ligation Auxiliary

2.1.1 Equipment

1. 3 Å molecular sieves.
2. Round bottom flask(s) (glass) of appropriate size.
3. Magnetic stir bar(s) of appropriate size.
4. Heated magnetic stir plate.
5. Aluminum-backed silica gel thin-layer chromatography (TLC) plates with embedded fluorophore (e.g., TLC Silica gel 60 F_{254}).
6. Pasteur pipettes and bulb.
7. Aluminum foil.
8. Medium-fritted glass filter funnel of appropriate size and rubber funnel adapter.
9. Side-arm flask.
10. Separatory funnel of appropriate size.
11. Filter paper.
12. Rotary evaporator.
13. Dry ice.
14. Rubber septa of appropriate size.
15. 18 gauge needles connected to nitrogen and vacuum lines via adapter.
16. Tygon vacuum tubing.
17. 18 gauge double tipped needle (cannula).
18. Silica gel 60 (70–230 mesh).
19. Washed sand.
20. Glass chromatography column of appropriate size, with glass frit.

2.1.2 Chemical Reagents

1. *N,N*-dimethylformamide (DMF).
2. *N*-hydroxyphthalimide.
3. Triethylamine.

4. 1,2-dibromoethane.
5. Ethyl acetate.
6. Hexane.
7. Milli-Q pure water or distilled water.
8. 1 N HCl.
9. $MgSO_4$, anhydrous.
10. Saturated NaCl (aqueous) or brine.
11. 95% (v/v) ethanol (aqueous).
12. DMF, dried over molecular sieves.
13. Sodium hydride, stored in dry glove box.
14. Triphenylmethanethiol.
15. N_2 gas.
16. Chloroform.
17. Hydrazine hydrate.
18. Basified water. Distilled water adjusted to pH 8 with 1 M NaOH.
19. $MgSO_4$, anhydrous.

2.1.3 Chemical Reagents for Mass Spectrometric Characterization of O-(2-(tritylthio)ethyl) Hydroxylamine (3)

1. DMF.
2. *N,N*-diisopropylethylamine (DIEA).
3. Iodoacetamide.
4. Acetonitrile (ACN).

2.1.4 Deuterated Solvents for Compound Characterization by NMR

1. Chloroform ($CDCl_3$).
2. DMF (DMF-D_7).
3. Dimethylsulfoxide (DMSO-D_6).

2.2 Solid-Phase Peptide Synthesis

2.2.1 Equipment

1. Reaction vessel.
2. Rubber vacuum tubing.
3. Solvent wash bottle(s) with spout cap.
4. Glass scintillation vial, 20 mL.
5. Benchtop shaker.
6. Refrigerated centrifuge and microcentrifuge.

2.2.2 Chemical Reagents

1. Rink amide resin (100–200 mesh size, 0.72 mmol/g).
2. DMF.
3. Methylene chloride (dichloromethane, DCM).
4. Methanol.

5. 9-fluorenylmethoxycarbonyl (Fmoc) deprotection solution: 20% (v/v) piperidine in DMF.
6. 4-methyltrityl (Mtt) group deprotection solution: 1% (v/v) triisopropylsilane (TIS), 1% (v/v) trifluoroacetic acid (TFA) in DCM.
7. DMSO.
8. DIEA.
9. *O*-(Benzotriazol-1-yl)-*N,N,N′,N′*-tetramethyluronium hexafluorophosphate (HBTU).
10. *N,N′*-diisopropylcarbodiimide (DIC).
11. Ethyl-2-cyano-2-(hydroxyimino)acetate (Oxyma).
12. 9-fluorenylmethoxycarbonyl (Fmoc) N^{α}-protected amino acids with standard, acid-labile protecting groups.
13. Fmoc-Lys(Mtt)-OH.
14. Acetic anhydride.
15. Di-*tert*-butyl dicarbonate (Boc_2O).
16. Bromoacetic acid.
17. N_2 gas.
18. Kaiser test solution 1: 4:1 (w/v) phenol/ethanol.
19. Kaiser test solution 2: 0.2 mM KCN in pyridine.
20. Kaiser test solution 3: 0.28 M ninhydrin in ethanol.
21. 60% (v/v) ethanol (aqueous).
22. Cleavage cocktail: 92.5:2.5:2.5:2.5 (v/v) TFA:TIS:H_2O: anisole.
23. Reagent K cleavage cocktail: 82.5:5.0:5.0:5.0:2.5 (v/v) TFA: thioanisole:H_2O:phenol:1,2-ethanedithiol.
24. Diethyl ether.

2.3 Recombinant Ub(1-75)MES and SUMO(1-91)MES α-Thioester Preparation

2.3.1 Equipment

1. Chitin resin.
2. Glass chromatography column, 50 mL.
3. Autoclave.
4. Glass culture flask, 2.8 L.
5. Sterilizing syringe filter (0.22 μm pore size).
6. Centrifuge bottles, 500 and 50 mL.
7. Refrigerated ultracentrifuge.
8. Ice.
9. Sonifier cell disruptor.
10. 0.45 μm syringe filter.
11. Nutator at 4 °C.

2.3.2 Chemical and Biological Reagents

1. *E. coli* BL21(DE3) cells.
2. pTXB1 vector.
3. Plasmids pUb(1-75) and pSUMO3(1-91). These plasmids were generated from the pTXB1 vector and contain truncated forms of the human ubiquitin and human SUMO-3 genes that are missing the terminal Gly residue (Gly76 for Ub and Gly92 for SUMO-3). The *ub(1-75)* and *sumo-3(1-91)* genes are fused in-frame and directly N-terminal to a mutant gyrase A intein from *Mycobacterium xenopi*.
4. Luria–Bertani medium, sterile.
5. 100 mg/mL ampicillin stock solution, sterile-filtered.
6. 1 M isopropyl β-D-1-thiogalactopyranoside (IPTG) stock solution (aqueous), protected from light.
7. Cell lysis buffer: Phosphate Buffered Saline (PBS), pH 7.2, containing 1 mM sodium 2-mercaptoethanesulfonate (MESNa).
8. Wash buffer: PBS, pH 7.75.
9. Thiolysis buffer: PBS, pH 7.75, containing 100 mM MESNa.

2.4 Expressed Protein Ligation and Auxiliary Removal

2.4.1 Equipment

1. Microcentrifuge tubes of appropriate size.
2. Liquid N_2.
3. 18 gauge needles connected to nitrogen and vacuum lines via needle adapter.
4. Tygon vacuum tubing.
5. Nutator at 25 °C.
6. Nutator at 37 °C.

2.4.2 Chemicals

1. Ligation buffer: 6 M guanidinium hydrochloride (Gn-HCl), 0.1 M Na_2HPO_4, 10 mM tris(2-carboxyethyl)phosphine (TCEP), pH 7.5
2. 0.1% formic acid (aqueous)
3. 1 M NaOH
4. Freshly activated Zn dust.
5. N_2 gas
6. 1 N HCl
7. Auxilary removal solution: 6 M Gn-HCl, pH 3.0, degassed
8. Zn wash solution: 6 M Gn-HCl, 50 mM Ethylenediaminetetraacetic acid (EDTA), pH 3.0

2.5 Protein and Peptide Purification and Characterization

1. C18 preparative reverse-phase high-performance liquid chromatography (RP-HPLC) column (10 micron particle size, 250 × 22 mm).
2. HPLC solvent A: H_2O, 0.1% (v/v) TFA.
3. HPLC solvent B: 90:10 (v/v) ACN: H_2O, 0.1% (v/v) TFA.
4. Collection tubes of appropriate size.
5. Electrospray ionization mass spectrometer (ESI-MS).
6. C18 LC-ESI-MS column (3.5 micron particle size, 100 × 2.1 mm).
7. LC-ESI-MS solvent A: 5:95 (v/v) ACN:H_2O, 1% (v/v) acetic acid.
8. LC-ESI-MS solvent B: ACN, 1% (v/v) acetic acid.

3 Methods

3.1 Synthesis of the S-trityl Protected Ligation Auxiliary, O-(2-(tritylthio)ethyl) hydroxylamine (3) (Fig. 2)

In this section we describe the synthesis of the ligation auxiliary, 2-aminooxyethanethiol, in the *S*-trityl protected form. The protected ligation auxiliary may be directly incorporated into any desired unprotected lysine side-chains on the solid-phase using the Zuckerman submonomer approach [13]. The facile three-step synthetic procedure we describe requires only a single chromatographic purification step, and may be undertaken at multigram scales with no significant changes in overall yield.

3.1.1 Synthesis of N-(2-bromoethoxy) Phthalimide (1)

1. Prepare a room temperature solution of *N*-hydroxyphthalimide (0.85 M in DMF, 41 mL, 34.9 mmol) in a round-bottom flask equipped with a magnetic stir-bar (*see* **Note 1**).
2. To this stirring solution, add triethylamine (10.7 mL, 76.7 mmol) dropwise.
3. To this stirring solution, add 1,2-dibromoethane (12.6 mL, 146.2 mmol) dropwise.
4. Stir the resulting mixture at room temperature. Keep protected from light by wrapping the flask with aluminum foil (*see* **Note 2**).

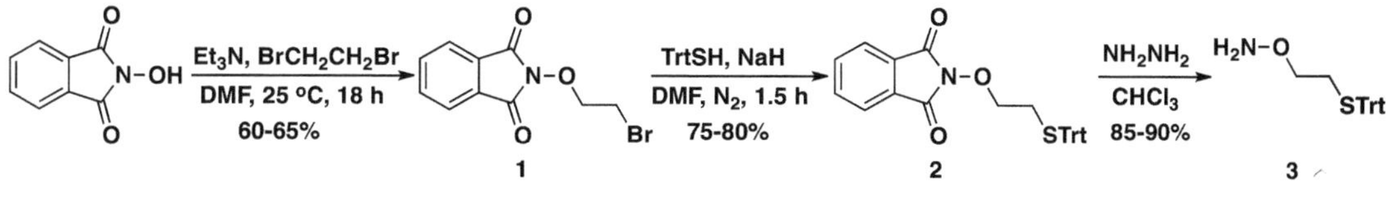

Fig. 2 Synthesis of the *S*-trityl-protected ligation auxiliary for incorporation into peptides on the solid phase. *Et_3N* Triethylamine, *Trt* Trityl group, *NaH* Sodium hydride, *DMF* *N,N*-dimethylformamide, *$CHCl_3$* Chloroform, *N_2* Nitrogen gas

5. Monitor reaction progress by TLC (*see* **Note 3**) in a 70:30 (v/v) mixture of hexane and ethyl acetate.
6. Upon disappearance of starting material (about 18 h), remove the solids formed by vacuum-filtration on a medium-fritted glass filter. The product remains dissolved in the DMF filtrate.
7. Precipitate the product by adding 350 mL water, then vacuum-filter on a medium-fritted glass filter and wash with 100 mL of ice-cold water.
8. Dissolve the precipitated solids in 200 mL ethyl acetate and extract with 1 N HCl (2 × 100 mL), water (1 × 100 mL), and saturated NaCl (1 × 100 mL).
9. Dry the resultant organic layer over anhydrous $MgSO_4$, and remove ethyl acetate in vacuo at 25 °C on a rotary evaporator.
10. Purify the crude product by recrystallization from 95% EtOH (aq) to give *N*-(2-bromoethoxy)phthalimide, **1**, as white, needle-like crystals (5.7 g, 64%).
11. Confirm product identity and purity by NMR and ESI-MS. ^{1}H NMR (300 MHz, $CDCl_3$): δ 7.81 (m, 4H), 4.47 (t, 2H, J = 6.88 Hz), 3.64 (t, 2H, J = 6.88 Hz). ^{13}C NMR (499 MHz, $CDCl_3$): δ 163.49, 134.81, 128.79, 123.78, 77.27, 26.89. ESI-MS: *m/z* $[M+Na]^+$ 291.96 Da.

3.1.2 Synthesis of N-(2-(tritylthio)ethoxy) Phthalimide (2)

This reaction is moisture-sensitive and should be performed under N_2 atmosphere. Glassware should be flame-fried under vacuum prior to use.

1. In a round-bottom flask equipped with a magnetic stir-bar, combine triphenylmethanethiol (1.23 g, 4.45 mmol) and sodium hydride (0.133 g, 5.54 mmol) (*see* **Note 4**).
2. To this mixture, add anhydrous DMF (18 mL) dropwise and continue to stir until the evolution of bubbles is no longer observed. The solution will become bright yellow in color as the deprotonated triphenylmethanethiolate sodium salt forms.
3. In a separate round-bottom flask equipped with a magnetic stir-bar, prepare a solution of *N*-(2-bromoethoxy)phthalimide (0.2 M in anhydrous DMF, 18 mL, 3.70 mmol).
4. To the stirring solution of *N*-(2-bromoethoxy)phthalimide, add the triphenylmethanethiolate solution by cannula. Monitor the reaction by TLC using a solvent system of 90:10 (v/v) hexane:ethyl acetate.
5. When no further reaction progress is observed, prepare and add a second identical solution of triphenylmethanethiolate (*see* **Note 5**).

6. After starting material has been consumed (about 1.5 h), quench the reaction with water until no further bubbling is observed (*see* **Note 6**).
7. Remove the precipitate by vacuum filtration and concentrate under high vacuum.
8. Purify the residue by silica gel column chromatography using a solvent system of 90:10 (v/v) hexane: ethyl acetate to give compound *N*-(2-(tritylthio)ethoxy)phthalimide, **2**, (1.43 g, 79%).
9. Confirm product identity and purity by NMR and ESI-MS. NMR: ^{1}H NMR (301 MHz, DMF-D_7): δ 7.89 (m, 4H), 7.44-7.16 (15H), 3.83 (t, 2H, J = 7.30 Hz), 2.65 (t, 2H, J = 7.30 Hz). ^{13}C NMR (499 MHz, DMF-D_7): δ 163.51, 144.90, 135.10, 129.72, 129.17, 128.34, 127.16, 123.48, 76.45, 67.23, 30.11. ESI-MS *m/z* $[M+Na]^+$ 488.13 Da.

3.1.3 Synthesis of O-(2-(tritylthio)ethyl) Hydroxylamine (3)

1. Prepare a room temperature solution of *N*-(2-(tritylthio)ethoxy)phthalimide, 2, (58.9 mM in $CHCl_3$, 27 mL, 1.59 mmol) in a 100 mL round-bottom flask equipped with a magnetic stir-bar.
2. To this stirring solution, add hydrazine hydrate (0.3 mL, 4.81 mmol) (*see* **Note 7**).
3. Allow the reaction to proceed for 2 h at room temperature with vigorous stirring. Monitor reaction progress by TLC using a solvent system of 80:20 (v/v) hexane: ethyl acetate.
4. After 2 h, remove solids formed in the reaction mixture by filtration on a medium-fritted glass filter.
5. Wash the $CHCl_3$ filtrate with basified water (pH 8.0, 3 × 30 mL) in a 250 mL separation flask, and dry over anhydrous $MgSO_4$. Remove solvent in vacuo to yield compound **3** (0.48 g, 90%).
6. Confirm product identity and purity by NMR and ESI-MS. NMR: ^{1}H NMR (301 MHz, $CDCl_3$): δ 7.41–7.03 (15H), 5.23 (s, 2H), 3.46 (t, 2H, J = 6.32 Hz), 2.36 (t, 2H, J = 6.32 Hz). ^{13}C NMR (500 MHz, DMSO-D_6): δ 144.49, 129.10, 128.01, 126.70, 72.59, 65.93, 31.36. ESI-MS: Due to the poor ionization of **3** during ESI-MS, it is best observed as its acetamide derivative following reaction with 2-iodoacetamide and *N,N*-diisopropylethylamine (DIEA) in DMF. The ion of the derivatized compound is *m/z* $[M+Na]^+$ 415.5 Da.

3.2 Solid-Phase Synthesis of Peptide Substrates for Ubiquitylation

In this section we describe the synthesis of short peptide substrates for ubiquitylation using solid-phase peptide synthesis (SPPS). Following peptide assembly using either manual or automated SPPS, the desired lysine site of ubiquitylation is selectively deprotected and the *S*-trityl protected auxiliary is introduced by the submonomer approach [13]. Selective deprotection of one or more targeted lysines may be accomplished by their incorporation as either Lys(Mtt) [14] or Lys(ivDde) [15, 16] side-chain protected forms. Nonubiquitylated lysines are incorporated with standard Lys(Boc) side-chain protection that precludes incorporation of the ligation auxiliary at these residues.

3.2.1 Synthesis of Ac-QKE-CONH$_2$ on Rink Amide AM Resin

The peptide Ac-QKE-CONH$_2$ may be manually synthesized on a 0.5 mmol scale employing standard 9-fluorenylmethoxycarbonyl (Fmoc)-based N$^{\alpha}$-deprotection chemistry [17] and custom glassware (Adams and Chittenden Scientific Glass, catalog number 942167). The synthesis vessel comprises a 25 or 50 mL cylindrical reservoir with a medium-fritted glass filter at the bottom equipped with a two-way stopcock and spout for attaching rubber tubing. This facilitates the application of either vacuum to draw out solvent, or nitrogen to mix solutions during deprotection/coupling, by switching the tubing as needed (*see* **Note 8**). Standard black vacuum rubber tubing (1/4″ × 5/8″ I.D. × O.D., 3/16″ wall thickness) is sufficient for both vacuum and nitrogen application. Starting with commercially available Rink amide resin (100–200 mesh size, 0.70 g, 0.72 mmol/g) couple each amino acid in fourfold molar excess based on initial resin loading (*see* **Note 9**).

1. Deprotect the Fmoc- group with 20% piperidine in DMF (v/v) (5 mL, 3 × 10 min).
2. Perform each coupling reaction for a minimum of 1 h and maximum of 4 h with a mixture of Fmoc-amino acid (2.0 mmol), *O*-(benzotriazol-1-yl)-*N,N,N′,N′*-tetramethyluronium hexafluorophosphate (HBTU, 1.97 mmol), and *N,N*-diisopropylethylamine (DIEA, 4.0 mmol) in 4 mL DMF. Coupling efficiency may be monitored by the Kaiser test [18] and additional couplings performed until a negative test is obtained (*see* **Note 10**). Overnight couplings, if required, should be performed under base-free conditions with a mixture of Fmoc-amino acid (2.0 mmol) *N,N′*-diisopropylcarbodiimide (DIC, 4.0 mmol) and ethyl-2-cyano-2-(hydroxyimino)acetate (Oxyma, 2.0 mmol) [19]. Employ Fmoc-Lys(Mtt)-OH at the desired site of ubiquitylation, as it contains the acid-sensitive 4-methyltrityl (Mtt) protecting group at the ε-NH$_2$ [14].
3. After completion of the peptide and following Fmoc deprotection, acetylate the N-terminal Gln by reaction with DIEA

(20-fold molar excess) and acetic anhydride (40-fold molar excess) for 3 × 15 min.

4. To confirm identity of the peptide, remove a 25–50 mg test portion of the resin and dry in vacuo after extensive washing with DMF, DCM and finally MeOH (*see* **Note 11**).
5. Cleave the peptide from the resin with 300 μL of freshly prepared cleavage cocktail for 1 h at room temperature.
6. Remove the resin beads by filtration and precipitate the peptide from the filtrate by the addition of 5–10 volumes of ice-cold dry ether (*see* **Note 12**).
7. Centrifuge briefly to pellet the precipitate and decant the ether layer. Wash the pellet twice with ice-cold diethyl ether and dry under a slow stream of nitrogen in the fume-hood.
8. Purify the crude peptide C18 analytical RP-HPLC with a gradient of 0–40% B over 30 min, and characterize by ESI-MS. ESI-MS: *m/z* $[M+H]^+$ 445.2 Da.

3.2.2 Synthesis of H_2N-KAKI-$CONH_2$ on Rink Amide AM Resin

The peptide H_2N-KAKI-$CONH_2$ is synthesized by a similar method as that for Ac-QKE-$CONH_2$ with key differences being a second Lys and free N-terminal amine. Toward this, the N-terminal Lys is protected with *tert*-butoxycarbonyl (Boc-) protecting groups on both the α- and ε-amines (*see* **Note 13**). The interior Lys, which is the desired site of ubiquitylation, is orthogonally protected with the Mtt protecting group.

1. Following peptide assembly, a test cleavage of the resin should be performed as in Subheading 3.2.1.
2. Purify the peptide by C18 analytical RP-HPLC with a gradient of 0–73% B over 30 min, and characterize by ESI-MS. ESI-MS: *m/z* $[M+H]^+$ 680.4 Da.

3.2.3 Introduction of O-(2-(tritylthio)ethyl) Hydroxylamine (3) in Peptide Substrates

1. Treat each resin-bound protected peptide, that is confirmed to have the correct sequence by test cleavage, with a solution of 1% TFA and 1% TIS in DCM (v/v) to remove the Mtt protecting group and expose the Lys ε-amine (*see* **Note 14**).
2. Neutralize the protonated ε-amine by washing with 10% DIEA in DMF (v/v).
3. Couple the Lys ε-amine with 8 equivalents of bromoacetic acid and 8 equivalents of DIC (based on initial resin-loading) in DMF for 45 min at room temperature. Repeat the coupling once.
4. Cleave a test portion of the resin to confirm complete coupling of bromoacetic acid to the side-chain.
5. Wash the bromoacetylated peptide resin extensively with DMF, DCM, and MeOH, then dry in vacuo.

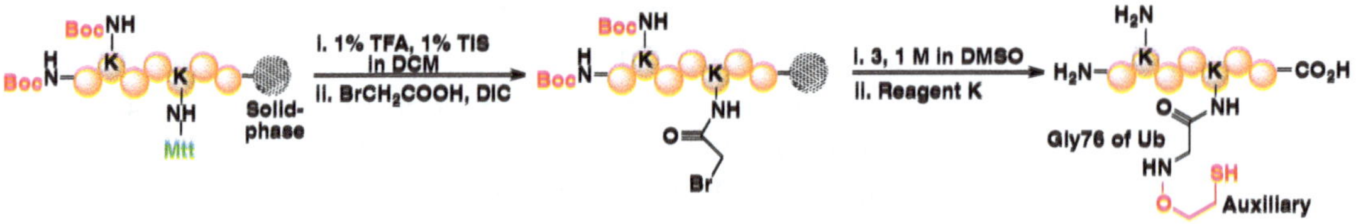

Fig. 3 Synthesis of peptides containing the ligation auxiliary 2-aminooxyethanethiol. *Fmoc* Fluorenylmethoxycarbonyl group, *Boc tert*-Butyloxycarbonyl group, *Mtt* 4-methyltrityl group, *TFA* Trifluoroacetic acid, *TIS* Triisopropylsilane, *DIC N,N'*-Diisopropylcarbodiimide

6. Transfer the dry peptidyl resin to a scintillation vial and suspend in solution containing 9 equivalents of *S*-trityl protected auxiliary **3** (1 M in DMSO) (*see* **Note 15**).
7. Gently shake the mixture for 24 h at room temperature on a benchtop shaker occluded from light.
8. Cleave a test portion of the resin to determine the degree of bromine displacement by **3** (Fig. 3).
9. When the test cleavage shows complete conversion of the bromoacetylated peptide, remove and recover the solution containing unreacted auxiliary. This solution can be lyophilized and stored at −80 °C for reuse (*see* **Note 16**). Wash the resin extensively with DMF.
10. Deprotect the N-terminal Fmoc of the KAKI peptide with 20% (v/v) piperidine in DMF (5 mL, 2 × 15 min).
11. Wash the resin extensively with DMF, DCM, and finally MeOH. Dry in vacuo.
12. Cleave each peptide from resin by reaction with Reagent K [20] cleavage cocktail (20 μL/mg resin) for 1.5 h at room temperature.
13. Remove the resin beads by filtration and precipitate the peptide from the filtrate by the addition of 5–10 volumes of ice-cold dry ether.
14. Centrifuge briefly to pellet the precipitate and decant the ether layer. Wash the pellet twice with ice-cold diethyl ether and dry under a slow stream of nitrogen in the fume-hood.
15. Dissolve the peptides in RP-HPLC buffer A and purify by C18 analytical RP-HPLC with a gradient of 0–40% B over 30 min.
16. Perform ESI-MS to verify identity and purity of the peptides. Ac-QK^{aux}E-CONH$_2$ (**4**) *m/z* [M+H]$^+$ 578.3 Da. H$_2$N-KA-K^{aux}I-CONH$_2$ (**5**) *m/z* [M+H]$^+$ 591.4 Da.

3.3 Intein-Mediated Generation of Ub(1-75)- and SUMO-3 (2-91)-α-thioesters

The plasmids pUb(1-75) and pSUMO3(1-91) contain truncated forms of the human ubiquitin and human SUMO-3 genes that are missing the terminal Gly residue (Gly76 for Ub and Gly92 for SUMO-3) [10]. The *ub(1-75)* and *sumo-3(1-91)* genes are fused in-frame and directly N-terminal to a mutant gyrase A intein from *Mycobacterium xenopi*. This mutant intein is only capable of catalyzing an N-to-S acyl shift of the amide bond between any N-terminal protein and itself, but cannot undertake the subsequent steps of intein-mediated ligation (pTXB1 vector, New England Biolabs). A C-terminal chitin-binding domain (CBD) present in the final fusion protein facilitates its purification from other *E. coli* proteins by selective binding to a chitin column.

1. Transform chemically competent ($CaCl_2$ method) *Escherichia coli* BL21(DE3) cells by heat shock with the plasmids pUb (1-75) and pSUMO3(1-91) (*see* **Note 17**).
2. Pick a single colony from Luria–Bertani (LB) agar plates containing 100 μg/mL of Ampicillin (Amp) and grow the transformed cells in small, 10 mL starter cultures in LB-Amp at 37 °C with shaking at 250 rpm for 14 h.
3. Inoculate six flasks containing 1 L LB-Amp medium with 10 mL started culture. Incubate at 37 °C with shaking at 250 rpm.
4. Once the OD_{600} of the culture reaches ~0.6–0.8, induce overexpression of the desired fusion proteins by adding 0.3 mM isopropyl β-D-1-thiogalactopyranoside (IPTG). Continue to culture the cells for an additional 6 h at 25 °C.
5. Harvest the cells by centrifugation at 7000 × *g* for 15 min. Discard the supernatant media.
6. Resuspend the cell pellet in lysis buffer consisting of phosphate buffered saline (PBS) at pH 7.2 containing 1 mM 2-mercaptoethanesulfonic acid sodium salt (MESNa) (*see* **Note 18**).
7. Lyse cells by pulsed sonication on ice, 3 × 5 min, with 5 min between each sonication step.
8. Centrifuged at 20,000 × *g* for 15 min to remove insoluble cellular debris.
9. Pass the lysate supernatant through a 0.45 μm syringe-filter and apply to a 30 mL chitin column preequilibrated with 10 column volumes (CV) of lysis buffer.
10. Bind proteins to the column over a period of 12 h at 4 °C with gentle shaking on a nutator.
11. Wash the column with 20 CV of lysis buffer followed by two CV of PBS, pH 7.75.
12. Cleave the Ub(1-75)-MESNa and SUMO-3(2-91)-MESNa thioesters from their respective intein-CBD fusions by incubation with 1.5 CV of PBS, pH 7.75 containing 100 mM MESNa for 72 h at 4 °C (Fig. 4, also *see* **Note 19**). We observe that the N-terminal Met of SUMO-3 (but not of Ub) is consistently

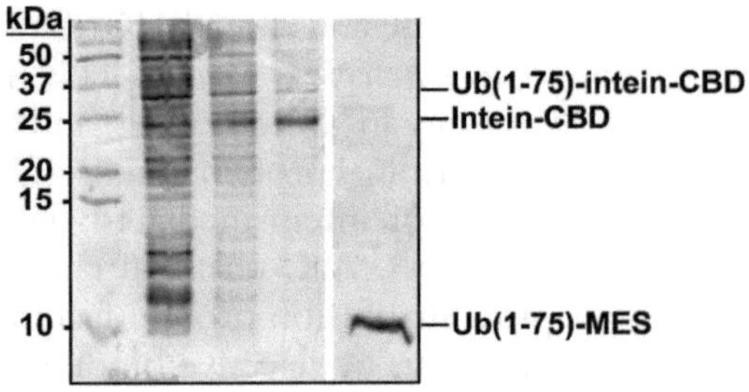

Fig. 4 Intein-mediated generation of Ub(1-75)-MES (6). Lane 1, Prestained protein molecular weight standards (NEB). Lane 2, cell lysate supernatant. Lane 3, column wash with PBS. Lane 4, residual protein on chitin beads after 72 h of treatment with intein-cleavage buffer. Lane 5, column eluate after 72 h of treatment with intein-cleavage buffer. *MES* 2-mercaptoethanesulfonic acid

processed by Met aminopeptidases in vivo, leading to the SUMO-3(2-91)-α-thioester product.

13. Purify the eluted Ub(1-75)- and SUMO-3(2-91)-α-thioesters by C18 preparative RP-HPLC employing a gradient of 30–60% B over 60 min. Identify fractions containing the desired thioesters by ESI-MS. (a) Ub(1-75)-MES (**6**) *m/z* $[M+H]^+$ 8632.8 Da. (b) SUMO-3(2-91)-MES (**7**) *m/z* $[M+H]^+$ 10,461.6 Da.

3.4 Expressed Protein Ligation of Auxiliary-Containing Peptides and Protein α-Thioesters

In this section we describe expressed protein ligation of peptides containing the ligation auxiliary, 2-aminooxyethanethiol, with protein α-thioesters. Due to our incorporation of the auxiliary into peptides using the submonomer approach, the terminal Gly of Ub and SUMO-3 are already attached to the target lysine. Therefore, C-terminally truncated α-thioesters missing Gly76 in Ub, Ub (1-75) -α-thioester, and Gly92 in SUMO-3, SUMO-3 (2-91)-α-thioester, are employed in the expressed protein ligation reaction.

1. Dissolve the purified auxiliary-conjugated peptides **4** or **5** (0.35 mg, 0.6 μmol) and protein α-thioesters **6** or **7** (1.0 mg, 0.12 μmol) separately dissolved in 120 μL of ligation buffer each. Combine the two solutions and gently shake at room temperature for 24 h.
2. To monitor the kinetics of ligation, remove 3 μL aliquots from the reaction mixture at specific time-points. Dilute the aliquots into 27 μL H_2O, 0.1% formic acid and analyze by LC-ESI-MS.
3. To monitor the extent of the S-to-N acyl shift (leading to the desired amide-linked product), remove 3 μL aliquots from the reaction mixture at specific time points. To each aliquot, add 3 μL 1 M NaOH and incubate on ice for 30 s. These conditions lead to the hydrolysis of nonrearranged thioester-linked ligation products. Quench the hydrolysis reaction with 25 μL

H_2O, 0.1% formic acid, and analyze by LC-ESI-MS employing a gradient of 0–50% D, 40 min. The ratio of peak heights in ESI-MS of the hydrolyzed Ub(1-75)-COOH product to the desired ligation product was considered an indicator of S-to-N acyl rearrangement.

4. Purify the final rearranged ligation products (*see* **Note 20**) by C18 analytical RP-HPLC employing a gradient of 30–60% B over 30 min. For each ligation, typical conversion to product is greater than 70%, with the major side product being the hydrolyzed Ub(1-75)-COOH, or SUMO-3(2-91)-COOH protein. Identify fractions containing the desired thioesters by ESI-MS. (a) $QK^{Ub(aux)}E$ *m/z* $[M+H]^+$ 9068.0 Da. (b) $KAK^{Ub(aux)}I$ *m/z* $[M+H]^+$ 9081.1 Da. (c) $QK^{Su(aux)}E$ *m/z* $[M+H]^+$ 10,896.8 Da. (d) $KAK^{Su(aux)}I$ *m/z* $[M+H]^+$ 10,909.9 Da.

3.5 Auxiliary Removal by Reductive Cleavage of the N–O Bond with Activated Zn

In this section we describe the facile removal of the ligation auxiliary from the final expressed protein ligation product upon treatment with activated Zn under acidic conditions. We have found this method to be highly compatible with proteins that may be refolded from the denatured state. Although denaturants are not strictly required when cleaving the N-O bond in peptides, the expressed protein ligation product with Ub or SUMO does require the inclusion of a denaturing agent to enable effective electron transfer from activated Zn to the N-O bond.

1. Freshly activate 5 mg of metallic Zn by rinsing with 1 N HCl, then extensively wash with water [21]. Dry the activated Zn in vacuo.
2. Add the activated Zn to 200 μL of degassed 6 M Gn-HCl, pH 3.0 containing 0.2 mg purified ligation product.
3. Degas the mixture by three freeze–thaw cycles under N_2. This step is necessary to reduce the generation of oxidized side products.
4. Allow the reduction to proceed at 37 °C under N_2 with gentle shaking for 24 h.
5. Briefly centrifuge the reaction at 15,000 × *g* to pellet the Zn dust and remove the supernatant containing reduced products.
6. Wash the pelleted Zn twice with 150 μL of 6 M Gn-HCl, 50 mM EDTA, pH 3.0.
7. Reduce the combined supernatant and washes with 10 mM TCEP at pH 7.5 (*see* **Note 21**) and purify by C18 analytical RP-HPLC with a gradient of 30–60% B over 30 min. Identify fractions containing the desired thioesters by ESI-MS. (a) $QK^{Ub}E$ *m/z* $[M+H]^+$ 8992.0 Da. (b) $KAK^{Ub}I$ *m/z* $[M+H]^+$ 9005.1 Da. (c) $QK^{Su}E$ *m/z* $[M+H]^+$ 10,820.8 Da. (d) $KAK^{Su}I$ *m/z* $[M+H]^+$ 10,833.9 Da.

4 Notes

1. Freshly distilled triethylamine stored over 3 Å molecular sieves gives the best yields. Triethylamine should be stored under argon and protected from light that can cause unwanted oxidation.
2. When working with brominated compounds, rigorously occluding light reduces their decomposition and hence the number of side-products observed.
3. Aluminum-backed silica gel plates are easier to use than glass-backed TLC plates as they require less solvent and can be easily cut to desired sizes. They can be bent to be slightly convex along the longitudinal axis to ensure an even solvent front.
4. Sodium hydride is very moisture sensitive and reacts violently with water. It should be stored in a glove box at all times and aliquots weighed out in capped scintillation vials as needed. All solvents to be used with sodium hydride should be dried over 3 Å molecular sieves for 48 h.
5. The steric bulk of the trityl group hinders nucleophilic attack by the thiolate and slows reaction kinetics. Due to undesirable competing reactions, such as disulfide formation, we have found that adding a second equivalent of freshly prepared thiolate gives higher yields than starting with a large excess of the thiolate.
6. A small excess of sodium hydride (1.25 equivalents) is employed to deprotonate the trityl thiol and we have never observed violent reactions upon quenching, however, the solvent used to quench sodium hydride (water or methanol may be used) must be added in a slow dropwise manner.
7. Hydrazine hydrate is corrosive and toxic and must be handled with care. Wear personal protective equipment including gloves and safety glasses. Work with this chemical in a functioning fume-hood.
8. In order to prevent cross-contamination of the vacuum and nitrogen delivery tubes, we employ a reaction vessel with one spout and switch the lines as needed. However, alternate versions of the synthesis vessel with three-way stopcocks and dedicated lines for nitrogen and vacuum may be used. Peptides containing orthogonally protected lysines may also be synthesized by automated microwave-assisted peptide synthesis using the 1-(4,4-Dimethyl-2,6-dioxocyclohexylidene)-3-methylbutyl (ivDde) protecting group [15], which is stable to heating at 75 °C for 5 min that is employed by commercial synthesizers (for example, the Liberty Blue synthesizer from CEM corporation).

9. Rink amide aminomethyl (AM) or 4-methylbenzhydrylamine (MBHA) solid-phase resins may be used with no significant difference for short peptides <10 amino acids long. For longer peptides, MBHA is recommended [22]. Wang resin preloaded with the first Fmoc-amino acid (EMD Millipore or Anaspec) should be employed if the C-terminal carboxylic acid form is desired instead of the C-terminal amide.
10. Although methods for quantitative ninhydrin have been reported [23], for routine tests of coupling efficiency the resin beads may be visually evaluated after heating at 95–100 °C for 5 min with ninhydrin reagents. A few blue beads (<10%) in a majority of yellow beads and a yellow solution is considered to be a successful coupling reaction, and it is acceptable to proceed with the next step of deprotection. Incomplete couplings, which require additional rounds of coupling, are judged by the presence of >10% blue beads and a bluish solution during the Kaiser test. If the Kaiser test remains unchanged after three rounds of coupling to the same deprotected amino acid, the resin is capped with acetic anhydride before proceeding to the next step of deprotection.
11. In order to conserve peptidyl-resin and increase final yields, test cleavages may be performed with as little as 10 mg of dry resin. However, care must be taken in the precipitation step to add diethyl ether dropwise as only a small amount of white precipitate is formed and may not be visible in a large volume of ether. The precipitate is centrifuged in a microfuge at 17,000 × g for 5 min and washed twice with ether, then taken up in 100 μL of HPLC buffer for analysis.
12. An alternate strategy for peptide cleavages is to directly evaporate the filtrate under a stream of nitrogen in the fume hood. This typically works well for small-scale cleavages and does not require careful addition of diethyl ether to generate small amounts of precipitates. However, the increased concentration of reactive cations, derived from the side-chain protecting groups employed, may generate covalent adducts on the final peptide. Therefore, this alternate procedure must be evaluated at the test-scale for each new peptide.
13. Although the Boc- group is useful for protecting the α-amine of an N-terminal Lys, it is labile to extended washing with 1% TFA that may be required to completely remove the Mtt-group. Therefore, an N-terminal Fmoc- protection strategy may also be employed. For short peptides where the Mtt-group is readily deprotected, an N-terminal Boc- group is sufficient.
14. One problem with deprotecting the Mtt- group is the persistence of yellow color on the resin even after several washes with

the 1% TFA, 1% TIS mixture in DCM. However, we find that 4–6 flow washes for 1 min each are sufficient to remove the Mtt group in most peptides, even if the beads remain somewhat yellow. A flow wash involves spraying solvent from a Nalgene wash bottle in a continuous stream on top of the resin while simultaneously draining solvent from the bottom. This ensures that the level of solvent on top of the resin remains largely constant and only a few millimeters above the resin bed. This prevents the creeping of reagents along the sides of the reaction vessel and leads to cleaner final products. Longer batch washes of the resin may lead to deprotection of Boc- protecting groups and should be undertaken cautiously, with the optimal duration empirically determined for each new peptide.

15. It is important to use completely dry resin and solvents for this reaction as the presence of water leads to displacement of –Br by –OH. Drying the resin overnight on a lyophilizer and the solvents over 3 Å molecular sieves for 48 h is sufficient to prevent unwanted reactions at this step.
16. The recovered auxiliary solution in DMSO can be reused up till four times without significant deterioration in displacement yields. However, the addition of 1 equivalent of DIEA is necessary after each use, in order to deprotonate the auxiliary hydrobromide salt. Dissolving the auxiliary at 1 M in DMSO requires gentle heating, and it is prone to precipitation at temperatures below 25 °C. If this is observed repeatedly, a 0.5 M solution of the auxiliary in DMSO may be employed with no appreciable reduction in yields.
17. Heat shock transformations were undertaken as described by Sambrook and Russell [24]. Briefly, chemically competent *E. coli* DH5α or BL21(DE3) cells stored at −80 °C were thawed on ice for 15 min. To each 40 μL aliquot of competent cells, 1 pg–100 ng of plasmid DNA (1–5 μL) was added and mixed by gentle pipetting. The cells were placed on wet ice for 2 min, followed by heat shock in a water bath at 42 °C for 45 s. The heat shocked cells were diluted with 950 μL of room temperature LB or SOC media and grown for an additional 1 h at 37 °C. Finally 200–300 μL of cells were plated on LB-Agar plates with the appropriate antibiotic for selection and grown for 12–14 h at 37 °C. Single well-resolved colonies on the plate were selected by using a sterile pipette tip and grown in liquid media for protein expression.
18. Tris-based buffers may be used for intein-mediated thioester production. However, in some instances the protein-α-thioester reacts with the primary amine of Tris to generate the C-terminal amide, which is inert toward native chemical ligation. This is indeed a problem for the SUMO-3

(2-91)-α-thioester but not for the Ub(1-75)-α-thioester. This can be preempted by employing phosphate buffer whenever possible.

19. Lower temperatures prevent hydrolysis of the protein-α-thioester, but also slow down the rate of transthioesterification of the target-intein fusion with MESNa. Hence, for proteins that do not readily cleave from the intein with 200 mM MESNa at 4 °C, a good compromise is to perform thioesterification at pH 7.5 and 25 °C.
20. Most ligations where one fragment is in at least fivefold excess appear to reach completion by 24 h, with some ligations complete in as little as 12 h. Therefore, it is acceptable to skip early time points and simply test the ligation mixture after incubation for one day.
21. Although activated Zn was reported to reduce peptide disulfide bonds [25], we have not observed this effect. This necessitates a quick reduction of ubiquitylated/sumoylated products with TCEP prior to purification. Treatment with 10 mM TCEP at pH 7.5 on ice for 30 min is usually sufficient to reduce disulfides. Extended treatment with TCEP at higher temperatures often results in desulfurization of Cys residues, which can be avoided by reducing the time of incubation or equivalents of TCEP employed.

Acknowledgments

C.C. is thankful for support from the Department of Chemistry and the University of Washington Royalty Research Fund, NIH/-NIGMS R01GM110430 and NSF-MCB 1715123. C.E.W. gratefully acknowledges support from an NSF GRFP (grant number DGH-1256082) and an ARCS Foundation fellowship.

References

1. Sharp PM, Li W-H (1987) Molecular evolution of ubiquitin genes. Trends Ecol Evol 2:328–332
2. Terrell J, Shih S, Dunn R, Hicke L (1998) A function for monoubiquitination in the internalization of a G protein-coupled receptor. Mol Cell 1:193–202
3. Meier F, Abeywardana T, Dhall A, Marotta NP, Varkey J, Langen R, Chatterjee C, Pratt MR (2012) Semisynthetic, site-specific ubiquitin modification of α-synuclein reveals differential effects on aggregation. J Am Chem Soc 134:5468–5471
4. Hicke L (2001) Protein regulation by monoubiquitin. Nat Rev Mol Cell Biol 2:195–201
5. Komander D, Rape M (2012) The ubiquitin code. Ann Rev Biochem 81:203–229
6. Hershko A, Ciechanover A (1998) The ubiquitin system. Ann Rev Biochem 67:425–479
7. Weller CE, Pilkerton ME, Chatterjee C (2014) Chemical strategies to understand the language of ubiquitin signaling. Biopolymers 101:144–155
8. Davies N, Lindsey GG (1994) Histone H2B (and H2A) ubiquitination allows normal

histone octamer and core particle reconstitution. Biochim Biophys Acta 1218:187–193
9. Weller CE, Dhall A, Ding F, Linares E, Whedon SD, Senger NA, Tyson EL, Bagert JD, Li X, Augusto O, Chatterjee C (2016) Aromatic thiol-mediated cleavage of N–O bonds enables chemical ubiquitylation of folded proteins. Nat Commun 7:12979
10. Weller CE, Huang W, Chatterjee C (2014) Facile synthesis of native and protease-resistant ubiquitylated peptides. ChemBioChem 15:1263–1267
11. Muir TW, Sondhi D, Cole PA (1998) Expressed protein ligation: a general method for protein engineering. Proc Nat Acad Sci U S A 95:6705–6710
12. Dawson PE, Muir TW, Clark-Lewis I, Kent SBH (1994) Synthesis of proteins by native chemical ligation. Science 266:776–778
13. Zuckermann RN, Kerr JM, Kent SBH, Moos WH (1992) Efficient method for the preparation of peptoids [Oligo(N-Substituted Glycines)] by submonomer solid-phase synthesis. J Am Chem Soc 114:10646–10647
14. Sax B, Dick F, Tanner R, Gosteli J (1992) 4-Methyltrityl (Mtt): a new protecting group for the side chain protection of Asn and Gln in solid-phase peptide synthesis. Pept Res 5:245–246
15. Chhabra SR, Hothi B, Evans DJ, White PD, Bycroft BW, Chan WC (1998) An appraisal of new variants of Dde amine protecting group for solid phase peptide synthesis. Tet Lett 39:1603–1606
16. Chatterjee C, McGinty RK, Pellois J-P, Muir TW (2007) Auxiliary-mediated site-specific peptide ubiquitylation. Angew Chem Int Ed Eng 46:2814–2818
17. Carpino LA, Han GY (1972) 9-Fluorenylmethoxycarbonyl Amino-Protecting Group. J Org Chem 37:3404–3409
18. Kaiser E, Colescott RL, Bossinger CD, Cook PI (1970) Color test for detection of free terminal amino groups in the solid-phase synthesis of peptides. Anal Biochem 34:595–598
19. Subiros-Funosas R, Prohens R, Barbas R, El-Faham A, Albericio F (2009) Oxyma: an efficient additive for peptide synthesis to replace the benzotriazole-based HOBt and HOAt with a lower risk of explosion. Chem-Eur J 15:9394–9403
20. King DS, Fields CG, Fields GB (1990) A cleavage method which minimizes side reactions following Fmoc solid phase peptide synthesis. Int J Peptide Protein Res 36:255–266
21. Canne LE, Bark SJ, Kent SBH (1996) Extending the applicability of native chemical ligation. J Am Chem Soc 118:5891–5896
22. Matsueda GR, Stewart JM (1981) A P-Methylbenzhydrylamine resin for improved solid-phase synthesis of peptide amides. Peptides 2:45–50
23. Sarin VK, Kent SB, Tam JP, Merrifield RB (1981) Quantitative monitoring of solid-phase peptide synthesis by the ninhydrin reaction. Anal Biochem 117:147–157
24. Sambrook J, Russell DW (2001) Molecular cloning: a laboratory manual, 3rd edn. Cold Spring Harbor Laboratory Press, Cold Spring Harbor
25. Erlandsson M, Hallbrink M (2005) Metallic zinc reduction of disulfide bonds between cysteine residues in peptides and proteins. Int J Pept Res Ther 11:261–265

Chapter 15

Investigating the Effects of O-GlcNAc Modifications in Parkinson's Disease Using Semisynthetic α-Synuclein

Ana Galesic and Matthew R. Pratt

Abstract

α-Synuclein is a small aggregation-prone protein associated with Parkinson's disease (PD). The protein's biochemical and biophysical properties can be heavily influenced by various types of posttranslational modification (PTMs) such as phosphorylation, ubiquitination, and glycosylation. To understand the site-specific effects of various PTMs have on the protein and its aggregation, obtaining a homogeneous sample of the protein of interest with the specific modification of interest is key. Expressed protein ligation (EPL) has emerged as robust tool for building synthetic proteins bearing site-specific modifications. Here, we outline our approach for building α-synuclein with site specific O-GlcNAc modifications, an intracellular subtype of glycosylation that has been linked to the inhibition of protein aggregation. More specifically, we provide specific protocols for the synthesis of α-synuclein bearing an O-GlcNAc modification at threonine 72, termed α-synuclein(gT72). However, this general approach utilizing two recombinant fragments and one synthetic peptide is applicable to other sites and types of modifications and should be transferable to various other protein targets, including aggregation prone proteins like tau and TDP-43.

Key words α-synuclein, Posttranslational modification, Synthetic protein chemistry, Expressed protein ligation, O-GlcNAc

1 Introduction

α-Synuclein is a small protein of 140 amino acids (14 kDa) [1] that is found at a fairly high concentration of 50 μM in presynaptic terminals [2]. This protein is composed of different domains: an N-terminal lysine rich domain that binds to membranes (residues 1–60), the nonamyloid component (NAC) domain that is hydrophobic and is the driving force in the aggregation process (residues 61–95), and an acidic C-terminal domain (residues 96–140) [3]. Monomeric α-synuclein is largely unstructured in solution but will bind to the membranes where the protein adopts an α-helical structure important for its vesicle trafficking function [4–7]. In a process that is still poorly understood, α-synuclein can also form β-sheet rich amyloid aggregates that are toxic to cells and

Miquel Vila-Perelló (ed.), *Expressed Protein Ligation: Methods and Protocols*, Methods in Molecular Biology, vol. 2133, https://doi.org/10.1007/978-1-0716-0434-2_15,

commonly found in the brains of patients suffering from Parkinson's disease and other synucleinopathies [8]. Although this aggregation process is understandably complicated, it can be broken down into a concentration-dependent two-step process: a nucleation step that yields small oligomers/seeds and a subsequent extension step that leads to the initial seeds being extended into larger, toxic fibrils [9].

α-Synuclein has been found to bear various posttranslational modifications (PTMs) in vivo, raising the possibility that these modifications may play either beneficial or deleterious roles in the initiation and/or progression of synucleinopathies [10, 11]. This includes O-GlcNAc modifications that have been found at nine different sites from a variety of in vivo proteomics studies (Fig. 1a) [12–16], including several within the NAC region, raising the question of whether they have an effect on the protein's aggregation properties. O-GlcNAcylation is the addition of the monosaccharide *N*-acetylglucosamine to serine and threonine residues of intracellular proteins (Fig. 1b) [17, 18]. Several lines of genetic and biochemical evidence support a role for O-GlcNAc in preventing neurodegeneration. Neuron-specific knockout of the enzyme responsible for adding O-GlcNAc, O-GlcNAc transferase, results in neurodegeneration and neuron cell death [19, 20], and the levels of this modification have been shown to be lower in Alzheimer's disease patients [21, 22]. Additionally, enzymatic O-GlcNAcylation of recombinant tau in vitro reduces its aggregation [23, 24], and increasing the amounts of O-GlcNAcylation slows neurodegeneration in a mouse model of Alzheimer's disease [23]. Recently, we have shown that site-specific O-GlcNAcylation at four different sites in the NAC domain of α-synuclein generally results in inhibition of its aggregation with interesting site-specific differences [25–27]. Importantly, this general inhibitory role for O-GlcNAcylation was confirmed by enzymatic modification of recombinant α-synuclein [28], but this resulted in a heterogeneous mixture of different O-GlcNAc modifications that were not separated.

Key to our ability to prepare site-specifically O-GlcNAcylated α-synuclein was the application of protein ligation techniques. More specifically, we took advantage of expressed protein ligation (EPL), an extension of native chemical ligation (NCL). NCL takes advantage of a reaction between a peptide thioester and peptide or protein bearing an N-terminal cysteine to create a native amide bond [29]. More specifically, the first step of an NCL reaction is a reversible transthioesterification between a C-terminal thioester of one peptide and an N-terminal cysteine residue of the other. Once this intermediate formed, an S-N acyl shift results in the native amide bond. While this technology can be used successfully for the generation of site-specifically modified proteins, the major drawback is the size limitation of solid phase peptide synthesis

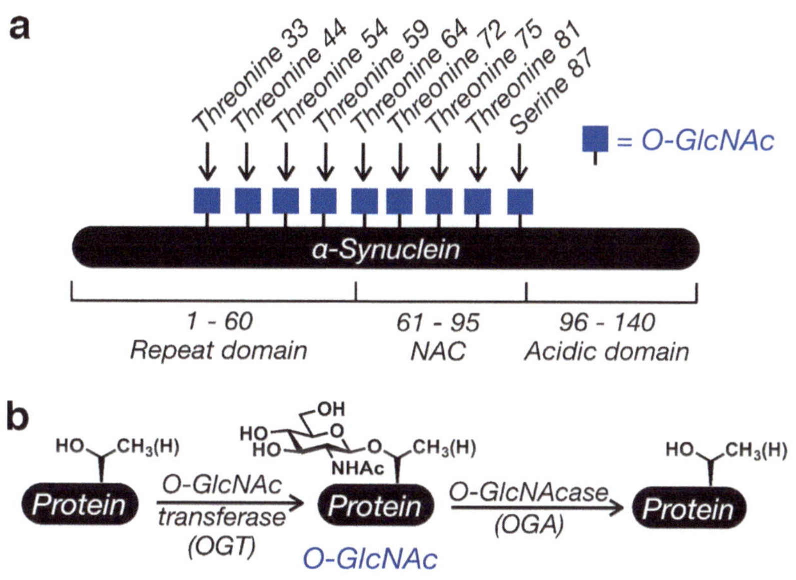

Fig. 1 O-GlcNAcylation of α-synuclein. (**a**) α-Synuclein has been found to be O-GlcNAcylated at nine different positions from in vivo proteomics studies. (**b**) O-GlcNAc is the reversible addition of the monosaccharide *N*-acetylglucosamine to serine and threonine residues of intracellular proteins

(SPPS). The solution to this problem, termed expressed protein ligation (EPL) [30], involves the generation of recombinant protein-thioesters, which occur naturally during protein splicing. Protein splicing is a posttranslational process where an intein fragment is removed from the sequence, which results in an amide bond formation between the two flanking segments or exteins. While the biological role of inteins is not clear, the splicing mechanism of inteins can be used to generate protein thioesters. Proteins of interest are recombinantly expressed as simple N-terminal fusion to an intein, which can be cleaved in the presence of exogenous thiols, yielding a recombinant C-terminal protein thioester. This protein thioester can then participate in a ligation reaction. Expressed protein ligation allows the incorporation of synthetic peptides that contain PTMs to be incorporated into proteins. Our route involves a protein thioester, a synthetic O-GlcNAc-modified peptide and a protein fragment generated from *E. coli*.

The following protocol describes the use of EPL to synthesize α-synuclein with a single O-GlcNAc modification at threonine 72, or α-synuclein(gT72), as schematized in Fig. 2a. Specifically, O-GlcNAc modified threonine [31] is first incorporated into a synthetic peptide thioester (A) using SPPS on the Dawson thioester resin. Next, peptide A undergoes a ligation reaction with recombinant fragment B, obtained by heterologous expression in *E. coli*. The N-terminal thiazolidine of the resulting product is then removed to give intermediate D, which readily undergoes a second

a

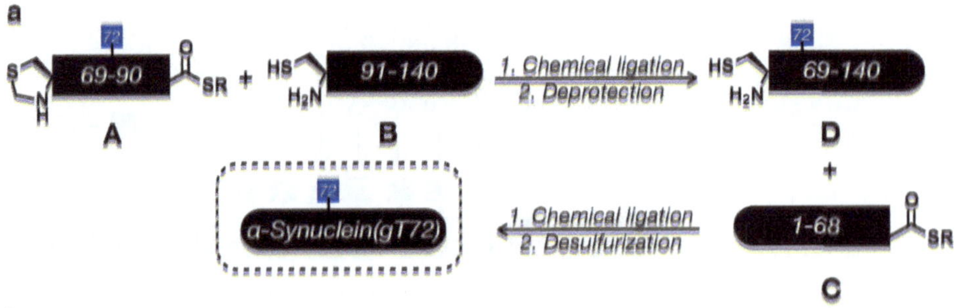

b

```
        10         20         30         40         50
MDVFMKGLSK AKEGVVAAAE KTKQGVAEAA GKTKEGVLYV GSKTKEGVVH
        60         70         80         90        100
GVATVAEKTK EQVTNVGGAV VTGVTAVAQK TVEGAGSIAA ATGFVKKDQL
       110        120        130        140
GKNEEGAPQE GILEDMPVDP DNEAYEMPSE EGYQDYEPEA
```

Fig. 2 Synthesis of O-GlcNAcylated α-synuclein. (**a**) Synthetic route of α-synuclein(gT72). (**b**) The primary sequence of human α-synuclein with the different O-GlcNAcylation sites (blue) that we have prepared and the ligation sites (red) that we have utilized

ligation reaction with protein thioester C to yield full-length O-GlcNAcylated α-synuclein. α-Synuclein contains no native cysteines residues. Therefore, the cysteines required for our synthesis can be transformed into the native alanine residues by a final desulfurization reaction. Importantly, we have applied this same general strategy to O-GlcNAc modifications at threonines 72, 75, and 81 or serine 87 by using the different ligation sites highlighted in Fig. 2b.

This protocol is representative of the synthesis of site-specific O-GlcNAcylated α-synuclein, but should be applicable toward the synthesis of other synthetic proteins bearing site specific PTMs. Specifically, we describe the incorporation of O-GlcNAc-modified threonine into peptides by using SPPS. Following the ligation reaction between the peptide (A) and C terminal fragment (B) of α-synuclein, expressed from *E. coli*, the ligation product 1 (D) is ligated to the N terminus of α-synuclein (C) that is expressed from *E. coli*. Finally, desulfurization of full length α-synuclein is conducted to convert cysteine to alanine (its native sequence) (Fig. 1b).

2 Materials

Prepare all solutions using ultrapure water (18 MΩ) H_2O at 25 °C. Dispose of waste appropriately.

2.1 Reagents

1. Growth Medium: LB (Luria–Bertani–Miller) broth [25 g/L]. This medium is used to generate recombinant proteins.

2. Growth Medium: TB (Terrific Broth) [50 g/L plus 8 mL of 50% glycerol]. This medium is used to generate recombinant proteins.
3. Antibiotic Stock Solution: 100 mg/mL ampicillin. Store at −20 °C. The final concentration in media is 100 μL/mL.
4. Antibiotic Stock Solution: 50 mg/mL kanamycin solutions. Store at −20 °C. The final concentration 50 μL/mL.
5. Lysis buffer for C-terminal fragment expression: 500 mM sodium chloride (NaCl), 100 mM Tris, 10 mM β-mercaptoethanol (βME, added fresh), 1 mM ethylenediaminetetraacetic acid (EDTA), pH 8.0. This buffer can be stored at 4 °C for ~2 months.
6. Lysis buffer for N-terminal fragment expression: 50 mM NaH_2PO_4, 300 mM NaCl, 5 mM imidazole, 2 mM tris(2-carboxyethyl)phosphine (TCEP) (freshly added), pH 8.0 with 2 mM PMSF (freshly added). This buffer can be stored at 4 °C for ~2 months.
7. Thiolysis buffer for protein thioester generation: 150 mM monosodium phosphate (NaH_2PO_4), 150 mM sodium 2-mercaptoethanesulfonate (MESNa), pH 7.4. This buffer should be prepared fresh.
8. Phenylmethylsulfonyl fluoride (PMSF) solution: 200 mM PMSF dissolved in iso-propanol. This stock solution should be prepared fresh.
9. Wash buffer for N-terminal fragment purification: 50 mM NaH_2PO_4, 300 mM NaCl, 20 mM imidazole, 2 mM TCEP (freshly added), pH 7.4. This buffer can be stored at 4 °C for ~2 months.
10. Elution buffer for N-terminal fragment purification: 50 mM NaH_2PO_4, 300 mM NaCl, 250 mM imidazole, 2 mM TCEP (freshly added), pH 7.8. This buffer can be stored at 4 °C for ~2 months without the addition of TCEP.
11. VA-061 stock solution for desulfurization: 200 mM Va-061 dissolved in methanol. This stock solution should be prepared fresh.
12. MPAA stock solution for protein ligation reactions: 250 mM 4-Mercaptophenylacetic acid (MPAA), 3M guanidine, 300 mM phosphate pH 7.4. This stock solution can be stored at −20 °C for ~1 month.
13. TCEP stock solution: 300 mM TCEP, 3 M guanidine, 300 mM phosphate pH 7.4. This stock solution can be stored at −20 °C for ~1 month.
14. Ligation buffer: 6 M guanidine, 300 mM phosphate, pH 7.4. This buffer should be prepared fresh.

15. Desulfurization buffer: 6 M guanidine, 300 mM phosphate, 300 mM TCEP, pH 7.4. This buffer should be prepared fresh.
16. HPLC buffer A: 0.1% trifluoroacetic acid (TFA) in water.
17. HPLC buffer B: 0.1% trifluoroacetic acid (TFA), 10% acetonitrile, 90% water.
18. Dichloromethane.
19. *N,N,N′,N′*-Tetramethyl-*O*-(1*H*-benzotriazol-1-yl)uronium hexafluorophosphate (HBTU).
20. *N,N*-Diisopropylethylamide (DIEA).
21. Fmoc-protected amino acids.
22. Boc-Thz-OH.
23. Hydrazine hydrate solution: 80% hydrazine hydrate v/v in methanol (MeOH).
24. p-Nitrophenyl chloroformate.
25. Cleavage cocktail solution: 95% trifluoroacetic acid, 2.5% triisopropylsilane, 2.5% water.
26. Peptide precipitation solution: diethyl ether.
27. Fmoc-deprotection solution: 20% v/v piperidine in DMF.
28. Dawson Dbz AM resin.
29. IPTG solution: 1 M isopropyl β-D-1-thiogalactopyranoside in H_2O.
30. Ethanethiol.
31. Tertbutylthiol.

2.2 Plasmids

1. pET42b vector: NdeI and SpeI restrictions sites for introducing C-terminal protein fragments (Fig. 2, fragments B).
2. pTBX1 vector: NdeI and Bpu10I restrictions sites for introducing N-terminal protein fragments as in-frame, 5′ genetic fusions to the AvaE intein (Fig. 2, fragment C).

2.3 Equipment

1. Temperature controlled incubator/shaker for protein expression.
2. Water bath temperature controlled incubator.
3. Temperature controlled centrifuge.
4. Rocking platform to gently introduce mixing into solution.
5. High performance liquid chromatography (HPLC) for separation of various species.
6. Mass Spectrometer for peptide and protein characterization.
7. Lyophilizer.
8. Bio-Spin disposable chromatography column.
9. Gas dispersion tube, large, fritted.

3 Methods

3.1 C-terminal Protein Fragment Expression (B, Fig. 2)

The C-terminal protein fragment is obtained by heterogeneous expression in *E. coli*. During this expression the initiator methionine is endogenously removed from the recombinant protein, resulting in an N-terminal cysteine residue that can react with metabolites in the *E. coli* host. After purification, these modifications are removed by methoxyamine treatment to generate a free N-terminal cysteine required for protein ligation.

1. Transform BL21 (DE3) chemically competent *E. coli* with pET42b plasmid encoding the protein fragment onto a kanamycin-containing LB plate.
2. Inoculate 50 mL LB media containing kanamycin (50 μL/mL) with a single colony to initiate a starter culture for recombinant expression. Incubate at 37 °C with constant agitation of 250 rpm for 12–16 h.
3. Inoculate 300 mL TB media with 3 mL of the starter culture, 50 μL/mL kanamycin and incubate at 37 °C till OD_{600} 0.7–1. Add final concentration of 0.5 mM isopropyl β-D-1-thiogalactopyranoside (IPTG) and incubate at room temperature (RT) with constant agitation of 250 rpm for 18 h.
4. Harvest bacteria by centrifugation at 6000 × *g*, 4 °C for 30 min. Discard the media.
5. Resuspend the bacterial pellet in 10 mL cold lysis buffer.
6. Lyse cells by incubating the sample at 80 °C for 10 min. Mix the contents of the tube every minute by inversion.
7. Incubate the sample at room temperature for 30 min.
8. Add freshly prepared phenylmethylsulfonyl fluoride (PMSF) to a final concentration of 2 mM and incubate on ice for 30 min.
9. Centrifuge at 7000 × *g*, 4 °C for 30 min.
10. Collect the supernatant by decanting and adjust its pH to 3.5 (*see* **Note 1**).
11. Incubate supernatant on ice for 30 min.
12. Centrifuge at 7000 × *g*, 4 °C for 30 min.
13. Collect the supernatant by decanting and dialyze it against 2 L of degassed 1% acetic acid at 4 °C overnight (*see* **Note 2**).
14. Centrifuge protein solution at 7000 × *g*, 4 °C for 30 min.
15. Reveal the N-terminal cysteine residue by adding methoxyamine hydrochloride to a final concentration of 250 mM and adjusting the pH to 4.0. Incubate at room temperature overnight on a rocking platform.

16. Purify protein fragment using reverse-phase liquid chromatography (HPLC). Identify the desired protein fragment by MS.
17. Lyophilize the C-terminal fragment and store as a solid at −20 °C.

3.2 O-GlcNAcylated Peptide (A, Fig. 2)

The O-GlcNAc-modified threonine is incorporated into a synthetic peptide that contains the thioester that is necessary for NCL reaction. The ligation sites are located at the junction of glycine alanine or two alanine residues (Fig. 2b), where one alanine is mutated to a cysteine for NCL reaction. The last residue on the N-terminal portion of the peptide is Boc-Thz-OH. This thiazolidine ring is opened by methoxyamine to generate cysteine that is later necessary for NCL.

1. Add Dawson Dbz AM resin, 0.1 mmol scale, to a 5 mL disposable reaction vessel with frit for SPPS (*see* **Note 3**).
2. Incubate resin in 3–4 mL of dichloromethane (DCM) for 20 min, wash once with dimethylformamide (DMF), and incubate resin in 3–4 mL of DMF for an additional 20 min (*see* **Note 4**).
3. Incubate 2 × 20 min in 3–4 mL of 20% piperidine in DMF.
4. Wash the resin with DMF (*see* **Note 4**).
5. Activate amino acid by dissolving 5 equivalents (eq) of the appropriate *N*-Fmoc-protected amino acid, 5 eq of HBTU, 5 eq of N,N-diisopropylethylamine (DIEA) in 3–4 mL of DMF. Incubate for 4–6 min.
6. Incubate resin and preactivated amino acid (from **step 6**) for 1 h (*see* **Note 5**).
7. Wash the resin with DMF and repeat the **steps 3–7** to iteratively couple the remaining amino acids in the peptide sequence.
8. Couple O-GlcNAcylated threonine or serine amino acids at the appropriate positions (*see* **Note 6**). Dissolve 2 eq of pentafluorophenyl (PFP) O-GlcNAcylated threonine or serine in 4 mL of DMF and add to the growing peptide in the reaction vessel. Incubate overnight.
9. Couple Boc-Thz-OH using the reagents in **step 5** as the final amino acid.
10. Incubate peptide with hydrazine hydrate (80% v/v in methanol (MeOH)) twice for 45 min to remove the *O*-acetate groups on the O-GlcNAc.
11. Activate Dawson linker with treatment of 5 eq p-nitrophenyl chloroformate in 3–4 mL of DCM for 1 h, followed by 5 eq of DIEA in 3–4 mL of DMF for 30 min.

12. Cleave peptide with 7 mL of cleavage cocktail (95:2.5:2.5 trifluoroacetic acid (TFA)/H_2O/triisopropylsilane) room temperature for 3 h.
13. Use a Bio-Spin disposable chromatography column to filter cocktail cleavage solution into a clean tube.
14. Precipitate crude peptide solution in cold diethyl ether (1:10 cleavage solution to diethyl ether) at −80 °C overnight.
15. Centrifuge at 7000 × *g*, 4 °C for 30 min and air dry the peptide pellet.
16. Completely dissolve peptide pellet in a mixture of HPLC buffers A and B and then lyophilize.
17. Dissolve the solid peptide in 8 mL of thiolysis buffer and incubate the mixture at room temperature for 1 h.
18. Purify the peptide on C18 semipreparative column and characterize the peaks by MS.
19. Lyophilize the peptide and store as a solid at −20 °C.

3.3 N-terminal Protein Fragment Expression (C, Fig. 2)

The C-terminal protein fragment is obtained by heterogeneous expression as an N-terminal fusion to the AvaE intein in *E. coli*. After an initial purification, the intein can be cleaved by the addition of exogenous thiols to yield the N-terminal protein fragment as a C-terminal thioester for subsequent ligation reactions.

1. Transform BL21 (DE3) chemically competent *E. coli* with pTBX1 plasmid encoding the protein fragment onto an ampicillin-containing LB plate (100 μL/mL).
2. Inoculate 50 mL LB media containing ampicillin (100 μL/mL) with a single colony to initiate a starter culture for recombinant expression. Incubate at 37 °C with constant agitation of 250 rpm for 12–16 h.
3. Inoculate 300 mL TB media with 3 mL of the starter culture, ampicillin (100 μL/mL) and incubate at 37 °C till OD_{600} 0.7–1. Add final concentration of 0.5 M isopropyl β-D-1-thiogalactopyranoside (IPTG) and incubate at room temperature (RT) with constant agitation of 250 rpm for 18 h.
4. Harvest bacteria by centrifugation at 6000 × *g*, 4 °C for 30 min. Discard the media.
5. Resuspend the cell pellets in 10 mL cold lysis buffer.
6. Lyse cells with a tip sonicator (35% amplitude, 30 s on/off, total time of 12 min) (*see* **Note 7**).
7. Centrifuge at 7000 × *g*, 4 °C for 30 min.
8. Collect the supernatant by decanting.
9. Add 1 mL of Ni-AGA Agarose resin to a Bio-Spin disposable chromatography column.

10. Rinse the Ni-AGA Agarose resin with water (10 mL) and then 2× 15 mL with wash buffer for N-terminal fragment purification.
11. Add the protein supernatant to the Ni-AGA agarose resin and incubate at 4 °C for 1 h.
12. Wash the matrix 3X with 10 mL of wash buffer for N-terminal fragment purification.
13. Elute the protein 3X with 4 mL of elution buffer for N-terminal fragment purification.
14. Combine the elution fractions and dialyze them against 1L of 0.5× Dulbecco's phosphate-buffered saline (DPBS) at 4 °C overnight (*see* **Note 2**).
15. Transfer protein solution to a clean tube and incubate the purified protein-intein fusion with 250 mM MESNa. This step will allow the intein to produce a protein thioester that can be intercepted with MESNa, resulting in the cleavage of the intein and generation of the N-terminal fragment thioester.
16. Adjust pH to 7.4 and incubate the reaction at RT overnight.
17. Purify protein on C4 semipreparative HPLC column and characterize protein peaks by MS.
18. Lyophilize the N-terminal protein fragment and store as a solid at −20 °C.

3.4 Ligation Reaction 1: O-GlcNAc Peptide (69–90) (A, Fig. 2) and C-terminal Fragment (91–140) (B, Fig. 2)

1. Completely dissolve the lyophilized C-terminal protein fragment (B, Fig. 2) (1.6 mM) in the ligation buffer by bath sonication and then add this solution to a fresh tube containing O-GlcNAc peptide fragment A (2 mM). Repeat bath sonication to dissolve both components.
2. Add MPAA and TCEP from the stock solutions to a final concentration of 30 mM each.
3. The reaction pH should be within a 7.0–7.4 range. If necessary, raise the pH with 1 M NaOH.
4. Place reaction tube on the rocking platform.
5. Monitor reaction by HPLC. Take 1 μL of the reaction mixture and add to 20 μL of TCEP stock solution. Mix and inject (21 μL) into HPLC analytical C18 column. Run gradient 0–70% B over 60 min and characterize peaks by MS.
6. When reaction is complete, add 250 mM methoxyamine, adjust the pH of the reaction to 4.0 and incubate on the rocking platform at room temperature overnight. This will remove the N-terminal thiazolidine protecting-group and reveal the N-terminal cysteine of ligation product one (D, Fig. 2).
7. Run HPLC, collect all peaks and identify the desired peak (fragment D) by MS.

8. Before the purification, add a small amount of solid TCEP (~1 mg per 100 μL) to the reaction and incubate the tube at room temperature on the rocking platform for 10 min. This will reduce any disulfide bonds that may have formed.
9. Purify reaction on HPLC C18 semi-prep column.
10. Lyophilize the ligation product (D) and store as a solid at −20 °C.

3.5 Ligation Reaction 2: N-Terminal Fragment (1–68) (C, Fig. 2) and Ligation-Product D (69–140) (Fig. 2)

1. Completely dissolve the ligation-product D (2 mM) and the N-terminal fragment C (8 mM) in ligation buffer by bath sonication.
2. Add MPAA and TCEP from the stocks to a final concentration of 30 mM each.
3. The reaction pH should be within a 7.0–7.4 range. If necessary, raise the pH with 1 M NaOH.
4. Place reaction tube on the rocking platform.
5. Monitor reaction by HPLC. Take 1 μL of the reaction mixture and add to 20 μL of TCEP stock solution. Mix and inject into HPLC analytical C18 column. Run gradient 0–70% B over 60 min and characterize peaks by MS.
6. When reaction is complete, add a small amount of solid TCEP (~1 mg per 100 μL) to the reaction and incubate the tube at room temperature on the rocking platform for 10 min. This will reduce any disulfide bonds that may have formed.
7. Purify reaction on HPLC C18 semi-prep column.
8. Lyophilize the ligation product and store as a solid at −20 °C.

3.6 Desulfurization of Ligation Product (1–140)

1. Dissolve ligation product (~0.75 mg/mL) in desulfurization buffer.
2. Transfer the solution to a round bottom flask and place the reaction in a water bath at 37 °C.
3. Add 2.5% v/v ethanethiol and 10% v/v tertbutylthiol and place the reaction flask under a nitrogen atmosphere by first evacuating the air under reduced pressure and replacing it with nitrogen gas.
4. Stir the reaction for 10 min at 37 °C.
5. Add VA-061 radical initiator to the reaction to give a final concentration of 2 mM.
6. Stir the reaction at 37 °C overnight.
7. Monitor the progression of the reaction by HPLC (0–70% over 60 min) and purify the final product on C4 semi-prep HPLC column.
8. Lyophilize the final product and store as a solid at −20 °C.

4 Notes

1. Lowering the pH to 3.5 results in the precipitation of endogenous *E. coli* proteins while the C-terminal protein fragment remains in solution.
2. To dialysis protein sample, use traditional dialysis membrane tubing with a molecular weight cutoff of 3500 Da.
3. For the peptide synthesis use reaction syringe with a filter frit.
4. Wash peptide by shaking the syringe in 3–4 mL of the appropriate solvent for 1 min, followed by extrusion of the solvent through the frit. Repeat this washing 4 times with additional DMF or DCM.
5. Couple first amino acid twice for 30 min (second time prepare fresh preactivation solution). Arginine and proline should be coupled twice for 45 min. Isoleucine, valine, and threonine should be coupled once for 1.5 h.
6. Do not add HBTU or DIEA to the PFP O-GlcNAcylated amino acid during preactivation step because PFP already acts as an activator of the carboxylate.
7. During tip sonication place the sample on the ice to prevent temperature increase of protein supernatant.

Acknowledgments

This research was supported by the National Institutes of Health (Grant R01GM114537 to M.R.P.).

References

1. Lashuel HA, Overk CR, Oueslati A, Masliah E (2013) The many faces of α-synuclein: from structure and toxicity to therapeutic target. Nat Rev Neurosci 14:38–48. https://doi.org/10.1038/nrn3406
2. Wilhelm BG, Mandad S, Truckenbrodt S, Kröhnert K, Schäfer C, Rammner B et al (2014) Composition of isolated synaptic boutons reveals the amounts of vesicle trafficking proteins. Science 344:1023–1028. https://doi.org/10.1126/science.1252884
3. Wang C, Zhao C, Li D, Tian Z, Lai Y, Diao J et al (2016) Versatile structures of α-synuclein. Front Mol Neurosci 9:48. https://doi.org/10.3389/fnmol.2016.00048
4. Jao CC, Hegde BG, Chen J, Haworth IS, Langen R (2008) Structure of membrane-bound alpha-synuclein from site-directed spin labeling and computational refinement. Proc Natl Acad Sci USA 105:19666–19671. https://doi.org/10.1073/pnas.0807826105
5. Varkey J, Isas JM, Mizuno N, Jensen MB, Bhatia VK, Jao CC et al (2010) Membrane curvature induction and tubulation are common features of synucleins and apolipoproteins. J Biol Chem 285:32486–32493. https://doi.org/10.1074/jbc.M110.139576
6. Mizuno N, Varkey J, Kegulian NC, Hegde BG, Cheng N, Langen R et al (2012) Remodeling of lipid vesicles into cylindrical micelles by α-synuclein in an extended α-helical conformation. J Biol Chem 287:29301–29311. https://doi.org/10.1074/jbc.M112.365817
7. Emanuele M, Chieregatti E (2015) Mechanisms of alpha-synuclein action on neurotransmission: cell-autonomous and non-cell

autonomous role. Biomolecules. 5:865–892. https://doi.org/10.3390/biom5020865
8. Fink AL (2006) The aggregation and fibrillation of α-synuclein. Acc Chem Res 39:628–634. https://doi.org/10.1021/ar050073t
9. Wood SJ, Wypych J, Steavenson S, Louis JC, Citron M, Biere AL (1999) alpha-synuclein fibrillogenesis is nucleation-dependent. Implications for the pathogenesis of Parkinson's disease. J Biol Chem 274:19509–19512. https://doi.org/10.1074/jbc.274.28.19509
10. Oueslati A, Fournier M, Lashuel HA (2010) Role of post-translational modifications in modulating the structure, function and toxicity of alpha-synuclein: implications for Parkinson's disease pathogenesis and therapies. Prog Brain Res. 183:115–145. https://doi.org/10.1016/S0079-6123(10)83007-9
11. Schmid AW, Fauvet B, Moniatte M, Lashuel HA (2013) Alpha-synuclein post-translational modifications as potential biomarkers for Parkinson disease and other synucleinopathies. Mol Cell Proteomics 12:3543–3558. https://doi.org/10.1074/mcp.R113.032730
12. Wang Z, Park K, Comer F, Hsieh-Wilson LC, Saudek CD, Hart GW (2009) Site-specific GlcNAcylation of human erythrocyte proteins potential biomarker(s) for diabetes. Diabetes 58:309–317. https://doi.org/10.2337/db08-0994
13. Wang Z, Udeshi ND, O'Malley M, Shabanowitz J, Hunt DF, Hart GW (2010) Enrichment and site mapping of O-linked N-acetylglucosamine by a combination of chemical/enzymatic tagging, photochemical cleavage, and electron transfer dissociation mass spectrometry. Mol Cell Proteomics 9:153–160. https://doi.org/10.1074/mcp.M900268-MCP200
14. Alfaro JF, Gong C-X, Monroe ME, Aldrich JT, Clauss TRW, Purvine SO et al (2012) Tandem mass spectrometry identifies many mouse brain O-GlcNAcylated proteins including EGF domain-specific O-GlcNAc transferase targets. Proc Natl Acad Sci USA 109:7280–7285. https://doi.org/10.1073/pnas.1200425109
15. Morris M, Knudsen GM, Maeda S, Trinidad JC, Ioanoviciu A, Burlingame AL et al (2015) Tau post-translational modifications in wild-type and human amyloid precursor protein transgenic mice. Nat Neurosci. 18:1183–1189. https://doi.org/10.1038/nn.4067
16. Wang S, Yang F, Petyuk VA, Shukla AK, Monroe ME, Gritsenko MA et al (2017) Quantitative proteomics identifies altered O-GlcNAcylation of structural, synaptic and memory-associated proteins in Alzheimer's disease. J Pathol 243:78–88. https://doi.org/10.1002/path.4929
17. Bond MR, Hanover JA (2015) A little sugar goes a long way: the cell biology of O-GlcNAc. J Cell Biol 208:869–880. https://doi.org/10.1083/jcb.201501101
18. Yang X, Qian K (2017) Protein O-GlcNAcylation: emerging mechanisms and functions. Nat Rev Mol Cell Biol 18:452–465. https://doi.org/10.1038/nrm.2017.22
19. O'Donnell N, Zachara NE, Hart GW, Marth JD (2004) Ogt-dependent X-chromosome-linked protein glycosylation is a requisite modification in somatic cell function and embryo viability. Mol Cell Biol 24:1680–1690
20. Wang AC, Jensen EH, Rexach JE, Vinters HV, Hsieh-Wilson LC (2016) Loss of O-GlcNAc glycosylation in forebrain excitatory neurons induces neurodegeneration. Proc Natl Acad Sci USA 113:15120–15125. https://doi.org/10.1073/pnas.1606899113
21. Liu F, Iqbal K, Grundke-Iqbal I, Hart G, Gong C (2004) O-GlcNAcylation regulates phosphorylation of tau: a mechanism involved in Alzheimer's disease. Proc Natl Acad Sci USA 101:10804–10809
22. Aguilar AL, Hou X, Wen L, Wang PG, Wu P (2017) A chemoenzymatic histology method for O-GlcNAc detection. ChemBioChem 18:2416–2421. https://doi.org/10.1002/cbic.201700515
23. Yuzwa SA, Shan X, Macauley MS, Clark T, Skorobogatko Y, Vosseller K et al (2012) Increasing O-GlcNAc slows neurodegeneration and stabilizes tau against aggregation. Nat Chem Biol 8:393–399. https://doi.org/10.1038/nchembio.797
24. Yuzwa SA, Cheung AH, Okon M, McIntosh LP, Vocadlo DJ (2014) O-GlcNAc modification of tau directly inhibits its aggregation without perturbing the conformational properties of tau monomers. J Mol Biol. 426:1736–1752. https://doi.org/10.1016/j.jmb.2014.01.004
25. Marotta NP, Lin YH, Lewis YE, Ambroso MR, Zaro BW, Roth MT et al (2015) O-GlcNAc modification blocks the aggregation and toxicity of the protein α-synuclein associated with Parkinson's disease. Nat Chem. 7:913–920. https://doi.org/10.1038/nchem.2361
26. Lewis YE, Galesic A, Levine PM, De Leon CA, Lamiri N, Brennan CK et al (2017) O-GlcNAcylation of α-synuclein at serine 87 reduces aggregation without affecting membrane binding. ACS Chem Biol

12:1020–1027. https://doi.org/10.1021/acschembio.7b00113

27. Levine PM, Galesic A, Balana AT, Mahul-Mellier A-L, Navarro MX, De Leon CA et al (2019) α-Synuclein O-GlcNAcylation alters aggregation and toxicity, revealing certain residues as potential inhibitors of Parkinson's disease. Proc Natl Acad Sci USA 116:1511–1519. https://doi.org/10.1073/pnas.1808845116
28. Zhang J, Lei H, Chen Y, Ma Y-T, Jiang F, Tan J et al (2017) Enzymatic O-GlcNAcylation of α-synuclein reduces aggregation and increases SDS-resistant soluble oligomers. Neurosci Lett 655:90–94. https://doi.org/10.1016/j.neulet.2017.06.034
29. Dawson P, Muir T, Clark-Lewis I, Kent S (1994) Synthesis of proteins by native chemical ligation. Science 266:776–779
30. Muir TW, Sondhi D, Cole PA (1998) Expressed protein ligation: a general method for protein engineering. Proc Natl Acad Sci USA 95:6705–6710
31. De Leon CA, Lang G, Saavedra MI, Pratt MR (2018) Simple and efficient preparation of O- and S-GlcNAcylated amino acids through InBr3-catalyzed synthesis of β-N-acetylglycosides from commercially available reagents. Org Lett 20:5032–5035. https://doi.org/10.1021/acs.orglett.8b02182

Chapter 16

Recombinant Expression of Cyclotides Using Expressed Protein Ligation

Maria Jose Campbell, Jingtan Su, and Julio A. Camarero

Abstract

Cyclotides are naturally occurring microproteins (≈30 residues long) present in several families of plants. All cyclotides share a unique head-to-tail circular knotted topology containing three disulfide bridges forming a cystine knot topology. Cyclotides possess high stability to chemical, physical, and biological degradation and have been reported to cross cellular membranes. In addition, naturally occurring and engineered cyclotides have shown to possess various pharmacologically relevant activities. These unique features make the cyclotide scaffold an excellent tool for the design of novel peptide-based therapeutics by using molecular evolution and/or peptide epitope grafting techniques. In this chapter, we provide protocols to recombinantly produce a natively folded cyclotide making use of a standard bacterial expression system in combination with an intein-mediated backbone cyclization with concomitant oxidative folding.

Key words Cyclotides, CCK motif, Intein, Protein splicing, GyrA intein

1 Introduction

Cyclotides are small globular microproteins (ranging from 28 to 37 residues) possessing a head-to-tail cyclized backbone topology and three interlocked disulfide bonds forming a cystine-knot (CCK) motif [1, 2] (Fig. 1). The CCK topology found in the cyclotide scaffold imparts high molecular rigidity to its backbone [3–5], making it exceptional stable toward physical, chemical, and biological degradation [6]. In fact, the use of cyclotide-containing plants in indigenous medicine first highlighted the fact that the peptides are resistant to boiling and are orally active [7].

The main features of cyclotides are a remarkable stability due to the CCK topology, a small size making them readily accessible to chemical synthesis [8], and an excellent tolerance to sequence variations. Naturally occurring cyclotides have shown to possess various pharmacologically relevant activities (*see* Refs. [1, 2] for recent reviews on the different biomedical applications of cyclotides). Cyclotides can also be engineered for the production of

Miquel Vila-Perelló (ed.), *Expressed Protein Ligation: Methods and Protocols*, Methods in Molecular Biology, vol. 2133, https://doi.org/10.1007/978-1-0716-0434-2_16,

compounds with novel biological activities to target extracellular [9–11] and intracellular [12] molecular targets, with some of them working in vivo in cancer animal models. Some of these novel cyclotides have been shown to be orally bioavailable [10] and able to cross cellular membranes efficiently [13, 14]. All these features make cyclotides a highly promising framework for peptide drug design [1, 2, 15].

Naturally occurring cyclotides are produced in plants from ribosomally produced precursor proteins [16–20]. The mechanism of excision of the cyclotide domains and ligation of the free N- and C-termini to produce the circular peptides involves two types of enzymes, papain-like cysteine proteases required in the N-terminal cleavage of cyclotide precursors [16] and asparaginyl endopeptidase-like ligases involved in the cyclization process [21–23].

Cyclotides can also be produced recombinantly using standard microbial expression systems by making use of modified protein splicing units [24–29]. This has made possible the production of genetically encoded libraries of cyclotides for rapid screening and selection of cyclotides with novel biological activities [25, 27].

We provide in this chapter a protocol to produce cyclotide MCoTI-I (Fig. 1) employing a standard *Escherichia coli*–based expression system in combination with the use of an intramolecular version of expressed protein ligation (EPL) [30–32]. Cyclotide

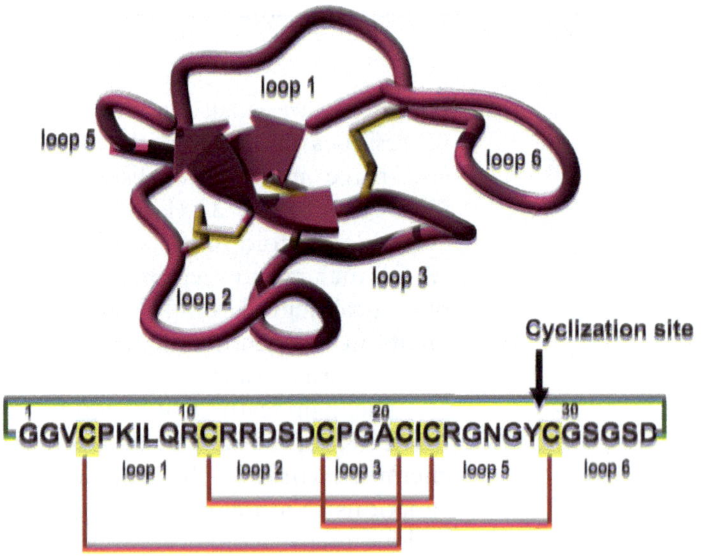

Fig. 1 Tertiary structure of cyclotide MCoTI-II (PDB code: 1IB9) [42] and the primary structure of related cyclotide MCoTI-I used in this study. The backbone cyclized peptide (connecting bond shown in green) is stabilized by the three disulfide bonds (shown in red). The cyclization site used in this chapter is indicated with a black arrow

MCoTI-I is a very potent trypsin inhibitor ($K_i \approx 20$ pM) that can be isolated from dormant seeds of *Momordica cochinchinensis*, a plant member of the *Cucurbitaceae* family [33]. Trypsin inhibitor cyclotides are very interesting candidates for the development of novel peptide-based therapeutics because they are not toxic to mammalian cells (up to concentrations of 100 μM) [12] and they can cross mammalian cell membranes [13, 14]. In addition, MCoTI-cyclotides are amenable to sequence modification through molecular evolution or by grafting of bioactive peptide epitopes allowing for the generation of cyclotides with novel biological functions [9, 12, 25].

Backbone cyclization of a polypeptide using EPL can be accomplished by placing a cysteine at the N-terminus of the target protein while the C-terminus is fused to an N-terminus of a modified Cys intein engineered to favor N-terminal cleavage (Fig. 2) [8, 34]. A cysteine can either be generated by introducing an upstream intein or by conventional proteolytic cleavage. The Cys residue can then react in an intramolecular fashion with an α-thioester generated by the downstream intein, thus providing a backbone cyclized polypeptide (Fig. 2). EPL has been used for the production of different disulfide-rich backbone cyclized polypeptides including sunflower trypsin inhibitor 1 (SFTI-1) [35], θ-defensins [36, 37], and cyclotides [12, 28, 29].

Heterologous production of cyclotide MCoTI-I will be accomplished employing a modified version of the *Mxe* GyrA intein [38]. This bacterial-derived mini-intein has a relatively small size (≈27 kDa) and shows high levels of expression in bacterial-based expression systems. This ensures a higher level of expression for the corresponding cyclotide linear precursor. Incorporation of a Met residue at the N-terminus of the cyclotide linear precursor sequence makes possible the generation of a N-terminal Cys residue by endogenous Met aminopeptidase (MAP) as the corresponding cyclotide-intein precursor protein is translated in vivo [39].

In this chapter we provide protocols for the cloning, expression, and purification of the naturally occurring MCoTI-I cyclotide. In cell production of folded MCoTI-I can be accomplished by expressing MCoTI-intein fusion protein 1 (Fig. 3). This construct contains an MCoTI-I linear precursor fused to the N-terminus of the *Mxe* GyrA intein. None of the additional native N-extein residues of the intein are used in this construct. To facilitate backbone cyclization we use the native Cys residue located at the beginning of loop 6 of MCoTI-II (Figs. 1 and 3). This loop contains a highly flexible peptide sequence and it is not required for folding or biological activity [3, 40]. Construct **1** also contains a chitin binding domain (CBD) fused at the C-terminus of the GyrA intein to facilitate purification. In-cell expression of cyclotide MCoTI-I using EPL-mediated backbone cyclization is achieved by transforming the plasmid encoding the cyclotide-precursor **1** into

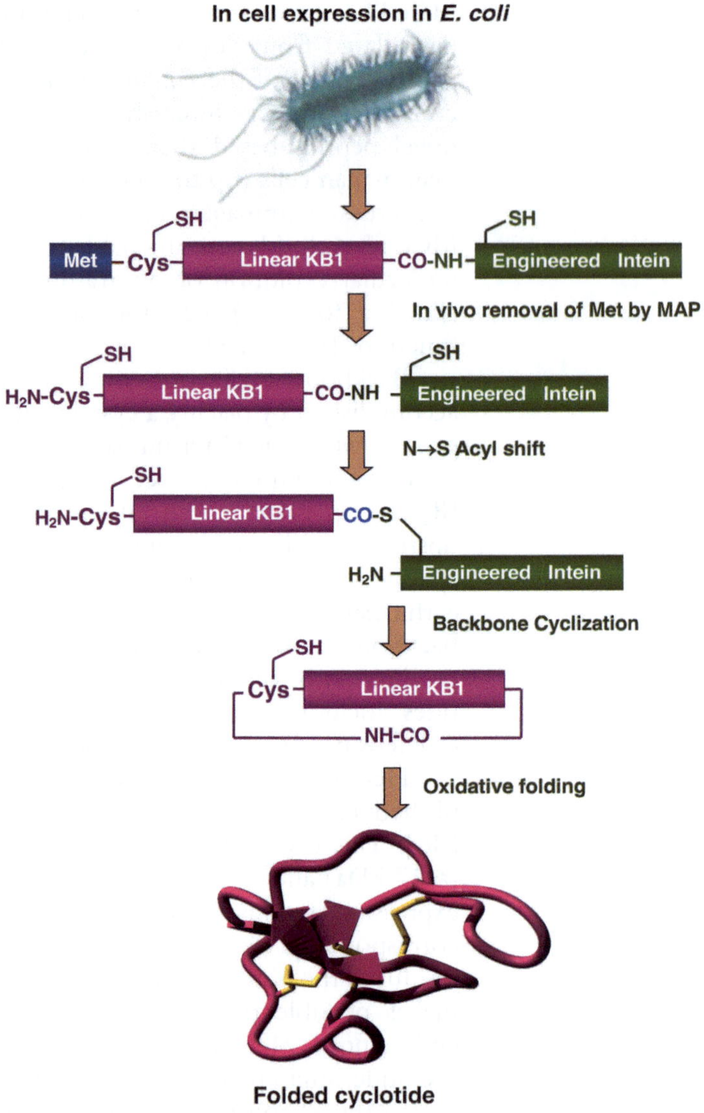

Fig. 2 In-cell expression of native folded cyclotide MCoTI-I using EPL-mediated backbone cyclization in bacterial cells

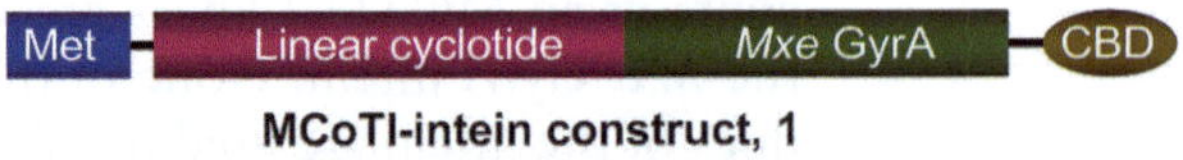

Fig. 3 Architecture of the intein precursor used for the expression of cyclotide MCoTI-I described in this protocol

Origami 2(DE3) cells to facilitate folding. Origami *E. coli* strains are K-12 derivatives that have mutations in both the thioredoxin reductase (trxB) and glutathione reductase (gor) genes, which greatly enhance disulfide bond formation in the *E. coli* cytoplasm [41].

2 Materials

All solutions are prepared using ultrapure water with a resistivity of 18 MΩ × cm at 25 °C and analytical grade reagents. All reagents and solutions are stored at room temperature, unless indicated otherwise.

2.1 Instruments

1. Water bath able to operate at 95 °C.
2. Tabletop microcentrifuge capable of operating at 18,000 × *g*.
3. Microbiology incubator set at 37 °C.
4. Temperature-controlled incubator Shaker.
5. Orbital shaker.
6. Polymerase chain reaction thermocycler.
7. Agarose gel electrophoresis unit.
8. Electrophoresis power pack able to operate up to 250 V.
9. UV–visible spectrophotometer.
10. Sonicator.
11. High-speed centrifuge.
12. SDS-PAGE electrophoresis apparatus.
13. Centrifuge tubes of 0.5 mL, 1.5 mL, 15 mL, and 30 mL of capacity.
14. 5 mL polypropylene columns.
15. Lyophilizer.
16. HPLC system equipped with gradient capability and UV-Vis detection.
17. C18 reverse phase HPLC columns.
18. Electrospray mass spectrometer (ES-MS or similar mass spectrometer).

2.2 Cloning of MCoTI-intein Construct 1

1. Expression plasmid pTXB-1 (New England Biolabs). This vector contains an engineered GyrA intein and a chitin-binding domain (CBD).
2. DNA ultramers encoding MCoTI-II (20 nmol scale, 5′-phosphorylated and purified by PAGE) (Table 1).
3. TE buffer: 10 mM Tris–HCl and 1 mM EDTA buffer at pH 8.0.
4. Annealing buffer: 20 mM Na_2HPO_4, 300 mM NaCl buffer at pH 7.4.
5. PCR Purification Kit (QIAGEN or similar).
6. Miniprep Kit (QIAGEN or similar).
7. Gel Extraction Kit (QIAGEN or similar).

Table 1
Forward (p5) and reverse (p3) 5′-phosphorylated oligonucleotides used to clone the different MCoTI-intein linear precursor into the pTXB1 expression plasmid

Linear	Oligonucleotide sequence	Precursor
1	p5	5′-TATGTGCGGTTCTGGTTCTGACGGTGGTGTTTGCCCGAAAATCCTGAAAA
	AATGCCGTCGTGACTCTGACTGCCCGGGTGC TTGCATCTGCCGTGGTAACGG	TTAC-3′
	p3	5′-GCAGTAACCGTTACCACGGCAGATGCAAGCACCCGGGCAGTCAGAGTCA
	CGACGGCATTTTTTCAGGA TTTTCGGGCAAACACCACCG TCAGAACCAGAA	CCGCACA-3′

DNA sequences were generated using optimal codons for expression in *E. coli*

8. PCR reagents: Vent DNA polymerase, Taq DNA polymerase, dNTP solution, 10× thermopol PCR buffer, 10× Taq DNA polymerase buffer (New England Biolabs Inc).
9. Chemical competent DH5α cells.
10. LB medium: Dissolve 25 g of LB broth in 1 L of pure H_2O and sterilized by autoclaving at 120 °C for 30 min.
11. LB medium-agar: Suspend 3.3 g of LB agar in 100 mL of pure H_2O and sterilized by autoclaving at 120 °C for 30 min. To prepare plates, allow LB medium-agar to cool to ≈50 °C, then add 0.1 mL of ampicillin stock solution (100 mg ampicillin/mL in H_2O, sterilized by filtration over a 45 μm filter), gently mix and pipet 20 mL into a sterile petri dish (100 mm diameter).
12. SOC Medium: 20 g of tryptone, 5 g yeast extract, 0.5 g NaCl, and 0.186 g KCl in water. Suspend ingredients into 980 mL of pure water and sterilize by autoclaving at 120 °C for 30 min. Dissolve 4.8 g $MgSO_4$, 3.603 g dextrose in 20 mL of pure H_2O and filter-sterilize over a 45 μm filter and add to the autoclaved medium.

2.3 Cyclotide Expression, Purification and Characterization

1. Chemical competent Origami2 (DE3) cells (EMD Millipore).
2. 1 M isopropyl-thio-β-D-galactopyranoside (IPTG) in H_2O. Sterilize by filtration over 45 μm filter and store at −20 °C.
3. Sterile conical bottom flask.
4. Lysis buffer: 0.1 mM ethylenediaminetetraacetic acid (EDTA), 1 mM phenylmethylsulfonyl fluoride (PMSF), 50 mM sodium phosphate, 250 mM NaCl buffer at pH 7.2 containing 10% glycerol.
5. Chitin agarose beads (New England Biolabs).
6. 100 mM phenylmethylsulfonyl fluoride (PMSF) in EtOH (better to prepare fresh before use).
7. 4×SDS-PAGE sample buffer: 1.5 mL of 1 M Tris–HCl buffer at pH 6.8, 3 mL of 1 M DTT (dithiothreitol) in pure H_2O, 0.6 g of sodium dodecyl sulfate (SDS), 30 mg of bromophenol blue, 2.4 mL of glycerol, bring final volume to 7.5 mL.
8. SDS-PAGE sample buffer: dilute 4 times 4×SDS-PAGE sample buffer in pure H_2O and add 20% 2-mercaptoethanol (by volume). Prepare fresh.
9. SDS-4-20% PAGE gels, 1× SDS running buffer.
10. Gel stain: Gelcode® Blue (Thermo scientific or similar).
11. N-hydroxy-succinimide ester (NHS)-activated sepharose beads.

12. Porcine pancreatic trypsin type IX-S (14,000 units/mg).
13. Coupling buffer: 200 mM sodium phosphate, 250 mM NaCl buffer at pH 6.0.
14. Washing buffer: 200 mM sodium acetate, 250 mM NaCl buffer at pH 4.5.
15. Column buffer: 0.1 mM ethylenediaminetetraacetic acid (EDTA), 50 mM sodium phosphate, 250 mM NaCl buffer at pH 7.2.
16. 8 M guanidium chloride in pure water for molecular biology (Sigma-Aldrich, USA).
17. Solid-phase extraction silica-C18 cartridge (820 mg of silica-C18, 55–105 μm particle size).
18. HPLC buffers. Buffer A: pure and filtered (over 45 μm filter) H_2O with 0.1% trifluoroacetic acid (TFA) (HPLC grade). Buffer B: 90% acetonitrile (HPLC grade) in pure and filtered (over 45 μm filter) H_2O with 0.1% TFA.

3 Methods

3.1 Construction of MCoTI-Intein Construct 1

This section allows for the production of an expression plasmid encoding the cyclotide-intein fusion precursor protein **1** (Figs. 2 and 3).

3.1.1 Obtaining the dsDNA Encoding Cyclotide MCoTI-I

1. Dissolve each ultramer shown in Table 1 in TE buffer to a concentration of 1 μg/μL with TE buffer.
2. Add 5 μL of a solution containing the upper DNA strand (P5) and 5 μL of a solution containing the lower strand (P3) to a 0.5 mL centrifuge tube containing 2.5 μL of 10× annealing buffer and 12 μL of pure H_2O.
3. Incubate sample from **step 2** on a preheated water bath to 95 °C for 15 min. After the allocated time, turn the power of the water bath off and allow the samples to slowly cool down until room temperature is reached. The cooling process should not take less than 60 min.
4. Purify the double strand DNA fragments using a PCR purification kit following the manufacturer's instructions.
5. Quantify double strand DNA fragments by UV-Vis spectroscopy (for a 1-cm pathlength, an optical density at 260 nm (OD_{260}) of 1.0 equals to a concentration of 50 μg/mL solution of dsDNA).

3.1.2 Preparation of Expression Plasmid pTXB1-MCoTI

1. Digest plasmid pTXB1 (New England Biolabs) with restriction enzymes *Nde*I and *Sap*I. Use a 0.5 mL centrifuge tube and add 5 μL of cut smart buffer (New England Biolabs), add enough pure sterile water to have a final volume reaction of 50 μL, add ≈10 μg of the corresponding dsDNA to be digested add finally add 1 μL (20 units) of restriction enzyme *Nde*I (New England Biolabs). Incubate at 37 °C for 3 h. Then, add 1 μL (20 units) of restriction enzyme *Sap*I (New England Biolabs) to the same tube and incubate at 37 °C for 3 h.
2. Purify double digested pTXB1 plasmid by using 0.8% agarose gel electrophoresis. Cut the band corresponding to the double digested DNA and purify it using the QIAquick Gel Extraction Kit (QIAGEN or similar), elute it with TE buffer and quantify it using UV–visible spectroscopy.
3. Ligate double digested pTXB1 and dsDNA encoding MCoTI-intein. Use a 0.5 mL centrifuge tube, add ≈100 ng of *Nde*I, *Sap*I-digested pTXB1, ≈50 ng of 5′-phosporylated dsDNA MCoTI-I, enough pure sterile H_2O to make a final reaction volume of 20 μL, 2 μL of 10× T4 DNA ligase buffer, 1 μL of 10 mM ATP and 1 μL (400 units) T4 DNA ligase. Incubate at 16 °C overnight.
4. Transform the ligation mixture into DH5α competent cells. ≈100 μL of chemical competent cells are thawed on ice and mixed with the ligation mixture (20 μL) for 30 min. Heat-shock cells at 42 °C for 45 s and then kept on ice for an extra 10 min. Add 900 μL of SOC medium and incubate at 37 °C for 1 h in an orbital shaker. Plate 100 μL on LB agar plate containing ampicillin (100 μg/mL) and incubate the plate at 37 °C overnight.
5. Pick up several colonies (most of the times 5 colonies should be enough) and inoculate each individual colony into separate 5 ml of LB media containing ampicillin (100 μg/mL). Incubate the tubes at 37 °C overnight in an orbital shaker.
6. Pellet down cells and extract DNA using the QIAprep Spin Miniprep Kit (QIAGEN) following the manufacturer's protocol and quantify plasmid using UV–visible spectroscopy.
7. Verify the presence of DNA encoding MCoTI-intein construct in each colony using *EcoR*I restriction enzyme as this restriction site is removed during cloning. Use a 0.5 mL centrifuge tube and add 2 μL of cut smart buffer (New England Biolabs), add enough pure sterile water to have a final volume reaction of 20 μL, add ≈100 ng of the corresponding plasmid to be digested, and finally add 1 μL (20 units) of restriction enzyme *EcoR*I (New England Biolabs). Incubate at 37 °C for 3 h. Analyze the reaction by 0.8% agarose gel electrophoresis. Positive clones should not be linearized by *EcoR*I digestion.

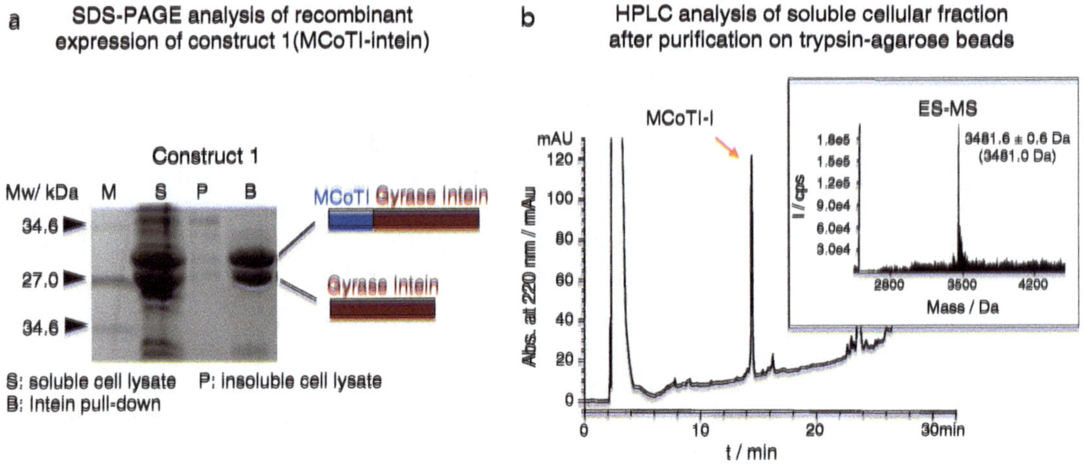

Fig. 4 Expression of cyclotide precursor 1 and characterization of cyclotide MCoTI-I produced in *E. coli* cells. (**a**) SDS-PAGE analysis of the recombinant expression of cyclotide precursors **1** in Origami2(DE3) cells for in-cell production of cyclotide MCoTI-I. The bands corresponding to precursor 1, and cleaved intein are marked with arrows. M stands for protein markers. (**b**) Analytical HPLC trace of the soluble cell extract of bacterial cells expressing precursor **1** after purification by affinity chromatography on a trypsin-sepharose column. Folded MCoTI-I is marked with an arrow. Mass spectrum (inset) of affinity purified MCoTI-I. The expected average molecular weight is shown in parentheses

3.2 Expression and Purification of MCoTI-Intein Precursor 1

This section allows for the expression of the cyclotide-intein protein construct **1** and to evaluate the expression level of **1** using SDS-PAGE as shown in Fig. 4.

3.2.1 Expression of Precursor Protein Encoding the MCoTI-Intein Construct 1

1. Transform chemical competent Origami2(DE3) cells with plasmid containing the DNA encoding MCoTI-intein construct **1** (plasmid pTXB1-MCoTI) (*see* **Note 1**). Transformed cells are plated on LB agar plates containing ampicillin (100 μg/mL) and incubated at 37 °C overnight as described is Subheading 3.1.2 (*see* **Note 2**).
2. All the colonies from 2 plates are resuspended in 2 mL of LB and inoculated in 1 L of LB containing ampicillin (100 μg/mL) in a 2.5 L flask.
3. Grow cells in an orbital shaker incubator at 37 °C for 2–3 h to reach mid-log phase (OD at 600 nm ≈ 0.5). Add IPTG to reach a final concentration of 0.3 mM. Adjust the temperature of the incubator to 25 °C and incubate cells in shaker for 16 h.
4. Pellet the cells down by centrifugation at 6000 × *g* for 15 min at 4 °C. Discard the supernatant and process the pellet immediately (*see* **Note 3**).

3.2.2 Determination of Protein Precursor 1 Expression Level

1. Resuspend cell pellet with 30 mL of Lysis buffer containing 5% glycerol. Lyse cells by sonication on ice using 25 s bursts spaced 30 s each (*see* **Note 4**). Repeat the cycle six times (*see* **Note 5**).

Take two 100 μL aliquots. In one of the samples add 33 μL 4× SDS-PAGE sample buffer containing 20% 2-mercaptoethanol heat it at 94 °C for 5 min (label it total cell lysate, T). For the other aliquot, separate the insoluble and soluble fractions by centrifugation at 18,000 × *g* in a microcentrifuge at 4 °C for 30 min. Take the supernatant fraction and resuspend the pellet in 100 μL of lysis buffer. Add 33 μL of 4× SDS-PAGE sample buffer containing 20% 2-mercaptoethanol to both fractions and heat them at 94 °C for 5 min (label them soluble and insoluble cell lysate samples, S and P, respectively). Save the samples for later SDS-PAGE analysis.

2. Separate the soluble cell lysate fraction by centrifugation at 15,000 × *g* for 20 min at 4 °C. Store the pellets at −80 °C in case they need to be reprocessed.
3. Transfer the soluble cell lysate fraction (≈30 mL) into a 50 mL centrifuge tube and add 1 mL of preequilibrated chitin-agarose beads by washing them with at least 10 column volumes of column buffer. Incubate with gentle rocking for 30 min at 4 °C.
4. Separate the beads from the supernatant by centrifugation at 3000 × *g* for 10 min at 4 °C. Take the supernatant and save it at 4 °C for later processing (Subheading 3.4). Separate the beads and wash them in a 5 mL polypropylene column (QIAGEN) with no less than 50 column volumes of column buffer containing 0.1% Triton X-100 and then rinsed and equilibrated with 100 bead-volumes of column buffer. Take ≈100 μL chitin-agarose beads into a 0.5 mL centrifuge tube, add 33 μL of 4× SDS-PAGE sample buffer containing 20% 2-mercaptoethanol heat it at 94 °C for 5 min (label it soluble cell lysate bound to chitin-agarose beads, B). Save the sample for later SDS-PAGE analysis.
5. Analyze the expression level of the precursor protein **1** using SDS-PAGE (Fig. 4a). Load 25 μL of samples labeled T, P, S, and B (see above) onto an SDS-4-20% PAGE gel. Run the samples at 125 V for about 1 h and 30 min in 1× SDS running buffer. Remove SDS with pure water and stain the gel with 20 mL GelCode® Blue reagent (Thermo Scientific) using the manufacturer's protocol (Fig. 4a). A bacterial clone with good expression level of precursor protein **1** should show two protein bands on the analysis by SDS-PAGE of the chitin beads fraction (Fig. 4a), where the upper and lower bands are the uncleaved and cleaved cyclotide-intein precursor. The ratio of cleaved–uncleaved proteins usually ranges from 6:4 to 7:3 for this cyclotide.

6. Discard any samples not showing intein precursor protein **1** or its cleaved product as they will not contain cyclotide to be purified.

3.3 Preparation of Trypsin-Immobilized Agarose Beads

This section allows for the production of trypsin-agarose beads that will be used for the purification of folded cyclotide MCoTI-I by using affinity chromatography (Subheading 3.4).

1. Wash ≈1 mL of NHS-activated sepharose (GE Healthcare) with 15 column volumes of ice-cold 1 mM HCl using a 5 mL polypropylene column (QIAGEN).
2. Equilibrate the column with 15 volumes of coupling buffer.
3. Dissolve 4 mg of porcine pancreatic trypsin in 500 μL of coupling buffer using gentle rocking
4. Add the trypsin solution to the equilibrated NHS-activated sepharose beads and incubate for 3 h with gentle rocking at room temperature.
5. Wash the sepharose beads with 10 volumes of coupling buffer containing 100 mM ethanolamine.
6. Incubate beads with 3 column volumes of coupling buffer containing 100 mM ethanolamine for 3 h with gentle rocking at room temperature.
7. Wash the sepharose beads with 50 column volumes of washing buffer and store a 4 °C until use (*see* **Note 6**). The loading of the trypsin-sepharose can be determined by incubating a small aliquot of the beads with a known amount of pure MCoTI-I and determining the amount of cyclotide captured on the beads using HPLC (see below).

3.4 Affinity Purification of Cyclotide MCoTI-I from Bacterial Soluble Cell Lysate

This section allows for the purification of the natively folded cyclotide MCoTI-I produced during the intein-induced backbone cyclization of the cyclotide precursor **1** by using by affinity chromatography with trypsin-agarose beads. This procedure will allow only to capture natively folded cyclotide as misfolded product/s will not be able to bind the trypsin immobilized onto the sepharose beads. The bound cyclotide can be simply eluted using a chaotropic salt to destabilize the complex between the cyclotide and trypsin.

1. Wash ≈500 μL of trypsin-sepharose beads with 10 column volumes of column buffer.
2. Incubate the washed beads with the soluble cell lysate flow-through from the chitin agarose beads purification step (Subheading 3.2.2, **step 4**) for 1 h at room temperature with gentle rocking.

3. Separate the supernatant by centrifugation at 3000 × *g* for 10 min at 4 °C.
4. Transfer the trypsin-beads to a 5 mL polypropylene column (QIAGEN) and wash the beads with 50 volumes of column buffer containing 0.1% Tween 20.
5. Wash trypsin-beads with 50 volumes of column buffer with no detergent added.
6. Elute bound cyclotide MCoTI-I with 3 volumes (3 × 500 μL) of 8 M guanidium hydrochloride at room temperature for 15 min by gravity.
7. Desalt the sample using a solid-phase extraction cartridge (Sep-Pak or similar) by following the manufacturer's protocol. Briefly, preswell the SepPak cartridge with 20 mL of 50% acetonitrile in water containing 0.1% TFA. Equilibrate cartridge with 50 mL of HPLC buffer A. Load sample into the cartridge using a plastic 20 mL syringe slowly with a flow rate of ≈1 mL/min. Collect the flow-through and repeat this step for efficient binding. Wash the cartridge with 20 mL of 5% acetonitrile in H_2O containing 0.1% TFA. Elute cyclotide MCoTI-I from the solid-phase extraction cartridge with 5 mL of HPLC buffer-B. Lyophilize to remove solvents.
8. Dissolve in 5 mL of HPLC buffer A. Analyze sample by HPLC using an isocratic of 0% buffer B for 2 min and then a linear gradient of 0–70% buffer B in 30 min. Use detection at 220 and 280 nm. Using these conditions the retention time of the cyclotide should be around 15 min (Fig. 4b). Collect the peak and analyze by mass spectrometry to confirm identity of cyclotide MCoTI-I (expected molecular weight: 3481.0 Da) (Fig. 4b).
9. Quantify the cyclotide using UV-Vis spectroscopy and a molar absorptivity at 280 nm of 2240 $M^{-1} \times cm^{-1}$. Around 40 μg of folded cyclotide should be obtained per liter of LB culture.

4 Notes

1. We recommend at this point to use at least 3–4 different colonies or clones for evaluating the expression levels of the MCoTI-I intein construct **1**.
2. When plating the transformed cells with pTXB1-MCoTI, it is better to aim for plates containing 100–200 colonies. Plates containing excessive number of colonies may facilitate the appearance of colonies not containing the plasmid due to the destruction of the antibiotic by the secreted beta-lactamase.

3. Cell pellets can be stored at −80 °C for no more than 2–3 weeks before being processed.
4. During sonication, be sure the temperature of the sample does not overheat.
5. A French press can be also used to lyse cells, depending on the availability.
6. The trypsin-sepharose column should not be stored for more than 2 weeks.

References

1. Gould A, Camarero JA (2017) Cyclotides: overview and biotechnological applications. ChemBioChem *18*:1350–1363
2. Wang CK, Craik DJ (2018) Designing macrocyclic disulfide-rich peptides for biotechnological applications. Nat Chem Biol *14*:417–427
3. Puttamadappa SS, Jagadish K, Shekhtman A, Camarero JA (2010) Backbone dynamics of cyclotide MCoTI-I free and complexed with trypsin. Angew Chem Int Ed Engl *49*:7030–7034
4. Puttamadappa SS, Jagadish K, Shekhtman A, Camarero JA (2011) Erratum in: backbone dynamics of cyclotide MCoTI-I free and complexed with trypsin. Angew Chem Int Ed Engl *50*:6948–6949
5. Daly NL, Thorstholm L, Greenwood KP, King GJ, Rosengren KJ, Heras B, Martin JL, Craik DJ (2013) Structural insights into the role of the cyclic backbone in a squash trypsin inhibitor. J Biol Chem *288*:36141–36148
6. Chaudhuri D, Aboye T, Camarero JA (2019) Using backbone-cyclized cys-rich polypeptides as molecular scaffolds to target protein-protein interactions. Biochem J *476*:67–83
7. Saether O, Craik DJ, Campbell ID, Sletten K, Juul J, Norman DG (1995) Elucidation of the primary and three-dimensional structure of the uterotonic polypeptide kalata B1. Biochemistry *34*:4147–4158
8. Li Y, Bi T, Camarero JA (2015) Chemical and biological production of cyclotides. Adv Bot Res *76*:271–303
9. Aboye TL, Ha H, Majumder S, Christ F, Debyser Z, Shekhtman A, Neamati N, Camarero JA (2012) Design of a novel cyclotide-based CXCR4 antagonist with anti-human immunodeficiency virus (HIV)-1 activity. J Med Chem *55*:10729–10734
10. Wong CT, Rowlands DK, Wong CH, Lo TW, Nguyen GK, Li HY, Tam JP (2012) Orally active peptidic bradykinin B1 receptor antagonists engineered from a cyclotide scaffold for inflammatory pain treatment. Angew Chem Int Ed Engl *51*:5620–5624
11. Chan LY, Gunasekera S, Henriques ST, Worth NF, Le SJ, Clark RJ, Campbell JH, Craik DJ, Daly NL (2011) Engineering pro-angiogenic peptides using stable, disulfide-rich cyclic scaffolds. Blood *118*:6709–6717
12. Ji Y, Majumder S, Millard M, Borra R, Bi T, Elnagar AY, Neamati N, Shekhtman A, Camarero JA (2013) In vivo activation of the p53 tumor suppressor pathway by an engineered cyclotide. J Am Chem Soc *135*:11623–11633
13. Contreras J, Elnagar AY, Hamm-Alvarez SF, Camarero JA (2011) Cellular uptake of cyclotide MCoTI-I follows multiple endocytic pathways. J Control Release *155*:134–143
14. Cascales L, Henriques ST, Kerr MC, Huang YH, Sweet MJ, Daly NL, Craik DJ (2011) Identification and characterization of a new family of cell-penetrating peptides: cyclic cell-penetrating peptides. J Biol Chem *286*:36932–36943
15. Camarero JA (2017) Cyclotides, a versatile ultrastable micro-protein scaffold for biotechnological applications. Bioorg Med Chem Lett *27*:5089–5099
16. Rehm FBH, Jackson MA, De Geyter E, Yap K, Gilding EK, Durek T, Craik DJ (2019) Papain-like cysteine proteases prepare plant cyclic peptide precursors for cyclization. Proc Natl Acad Sci U S A *116*:7831–7836
17. Mylne JS, Chan LY, Chanson AH, Daly NL, Schaefer H, Bailey TL, Nguyencong P, Cascales L, Craik DJ (2012) Cyclic peptides arising by evolutionary parallelism via asparaginyl-endopeptidase-mediated biosynthesis. Plant Cell *24*:2765–2778
18. Poth AG, Mylne JS, Grassl J, Lyons RE, Millar AH, Colgrave ML, Craik DJ (2012) Cyclotides associate with leaf vasculature and are the products of a novel precursor in petunia (Solanaceae). J Biol Chem *287*:27033–27046

19. Poth AG, Colgrave ML, Lyons RE, Daly NL, Craik DJ (2011) Discovery of an unusual biosynthetic origin for circular proteins in legumes. Proc Natl Acad Sci U S A *108*:10127–10132
20. Jennings C, West J, Waine C, Craik D, Anderson M (2001) Biosynthesis and insecticidal properties of plant cyclotides: the cyclic knotted proteins from Oldenlandia affinis. Proc Natl Acad Sci U S A *98*:10614–10619
21. Gillon AD, Saska I, Jennings CV, Guarino RF, Craik DJ, Anderson MA (2008) Biosynthesis of circular proteins in plants. Plant J *53*:505–515
22. Saska I, Gillon AD, Hatsugai N, Dietzgen RG, Hara-Nishimura I, Anderson MA, Craik DJ (2007) An asparaginyl endopeptidase mediates in vivo protein backbone cyclization. J Biol Chem *282*:29721–29728
23. Nguyen GK, Wang S, Qiu Y, Hemu X, Lian Y, Tam JP (2014) Butelase 1 is an Asx-specific ligase enabling peptide macrocyclization and synthesis. Nat Chem Biol *10*:732–738
24. Jagadish K, Camarero JA (2017) Recombinant expression of cyclotides using split inteins. Methods Mol Biol *1495*:41–55
25. Jagadish K, Gould A, Borra R, Majumder S, Mushtaq Z, Shekhtman A, Camarero JA (2015) Recombinant expression and phenotypic screening of a bioactive cyclotide against alpha-synuclein-induced cytotoxicity in Baker's yeast. Angew Chem Int Ed Engl *54*:8390–8394
26. Jagadish K, Borra R, Lacey V, Majumder S, Shekhtman A, Wang L, Camarero JA (2013) Expression of fluorescent cyclotides using protein trans-splicing for easy monitoring of cyclotide-protein interactions. Angew Chem Int Ed Engl *52*:3126–3131
27. Austin J, Wang W, Puttamadappa S, Shekhtman A, Camarero JA (2009) Biosynthesis and biological screening of a genetically encoded library based on the cyclotide MCoTI-I. ChemBioChem *10*:2663–2670
28. Camarero JA, Kimura RH, Woo YH, Shekhtman A, Cantor J (2007) Biosynthesis of a fully functional cyclotide inside living bacterial cells. ChemBioChem *8*:1363–1366
29. Kimura RH, Tran AT, Camarero JA (2006) Biosynthesis of the cyclotide Kalata B1 by using protein splicing. Angew Chem Int Ed Engl *45*:973–976
30. Sarmiento C, Camarero JA (2019) Biotechnological applications of protein splicing. Curr Protein Pept Sci *20*:408–424
31. Conibear AC, Watson EE, Payne RJ, Becker CFW (2018) Native chemical ligation in protein synthesis and semi-synthesis. Chem Soc Rev *47*:9046–9068
32. Berrade L, Camarero JA (2009) Expressed protein ligation: a resourceful tool to study protein structure and function. Cell Mol Life Sci *66*:3909–3922
33. Hernandez JF, Gagnon J, Chiche L, Nguyen TM, Andrieu JP, Heitz A, Trinh Hong T, Pham TT, Le Nguyen D (2000) Squash trypsin inhibitors from Momordica cochinchinensis exhibit an atypical macrocyclic structure. Biochemistry *39*:5722–5730
34. Borra R, Camarero JA (2013) Recombinant expression of backbone-cyclized polypeptides. Biopolymers *100*:502–509
35. Austin J, Kimura RH, Woo YH, Camarero JA (2010) In vivo biosynthesis of an Ala-scan library based on the cyclic peptide SFTI-1. Amino Acids *38*:1313–1322
36. Gould A, Li Y, Majumder S, Garcia AE, Carlsson P, Shekhtman A, Camarero JA (2012) Recombinant production of rhesus theta-defensin-1 (RTD-1) using a bacterial expression system. Mol Biosyst *8*:1359–1365
37. Conibear AC, Wang CK, Bi T, Rosengren KJ, Camarero JA, Craik DJ (2014) Insights into the molecular flexibility of theta-defensins by NMR relaxation analysis. J Phys Chem B *118*:14257–14266
38. Klabunde T, Sharma S, Telenti A, Jacobs WR, Sacchettini JC (1998) Crystal structure of GyrA intein from Mycobacterium xenopi reveals structural basis of protein splicing. Nat Struct Biol *5*:31–36
39. Hirel PH, Schmitter JM, Dessen P, Fayat G, Blanquet S (1989) Extent of N-terminal methionine excision from Escherichia coli proteins is governed by the side-chain length of the penultimate amino-acid. Proc Natl Acad Sci U S A *86*:8247–8251
40. Avrutina O, Schmoldt HU, Gabrijelcic-Geiger-D, Le Nguyen D, Sommerhoff CP, Diederichsen U, Kolmar H (2005) Trypsin inhibition by macrocyclic and open-chain variants of the squash inhibitor MCoTI-II. Biol Chem *386*:1301–1306
41. Bessette PH, Aslund F, Beckwith J, Georgiou G (1999) Efficient folding of proteins with multiple disulfide bonds in the Escherichia coli cytoplasm. Proc Natl Acad Sci U S A *96*:13703–13708
42. Felizmenio-Quimio ME, Daly NL, Craik DJ (2001) Circular proteins in plants: solution structure of a novel macrocyclic trypsin inhibitor from Momordica cochinchinensis. J Biol Chem *276*:22875–22882

Chapter 17

Highly Precise Protein Semisynthesis through Ligation–Desulfurization Chemistry in Combination with Phenacyl Protection of Native Cysteines

Somnath Mukherjee, Maria Matveenko, and Christian F. W. Becker

Abstract

Semisynthesis of proteins via expressed protein ligation is a powerful tool to furnish full-length proteins carrying site-specific (posttranslational) modifications. The development of various β-mercapto amino acid building blocks coupled with ligation–desulfurization chemistry enabled further advances in this methodology by alleviating the need for cysteine residues at the desired ligation sites. However, this expansion in the availability of viable ligation sites is sometimes counterbalanced by the inadvertent desulfurization of unprotected native cysteines, which might be of structural and/or functional importance. Here, we provide a detailed protocol for using the cysteine-selective protecting group phenacyl (PAc) to achieve precise protein semisynthesis preserving native cysteine residues. The PAc group can be easily installed on cysteine(s) within recombinantly produced protein thioesters, withstands standard ligation, desulfurization and reversed phase HPLC conditions, and can be smoothly removed. We have previously demonstrated the utility of this protecting group through the semisynthesis of two model proteins, human small heat shock protein Hsp27 and Prion protein, in which one or two native cysteines, respectively, were maintained through the ligation–desulfurization sequence.

Key words Posttranslational modifications, Protein protecting group, Cysteine-selective protection, Phenacyl, Expressed protein ligation, Ligation–desulfurization, Radical desulfurization, Traceless protein semisynthesis

1 Introduction

Posttranslational modification (PTM) of proteins represents a vast and intriguing area of eukaryotic cell biology critical for precisely orchestrating cellular functions as well as disease etiology. The majority of these modifications, including acetylation, glycosylation, methylation, and ubiquitination, are typically carried out by specific enzymes, involving covalent attachment of various molecules or moieties onto specific residues of target proteins.

Somnath Mukherjee and Maria Matveenko contributed equally to this work.

Miquel Vila-Perelló (ed.), *Expressed Protein Ligation: Methods and Protocols*, Methods in Molecular Biology, vol. 2133, https://doi.org/10.1007/978-1-0716-0434-2_17,

In addition to rather extensively studied enzymatic PTMs, another class of protein modifications, known as nonenzymatic PTMs (nPTMs), has been increasingly recognized to have important biological functions [1–4]. These covalent modifications typically occur on nucleophilic side chains of amino acid residues such as Arg, Lys, His, and Cys, through reactions with small electrophilic molecules and without enzymatic assistance. The resulting modifications are often irreversible and may be considered biomarkers of cellular stress and various pathological conditions [5].

In order to determine the precise correlation between the pathological outcomes of certain PTMs at specific residue(s) of a protein of interest, it is critical to generate homogeneous protein specimen containing such site-specific modifications. Available molecular biology technologies do not yet offer methodology to incorporate all pathologically relevant PTMs in a site-specific way into a protein of interest. As a result, semisynthesis remains to be the only way to introduce certain PTMs site-specifically.

Although there has been a tremendous surge in the repertoire of available strategies to ligate a synthetic segment carrying single/multiple PTM(s) with a recombinantly produced counterpart and thus to synthesize a homogeneously modified full length protein, native chemical ligation (NCL) [6] and expressed protein ligation (EPL) [7] remain to be the most versatile and powerful methods for this purpose. Since the development of NCL by Kent and coworkers, several advances in this methodology have been made to address the low natural abundance of cysteine (1.7% [8]), a critical residue for such ligations to take place. As such, the ligation-desulfurization strategy, pioneered by Yan and Dawson [9] and Wan and Danishefsky [10], utilizes suitably positioned alanine (s) as ligation site(s) by using the corresponding nonnative cysteine residues for the NCL reaction and subsequently reverting these to alanine via desulfurization. In addition to cysteine, various other β-mercapto amino acid building blocks have been developed to expand the scope of NCL [11–21]. Although such building blocks broaden the choice of viable ligation site(s), with the exception of a few such as β-mercapto-aspartate or selenocysteine [22–24], they do not allow a "traceless" semisynthetic approach, in which all the introduced β-mercapto residues are desulfurized and the native cysteines remain intact. Cysteine residues often play important roles in structure and activity of proteins, and thus should be kept from being inadvertently mutated to alanine under desulfurization conditions. Therefore, it is pertinent to expand the range of protecting groups that allow for protein semisynthesis via the traceless approach.

Here, we describe a detailed protocol for utilizing the cysteine-selective and easy to handle phenacyl (PAc) protecting group in the traceless semisyntheses of 22.8 kDa human small heat shock protein Hsp27 containing one native cysteine (Fig. 2), while the analogous

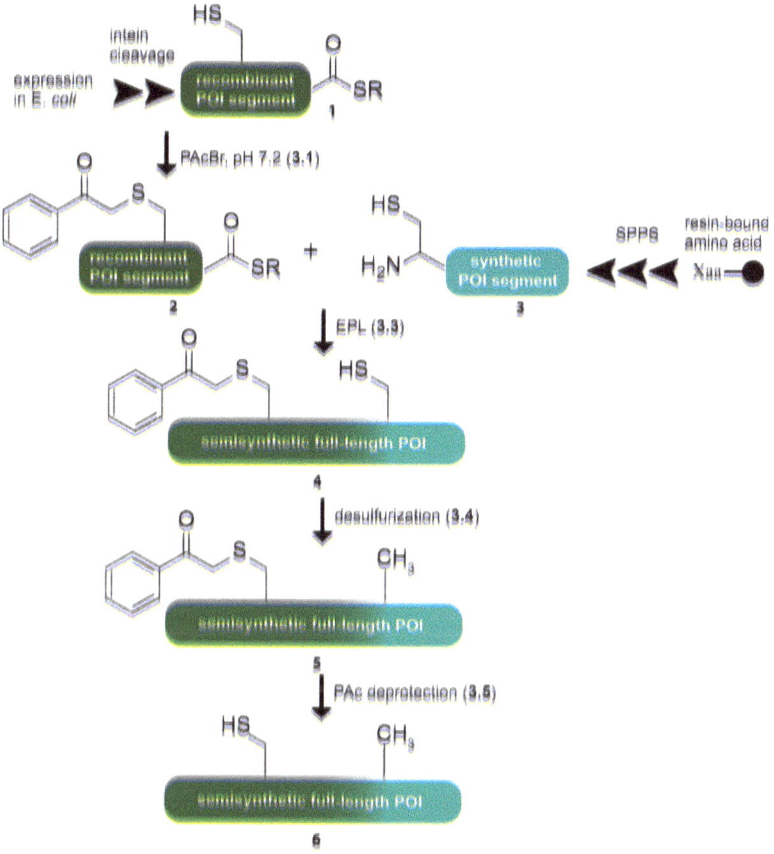

Fig. 1 MS data obtained for Hsp27(1–172)-thioester undergoing PAc protection, followed by native chemical ligation, radical desulfurization, then PAc deprotection to obtain full length semisynthetic Hsp27 retaining its native cysteine residue [25]

protocol has also been successfully applied in our laboratories to generate Prion protein containing two native cysteines [25–28]. Importantly, this protecting group could be easily installed on cysteine residues in recombinantly produced protein thioester segments, withstood standard ligation-desulfurization conditions as well as rounds of RP-HPLC purifications, and could be easily removed at the end of the synthetic sequence (Figs. 1 and 2). The overall operational simplicity of this protocol offers a useful option to the general scientific community engaged in protein semisynthesis.

2 Materials

All aqueous buffers and solutions should be prepared with Milli-Q or equivalent purified water (ddH_2O) (total organic carbon: <2 ppb). Use all reagents and organic solvents of the highest

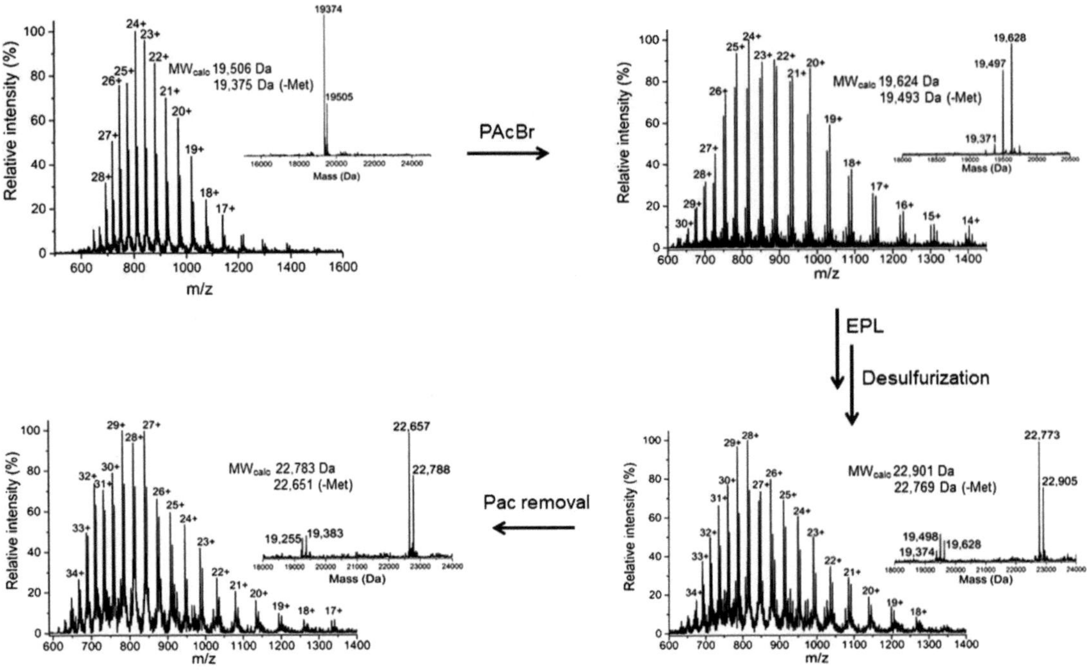

Fig. 2 Overview of traceless semisynthesis highlighting the use of the PAc group for temporary protection of a native cysteine residue within a recombinantly produced protein segment during desulfurization of a nonnative cysteine at the ligation site into the native alanine in a protein of interest (POI). In the protocol described herein, the semisynthesis of Hsp27 is used as an example. The 205-amino acid Hsp27 protein was generated via EPL between the recombinant Hsp27(1–172) segment having a C-terminal thioester and the shorter synthetic segment Hsp27(173–205) modified at the N-terminus as Ala173Cys. The use of the PAc strategy traceless approach allowed to prevent the undesired desulfurization of Cys137, while reverting to residue 173 to the native Ala via desulfurization. For more details of this previously published work *see* Ref. [25]

available quality. As most of the reactions described herein critically rely on thiols and/or phosphine reagents prone to oxidation with molecular oxygen, diligently follow all instructions to degas the buffers (*see* **Note 1**) and prepare them freshly in small batches (typically 2–50 mL). Follow all waste disposal guidelines and safety regulations throughout the workflow.

2.1 Phenacyl (PAc) Protection of Cysteine(s) in Recombinantly Produced Protein α-Thioesters

1. RP-HPLC-purified and lyophilized protein α-thioester **1** (*see* Fig. 2 and **Notes 2–5**). Using a solution of a protein α-thioester segment of known concentration may also be suitable (*see* **Note 6**).
2. Protection buffer: 6 M Gdn·HCl in 0.4 M sodium phosphate buffer, pH 7.15 (*see* **Note 7**). To prepare 50 mL of this buffer, weigh 28.7 g of Gdn·HCl into a graduated vessel such as a 50 mL Falcon tube, then make up to 50 mL with required amount of 0.4 M Na_2HPO_4/0.4 M NaH_2PO_4 (*see* **Note 8**) to reach pH 7.15, as monitored by a pH meter, with stirring. Alternatively, weigh out 1.87 g of Na_2HPO_4, 0.82 g of

NaH_2PO_4 and 28.7 g of Gdn·HCl, add 25–30 mL of Milli-Q water (ddH_2O). In case the pH of this buffer needs adjustment, titrate this slurry with NaOH solution or phosphoric acid to reach pH 7.15 and fill up to 50 mL. This buffer can be stored at room temperature for several weeks, but must be degassed immediately prior to use in the PAc protection reaction.

3. Stock solution of phenacyl bromide (PAcBr) (*see* **Note 9**): 0.1 M solution of PAcBr in anhydrous DMF (*see* **Note 10**). To prepare, for example, 250 μL of this solution, weigh 5 mg of PAcBr into a 1.5 or 2 mL Eppendorf tube, and 250 μL of anhydrous DMF. Close the lid immediately then dissolve PAcBr by vortexing. For convenience, any small amount (5–15 mg) of PAcBr can be dissolved in the adjusted volume of DMF required to achieve the same PAcBr:DMF w/v ratio as above. This stock solution should be either made fresh, or stored under argon (by blowing argon gently on top of the solution then closing the lid and sealing it with Parafilm) at 7–8 °C (in the fridge) and used within 1–2 weeks.
4. Temperature controlled shakers with 2 mL Eppendorf tube holder block for incubation.
5. Calibrated pH meter with a micro tip.

2.2 Removal of Excess Small Molecule Reagents (PAcBr and Desulfurization Reagents)

1. A centrifugal filter unit (molecular weight cutoff 10K or as required) (*see* **Note 11**) of suitable/available size, and an equivalent tube for counterbalance during the centrifugation step.
2. Purification buffer A: 6 M Gdn·HCl pH ~6.5 (*see* **Note 12**). This purification buffer can be stored at room temperature for several weeks. To prepare 50 mL of this buffer, weigh 28.7 g of Gdn·HCl into a graduated vessel such as a 50 mL Falcon tube, then add <20 mL ddH_2O and adjust pH to 6.5 using 1 M NaOH or 1 M HCl, making up to 50 mL with additional ddH_2O if required.
3. Purification buffer B: 6 M Gdn·HCl, pH ~4.5 (*see* **Note 13**). Prepare and handle as directly above (Subheading 2.2, **item 2**), except for adjusting the pH to 4.5.
4. Centrifuge using rotors capable of holding appropriate centrifugal filter units and being temperature controlled.
5. Calibrated pH meter with a micro tip.

2.3 Expressed Protein Ligation

1. Ligation buffer: 6 M Gdn·HCl, 0.2 M sodium phosphate buffer, 0.25 M sodium 2-mercaptoethanesulfonate (MESNA), 50 mM tris(2-carboxyethyl)phosphine (TCEP), pH 7.8. This buffer must be made and used on the day of setting up the ligation reaction. To prepare 2 mL of this buffer, weigh 1.15 g of Gdn·HCl into a graduated vessel such as a

15 mL Falcon tube, then add <1 mL of 0.2 M sodium phosphate buffer pH ~7.8 (prepared as per **Note 8**, and *see* **Note 14**). To the resulting slurry, add 82.1 mg of MESNA (MW 164.2) and 28.7 mg of TCEP hydrochloride (MW 286.7) (*see* **Note 15**), in any order. These acidic reagents will cause the pH to drop significantly. Carefully (dropwise and with vortexing in between), adjust the pH to 7.8 using at first 5 M NaOH (aq), then 1 M NaOH (aq). All the above components will make up almost 2 mL already, but if necessary use further 0.2 M sodium phosphate buffer pH ~7.8 to make up to 2 mL. Degas the solution for at least 10 min.

2. Synthetic peptide segment carrying N-terminal cysteine, **3** (*see* Fig. 2, **Note 16**), RP-HPLC purified and lyophilized, and available as a fluffy powder: weigh out a required amount of this segment.
3. Temperature controlled shakers with 2 mL Eppendorf tube holder block for the incubation.
4. Calibrated pH meter with a micro tip.
5. Analytical balance suitable for weighing amounts of less than 1 mg.

2.4 Desulfurization

1. Desulfurization buffer: 6 M Gdn·HCl, 0.2 M sodium phosphate buffer, 0.25 M TCEP, 20 mM 2,2′-azobis(2-methylpropionamidine) dihydrochloride (V-50), pH 6.6. This buffer must be made and used on the day of setting up the desulfurization reaction. To prepare 2 mL of this buffer, weigh 1.15 g of Gdn·HCl into a graduated vessel such as a 15 mL Falcon tube, then add <1 mL of 0.2 M sodium phosphate buffer pH ~7 (prepared as per **Note 8**, and *see* **Note 14**). To the resulting slurry, add 143 mg of TCEP hydrochloride, which will cause the pH to drop significantly. Carefully (dropwise and with vortexing in between), adjust the pH to 6.6 using at first 5 M NaOH (aq), then 1 or 2 M NaOH (aq). All the above components will make up almost 2 mL already, but if necessary use further 0.2 M sodium phosphate buffer pH ~7 to make up to 2 mL, ensuring that the final pH is 6.6. Degas the resulting solution for at least 10 min. Subsequently, add 11 mg of V-50 to this buffer then vortex extensively to dissolve the reagent. Degas the resulting buffer for further 2 min.
2. Temperature controlled shakers with 2 mL Eppendorf tube holder block for the incubation.
3. Calibrated pH meter with a micro tip.

2.5 Removal of the PAc Group

1. Deprotection buffer: 6 M Gdn·HCl, 3-mercaptopropionic acid (MPA, 15% v/v, *see* **Note 17**), pH ~1 (unadjusted). This buffer is typically made and used on the day of setting up the deprotection reaction, but may be used within several days of being made and stored at room temperature. To prepare, for example, 2 mL of this buffer, weigh 1.15 g of Gdn·HCl into a graduated vessel such as a 15 mL Falcon tube, then add 300 μL of MPA and make up to 2 mL with ddH_2O.
2. Zinc powder: weigh out zinc powder (37 mg/mL of the deprotection reaction). For example, for a reaction with a total volume of 500 μL, weigh out 19 mg of zinc powder.
3. Temperature controlled shaker with 2 mL Eppendorf tube holder block for incubation.
4. Tabletop centrifuge.

2.6 RP-HPLC Purification of the Final Product

1. RP-HPLC system equipped with dual wavelength detectors (214 and 280 nm).
2. C4 RP-HPLC columns (e.g., 300-5-C4, 150 × 4.6 mm, 5 μm particle size; or larger columns as desirable/available) (*see* **Note 18**).
3. Solvents used: Solvent A [ddH_2O + 0.1% trifluoroacetic acid (TFA)], Solvent B [acetonitrile (ACN) + 0.08% TFA] (*see* **Note 19**).
4. Lyophilizer.

3 Methods

3.1 Phenacyl (PAc) Protection of Cysteine(s) of Recombinantly Produced α-Thioester

This section provides a method to protect the native cysteine (s) within the protein thioester segment (*see* **Notes 2–5** for comments on the thioester segment) using the PAc protecting group.

1. Degas an amount of the protection buffer in excess of that required for the protection reaction below.
2. Weigh out (*see* **Note 6**) protein α-thioester **1** (0.26 μmol) into a 1.5 or 2 mL Eppendorf tube, add the degassed protection buffer (500 μL), and dissolve the thioester segment by several rounds of vortexing/centrifuging (*see* **Note 20**).
3. Using a micropipette, add the solution of PAcBr in anhydrous DMF (6.5 μL, 0.65 μmol, 2.5 equivalents, *see* **Note 21**) to the solution of **1**. Close the lid and immediately vortex the resulting solution.
4. Flush the Eppendorf tube with argon and seal it with Parafilm.
5. Incubate the reaction mixture at 25 °C, 400 rpm for 1 h.

6. To check the outcome, take 5 μL aliquot from the reaction mixture and dilute it into 20 μL of 1:1 ACN:ddH_2O containing 0.1% TFA, and use 20 μL of the resulting solution for LC-MS analysis (*see* **Note 22**).
7. Record the LC-MS spectrum and analyze the outcome of the reaction. If the reaction is not complete, keep shaking the reaction mixture longer and repeat **step 6** (at 30 min then 1 h intervals) until the reaction is complete.

3.2 Removal of Excess PAcBr

This section provides a method to remove any unreacted PAcBr (remaining from the previous step) from the solution of the crude PAc-protected thioester segment, so that PAcBr does not interfere with the subsequent ligation reaction by reacting with the N-terminal cysteine of the synthetic peptide segment.

1. To do so, exchange the protection buffer to purification buffer A, using centrifugal filter units of the appropriate size (*see* Subheading 2 and **Note 11**). Use as much of purification buffer A as needed to transfer all of the protection reaction mixture into a centrifugal filter unit during this process to avoid product loss, and exchange the buffer 2–3 times by adding more purification buffer and spinning the solution down to a small volume.
2. To render the PAc-protected thioester ready for the next step, exchange the buffer to degassed ligation buffer, reducing the volume of this last solution to ~600 μL.

3.3 Expressed Protein Ligation (see Note 23)

This section provides a method to ligate the recombinant segment carrying a C-terminal α-thioester with the synthetic peptide having an N-terminal cysteine, so as to generate the full length protein.

1. Transfer the solution of PAc-protected thioester segment above (~600 μL) from the centrifugal filter unit to a 2 mL Eppendorf tube, using extra 100 μL of degassed ligation buffer for transferring (to wash centrifugal filter unit) to prevent product loss.
2. Optionally, carefully degas the resulting solution in the Eppendorf tube (*see* **Note 24**).
3. To this solution of crude PAc-protected thioester **2** in ligation buffer (here, total volume 700 μL), add the peptide segment **3** as a solid (0.77 μmol, 3 equivalents). Close the lid immediately and mix the reagents thoroughly by vortexing then briefly centrifuge to remove the reaction mixture from the bottom surface of the lid of the Eppendorf tube.
4. Flush the reaction with argon, close the lid again and seal it with Parafilm.
5. Incubate the reaction at 37 °C, shake at 400 rpm for 48 h.

6. To check the outcome, take 5 μL aliquot from the reaction mixture and dilute it with 20 μL of 1:1 ACN:ddH_2O containing 0.1% TFA to use 20 μL of this solution for LC-MS analysis.
7. Record the LC-MS spectrum and analyze the outcome of the reaction (*see* **Note 25**).

3.4 Desulfurization

This section provides a method to desulfurize the introduced cysteine residue at the ligation site (introduced for the EPL to occur) to alanine, thus realizing a "traceless" semisynthesis at Ala ligation sites.

1. Dilute (2.7×) the above crude ligation mixture (700 μL) with desulfurization buffer (1.2 mL), resulting in 1.9 mL of reaction mixture with the final concentration of TCEP of 0.15 M, and final concentration of V-50 of 13 mM.
2. Flush the reaction tube with argon, close the lid, seal with Parafilm, and vortex.
3. Incubate the tube at 40 °C, 400 rpm for 16 h.
4. To check the outcome, take 5 μL aliquot from the reaction mixture and dilute it with 20 μL of 1:1 ACN:ddH_2O containing 0.1% TFA. If reaction is not complete, stirring longer typically does not improve the conversion, with the quality of the product decreasing over time. Issues, when occurring, may involve interaction of the reagent TCEP with the substrate and/or product. Thus, if optimization is required, the first thing to try is to reduce the amount of TCEP in the reaction.
5. To remove excess desulfurization reagents, exchange the reaction buffer to the purification buffer A, twice, using centrifugal filter units.
6. To acidify the solution of the crude product in preparation for PAc removal, exchange the buffer to the purification buffer B, using centrifugal filter units.
7. Reduce the volume of the crude mixture to at least as low as 600 μL (*see* **Note 26**).

3.5 Removal of the PAc Group

This section provides a method to remove the PAc protection group(s) from the native cysteine(s) and to reveal the free full length protein product.

1. Transfer the solution obtained above (here, 600 μL) to a 2 mL Eppendorf tube, then add to it deprotection buffer (600 μL), and mix by vortexing (producing the total volume of 1.2 mL).
2. Treat the resulting solution with zinc powder (45 mg, 37 mg/mL) directly, mix by vortexing, then open the lid (*see* **Note 27**).

3. Incubate the reaction mixture at 25 °C for 1 h, while keeping the lid of the Eppendorf tube slightly open (*see* **Note 27**), with occasional vortexing (always close the lid prior to vortexing).
4. To check the outcome, centrifuge the reaction mixture then take 5 μL of the supernatant and dilute it with 20 μL of 1:1 ACN:ddH_2O containing 0.1% TFA, and use 20 μL of this solution for LC-MS analysis.
5. Record the LC-MS spectrum and analyze the outcome of the reaction. If the reaction is not complete, stir longer and repeat **step 5**. If the reaction is not complete after 2–3 h, it will likely not work even with longer stirring or additional zinc powder. In such cases, the alternative deprotection procedure should be used as described in **Note 28**.

3.6 RP-HPLC Purification of the Final Product

1. If the reaction has gone to completion, centrifuge the crude reaction mixture (*see* **Note 29**) then purify the supernatant via RP-HPLC using, for example, a 5–65% linear gradient of Solvent B (ACN + 0.08% TFA) in Solvent A (ddH_2O + 0.1% TFA) over 30 min at 25 °C and at a flow rate of 1 mL/min might be used. Purification gradients will need to be optimized for each protein. As injectable amount is limited by the size of the column, crude reaction mixtures may be split for several runs of purification.
2. Analyze the fractions from the HPLC run(s) by MS.
3. Pool the fractions containing pure desired product and lyophilize. For Hsp27, the desired product **6** (*see* Fig. 2) was obtained in 35% isolated yield over 4 steps (*see* Ref. [25]).
4. Store the lyophilized product, which is typically obtained as a fluffy solid, at −20 °C.

4 Notes

1. Throughout the procedures, degassing a buffer/solution means purging it by bubbling N_2 or Ar through the solution via a needle long enough to reach the bottom of the solution, using a moderate flow of gas, continuing to purge for ~10 min in case of ~2 mL volumes, or ~30 min in case of ~50 mL volumes.
2. Protein C-terminal thioesters are susceptible to hydrolysis in neutral and even slightly alkaline solutions. Always follow the suggested pH to avoid hydrolysis. For long-term storage, thioester segments should be RP-HPLC purified, lyophilized and kept at −20 °C.
3. During the long reaction time (~48 h) required for the thiol-mediated intein cleavage and purification of the thioester by

affinity chromatography, sometimes adducts form in which cysteine residues have been oxidized to produce mixed disulfides with the thiol reagents (e.g., MESNA). Such cysteines are therefore not available for the reaction with PAcBr, and should be reduced. As a matter of routine, we recommend adding a small amount of TCEP to the crude thioester fragment solution prior to purification by RP-HPLC.

4. Since recombinantly generated protein thioesters are typically not full-length functional proteins, they are often not folded and can be handled under denaturing reaction conditions as described here. The generated full length semisynthetic proteins are typically purified by RP-HPLC (removing all buffer additives such as thiol, TCEP, and guanidine hydrochloride) and lyophilized. The final folding process is highly protein dependent. In case the protein cannot be purified by RP-HPLC, alternative purification methods (e.g., size exclusion chromatography, ion-exchange chromatography) can be considered.
5. The protocol of generating a recombinant protein thioester starts with commonly practiced expression and purification steps. The protein segment of interest is N-terminally fused to an intein, such as the *Mxe* GyrA intein, that also carries an affinity tag for purification via, for example, IMAC and/or chitin affinity chromatography. Expression and purification is followed by subjecting the purified construct to "thiolysis" with an externally added thiol (e.g., MESNA) to cleave the intein and generate the protein thioester. The resulting cleaved thioester is then purified by RP-HPLC or affinity chromatography, taking care as per **Notes 2** and **3**.
6. If the thioester segment is only (or more conveniently) available as a solution of known concentration, calculate the required volume of the solution and exchange the buffer to the protection buffer, followed by thoroughly degassing the resulting solution. *See* **Note 3** above for potential problems with oxidized cysteines within thioester fragments. Should such oxidation occur, treat the solution of the thioester with a few crystals of TCEP followed by buffer exchange to the protection buffer. It is important for the protection step that all TCEP is removed prior to reaction with PAcBr.
7. The pH of the protection buffer is particularly important, as side reactions can start to occur at higher or lower pH ranges. pH of approximately 7.2 was considered optimum both by us and others [28] but was not optimized extensively and the specific value of 7.15 was used throughout experiments for consistency. In general, pH near 7 should be suitable for this reaction.

8. A phosphate buffer is made by mixing and adjusting the proportions of the more basic solution of Na_2HPO_4 and more acidic solution of NaH_2PO_4, both having the desired concentration of the buffer, to reach the required pH, within the required volume. More guidance on this matter can be found in collections of standard biochemistry laboratory techniques.
9. Phenacyl bromide is a lachrymator (substance causing eye irritation and tears); use sufficient safety protection, and always weigh it out inside a well-ventilated fume hood. The reagent was purchased from TCI-Europe (Zwijndrecht, Belgium) and stored under argon.
10. Anhydrous DMF (water content <0.005%) here means the type available from standard chemical suppliers in sealed bottles (bottles equipped with a sealed metal cap having a needle-penetrable septum just below the metal cap, so as to protect the content from moisture/oxygen) under inert gas, handled via a syringe with a needle to reach through the penetrable seal into the stock bottle. For convenience, an excess amount can be taken out via a syringe into, for example, a Falcon tube, then subsequently aliquoted via a micropipette to add the required amount to the reaction, while the remainder of this solvent should be discarded or used for other purposes.
11. During buffer exchange steps, make sure to use centrifugal filtration tubes with appropriate MW cutoff to avoid the desired product flowing through the filter. Commercial vendors usually supply such tubes with the following MW cutoff limits: 3 kDa, 10 kDa, 30 kDa, 50 kDa, 100 kDa. We recommend to use a filter size 2–3 times smaller than the molecular weight of the protein of interest, for example, if you are working with a 30 kDa protein, use a 10 kDa filter equipped tube. Centrifugal filtration tubes with volumes of 2–15 mL have been used to conduct the work described herein. Centrifugation should be done at reduced temperatures, typically 4–6 °C. These steps can be rather time-consuming, so allow sufficient time.
12. The pH of purification buffer A does not need to be exactly 6.5: any pH around 6.5 (i.e., between 6 and 7) is suitable, as long as it is not more alkaline. This is critical for the step of excess PAcBr removal following PAc protection, as thioesters are prone to hydrolysis under even mildly alkaline conditions.
13. The pH of purification buffer B does not need to be exactly 4.5, any pH around 4.5 (i.e., between 4 and 5) is suitable. This range was chosen as a compromise between acidifying the solution in preparation for PAc removal, and protecting the membrane of the centrifugal filter units.

14. The concentration of this already weak buffer reduces further in the final buffer mixture, without consequence.
15. TCEP tends to gradually become oxidized (and thus no longer active), so it is really important to gently flush the bottle with argon/nitrogen gas after use and seal the bottle with Parafilm.
16. In order to achieve a practical semisynthesis of a protein of interest via single or multiple ligations, it is crucial to plan the semisynthesis wisely. For example, a plan involving a synthetic peptide segment longer than 40 amino acids would be practically challenging. This is because synthesizing such long peptides with reasonable yield and purity is cumbersome. Additional "sequence-specific" difficulties may arise due to hydrophobicity, aggregation propensity, and so on of the peptide.
17. MPA has a faint but unpleasant odor so handle it only inside a fume-hood.
18. The choice RP-HPLC column size (i.e., whether analytical or semipreparative) as well as solvent gradient for a successful purification of the ligation product needs to be optimized according to the available systems and reaction amounts.
19. TFA is an extremely corrosive material and must be used with caution during the preparation of buffers A and B for RP-HPLC.
20. As RP-HPLC purified lyophilized substrates such as thioester and peptides are acidic in nature, pH of the reaction buffer may drop slightly upon their addition. In all the reactions described herein, this potential drop was inconsequential and pH was not readjusted. In general, as best practice, readjusting the pH of a reaction mixture containing substrates should be avoided if possible. If the protection or the ligation reaction is sluggish, consider checking the pH of the reaction mixture. If this pH is too low (lower than that of the reaction buffer before the addition of the reactants), pH of the reaction buffer should be increased prior to the addition of the reactants.
21. If the thioester segment contains multiple native cysteines, 2.5 equivalents of PAcBr per cysteine should be used for complete protection.
22. Every time the reaction vessel needs to be opened for analysis, it should be flushed with argon again before being closed.
23. Conditions for the EPL, such as relative amount of substrates, temperature, pH and reaction time, depend on the particular peptide/protein substrates at hand, and may need to be optimized to enhance reaction yield/product quality. In addition, different thiol additives may be used, keeping in mind that aromatic ligation additives such as 4-mercaptophenylacetic

acid (MPAA) cannot be used for the *one-pot* ligation-desulfurization procedure and require an extra step of RP-HPLC purification following ligation (prior to desulfurization) to remove all the MPAA (attempts to partially remove MPAA by extraction with diethyl ether will not be sufficient). Aromatic ligation additives such as MPAA inhibit the subsequent desulfurization reaction by quenching the radicals required for the process.

24. Any loss of buffer volume due to evaporation in this step is potentially beneficial for the ligation reaction, which is intermolecular and thus requires a high concentration of the ligating substrates.

25. For new reactions, this step should be done at 2 h intervals until the total of 8 h reaction time, then after 16 h, then after 48 h reaction time. Stirring such a ligation for much longer usually does not improve the outcome and other factors need to be optimized to improve the result.

26. Ideally, the volume of the obtained solution of PAc-protected final product should be as low as possible to allow using as much of the deprotection buffer as possible in the next step of the synthetic sequence without diluting the reaction too much. While the dilution factor is not critical to the reaction (because the amount of zinc can be adjusted accordingly), too high volume is not practical for the subsequent RP-HPLC purification step which may be limited in terms of injectable amount.

27. While cleaving the PAc group with Zn/MPA, the reaction tube should be kept open as hydrogen gas is generated and will cause overpressure otherwise.

28. In a related protocol, MPA can be replaced with acetic acid, coupled with a higher dose of zinc powder [26]. This alternative procedure was successfully used in our laboratories for the synthesis of Prion protein containing two cysteines [25]. The alternative deprotection procedure is as follows: the crude solution of desulfurized, PAc-protected product in purification buffer B (obtained by buffer exchange as in Subheading 3.4, **steps 6** and 7) is diluted (2×) with the equivalent amount of buffer B. The resulting solution is further diluted (1.85×) by acetic acid to produce a ~45% v/v solution of acetic acid in buffer B, then treated with zinc powder (180 mg/mL), and stirred at 25 °C for 1 h, followed by purification by RP-HPLC as described in Subheading 3.6.

29. The crude reaction mixtures should be thoroughly centrifuged prior to purification by RP-HPLC to remove all solid residues and prevent clogging of columns. This should be done immediately before each RP-HPLC run if several runs are needed to purify the same mixture.

Acknowledgments

This work was supported by the Austrian Academy of Sciences by an APART fellowship to M.M.

References

1. Harmel R, Fiedler D (2018) Features and regulation of non-enzymatic post-translational modifications. Nat Chem Biol 14:244. https://doi.org/10.1038/nchembio.2575
2. Matveenko M, Cichero E, Fossa P, Becker CFW (2016) Impaired chaperone activity of human heat shock protein Hsp27 site-specifically modified with Argpyrimidine. Angew Chem Int Ed 55(38):11397–11402. https://doi.org/10.1002/anie.201605366
3. Vistoli G, De Maddis D, Cipak A, Zarkovic N, Carini M, Aldini G (2013) Advanced glycoxidation and lipoxidation end products (AGEs and ALEs): an overview of their mechanisms of formation. Free Radical Res 47 (Suppl1):3–27. https://doi.org/10.3109/10715762.2013.815348
4. Moellering RE, Cravatt BF (2013) Functional lysine modification by an intrinsically reactive primary glycolytic metabolite. Science 341 (6145):549. https://doi.org/10.1126/science.1238327
5. Negre-Salvayre A, Coatrieux C, Ingueneau C, Salvayre R (2008) Advanced lipid peroxidation end products in oxidative damage to proteins. Potential role in diseases and therapeutic prospects for the inhibitors. Br J Pharmacol 153 (1):6–20. https://doi.org/10.1038/sj.bjp.0707395
6. Dawson PE, Muir TW, Clark-Lewis I, Kent SB (1994) Synthesis of proteins by native chemical ligation. Science 266(5186):776. https://doi.org/10.1126/science.7973629
7. Muir TW, Sondhi D, Cole PA (1998) Expressed protein ligation: A general method for protein engineering. Proc Natl Acad Sci USA 95(12):6705. https://doi.org/10.1073/pnas.95.12.6705
8. Nilsson BL, Soellner MB, Raines RT (2005) Chemical synthesis of proteins. Annu Rev Biophys Biomol Struct 34(1):91–118. https://doi.org/10.1146/annurev.biophys.34.040204.144700
9. Yan LZ, Dawson PE (2001) Synthesis of peptides and proteins without cysteine residues by native chemical ligation combined with desulfurization. J Am Chem Soc 123(4):526–533. https://doi.org/10.1021/ja003265m
10. Wan Q, Danishefsky SJ (2007) Free-radical-based, specific desulfurization of cysteine: a powerful advance in the synthesis of polypeptides and glycopolypeptides. Angew Chem Int Ed 46(48):9248–9252. https://doi.org/10.1002/anie.200704195
11. Crich D, Banerjee A (2007) Native chemical ligation at phenylalanine. J Am Chem Soc 129 (33):10064–10065. https://doi.org/10.1021/ja072804l
12. Chen J, Wan Q, Yuan Y, Zhu J, Danishefsky SJ (2008) Native chemical ligation at valine: a contribution to peptide and glycopeptide synthesis. Angew Chem Int Ed 47 (44):8521–8524. https://doi.org/10.1002/anie.200803523
13. Haase C, Rohde H, Seitz O (2008) Native chemical ligation at valine. Angew Chem Int Ed 47(36):6807–6810. https://doi.org/10.1002/anie.200801590
14. Ajish Kumar KS, Haj-Yahya M, Olschewski D, Lashuel HA, Brik A (2009) Highly efficient and chemoselective peptide ubiquitylation. Angew Chem Int Ed 48(43):8090–8094. https://doi.org/10.1002/anie.200902936
15. Yang R, Pasunooti KK, Li F, Liu X-W, Liu C-F (2009) Dual native chemical ligation at lysine. J Am Chem Soc 131(38):13592–13593. https://doi.org/10.1021/ja905491p
16. Chen J, Wang P, Zhu J, Wan Q, Danishefsky SJ (2010) A program for ligation at threonine sites: application to the controlled total synthesis of glycopeptides. Tetrahedron 66 (13):2277–2283. https://doi.org/10.1016/j.tet.2010.01.067
17. Hojo H, Ozawa C, Katayama H, Ueki A, Nakahara Y, Nakahara Y (2010) The Mercaptomethyl group facilitates an efficient one-pot ligation at Xaa-Ser/Thr for (glyco)peptide synthesis. Angew Chem Int Ed 49 (31):5318–5321. https://doi.org/10.1002/anie.201000384
18. Tan Z, Shang S, Danishefsky SJ (2010) Insights into the finer issues of native chemical ligation: an approach to cascade ligations. Angew Chem Int Ed 49(49):9500–9503. https://doi.org/10.1002/anie.201005513
19. Ding H, Shigenaga A, Sato K, Morishita K, Otaka A (2011) Dual kinetically controlled

native chemical ligation using a combination of sulfanylproline and sulfanylethylanilide peptide. Org Lett 13(20):5588–5591. https://doi.org/10.1021/ol202316v
20. Shang S, Tan Z, Dong S, Danishefsky SJ (2011) An advance in proline ligation. J Am Chem Soc 133(28):10784–10786. https://doi.org/10.1021/ja204277b
21. Siman P, Karthikeyan SV, Brik A (2012) Native chemical ligation at glutamine. Org Lett 14 (6):1520–1523. https://doi.org/10.1021/ol300254y
22. Thompson RE, Chan B, Radom L, Jolliffe KA, Payne RJ (2013) Chemoselective peptide ligation–desulfurization at aspartate. Angew Chem Int Ed 52(37):9723–9727. https://doi.org/10.1002/anie.201304793
23. Araman C, Thompson RE, Wang S, Hackl S, Payne RJ, Becker CFW (2017) Semisynthetic prion protein (PrP) variants carrying glycan mimics at position 181 and 197 do not form fibrils. Chem Sci 8(9):6626–6632. https://doi.org/10.1039/C7SC02719B
24. Conibear AC, Watson EE, Payne RJ, Becker CFW (2018) Native chemical ligation in protein synthesis and semi-synthesis. Chem Soc Rev 47(24):9046–9068. https://doi.org/10.1039/C8CS00573G
25. Matveenko M, Hackl S, Becker CFW (2018) Utility of the phenacyl protecting group in traceless protein semisynthesis through ligation–desulfurization chemistry. ChemistryOpen 7(1):106–110. https://doi.org/10.1002/open.201700180
26. Katayama H, Hojo H (2013) The phenacyl group as an efficient thiol protecting group in a peptide condensation reaction by the thioester method. Org Biomol Chem 11 (26):4405–4413. https://doi.org/10.1039/C3OB40644J
27. Tang G, Ji T, Hu A-F, Zhao Y-F (2008) Novel N,S-phenacyl protecting group and its application for peptide -synthesis. Synlett 2008 (12):1907–1909. https://doi.org/10.1055/s-2008-1077887
28. Kawakami T, Yoshikawa R, Fujiyoshi Y, Mishima Y, Hojo H, Tajima S, Suetake I (2015) Synthesis of histone proteins by CPE ligation using a recombinant peptide as the C-terminal building block. J Biochem 158 (5):403–411. https://doi.org/10.1093/jb/mvv056

Index

Miquel Vila-Perelló (ed.), *Expressed Protein Ligation: Methods and Protocols*, Methods in Molecular Biology, vol. 2133, https://doi.org/10.1007/978-1-0716-0434-2,

MIX
Papier aus verantwortungsvollen Quellen
Paper from responsible sources
FSC® C105338

If you have any concerns about our products,
you can contact us on
ProductSafety@springernature.com

In case Publisher is established outside the EU,
the EU authorized representative is:
Springer Nature Customer Service Center GmbH
Europaplatz 3, 69115 Heidelberg, Germany

Printed by Libri Plureos GmbH
in Hamburg, Germany